# Lecture Notes in Physics





Springer-Verlag Berlin Heidelberg GmbH

## The Editorial Policy for Proceedings

The series Lecture Notes in Physics reports new developments in physical research and teaching – quickly, informally, and at a high level. The proceedings to be considered for publication in this series should be limited to only a few areas of research, and these should be closely related to each other. The contributions should be of a high standard and should avoid lengthy redraftings of papers already published or about to be published elsewhere. As a whole, the proceedings should aim for a balanced presentation of the theme of the conference including a description of the techniques used and enough motivation for a broad readership. It should not be assumed that the published proceedings must reflect the conference in its entirety. (A listing or abstracts of papers presented at the meeting but not included in the proceedings could be added as an appendix.)
When applying for publication in the series Lecture Notes in Physics the volume's editor(s) should submit sufficient material to enable the series editors and their referees to make a fairly accurate evaluation (e.g. a complete list of speakers and titles of papers to be presented and abstracts). If, based on this information, the proceedings are (tentatively) accepted, the volume's editor(s), whose name(s) will appear on the title pages, should select the papers suitable for publication and have them refereed (as for a journal) when appropriate. As a rule discussions will not be accepted. The series editors and Springer-Verlag will normally not interfere with the detailed editing except in fairly obvious cases or on technical matters.
Final acceptance is expressed by the series editor in charge, in consultation with Springer-Verlag only after receiving the complete manuscript. It might help to send a copy of the authors' manuscripts in advance to the editor in charge to discuss possible revisions with him. As a general rule, the series editor will confirm his tentative acceptance if the final manuscript corresponds to the original concept discussed, if the quality of the contribution meets the requirements of the series, and if the final size of the manuscript does not greatly exceed the number of pages originally agreed upon. The manuscript should be forwarded to Springer-Verlag shortly after the meeting. In cases of extreme delay (more than six months after the conference) the series editors will check once more the timeliness of the papers. Therefore, the volume's editor(s) should establish strict deadlines, or collect the articles during the conference and have them revised on the spot. If a delay is unavoidable, one should encourage the authors to update their contributions if appropriate. The editors of proceedings are strongly advised to inform contributors about these points at an early stage.
The final manuscript should contain a table of contents and an informative introduction accessible also to readers not particularly familiar with the topic of the conference. The contributions should be in English. The volume's editor(s) should check the contributions for the correct use of language. At Springer-Verlag only the prefaces will be checked by a copy-editor for language and style. Grave linguistic or technical shortcomings may lead to the rejection of contributions by the series editors. A conference report should not exceed a total of 500 pages. Keeping the size within this bound should be achieved by a stricter selection of articles and not by imposing an upper limit to the length of the individual papers. Editors receive jointly 30 complimentary copies of their book. They are entitled to purchase further copies of their book at a reduced rate. As a rule no reprints of individual contributions can be supplied. No royalty is paid on Lecture Notes in Physics volumes. Commitment to publish is made by letter of interest rather than by signing a formal contract. Springer-Verlag secures the copyright for each volume.

## The Production Process

The books are hardbound, and the publisher will select quality paper appropriate to the needs of the author(s). Publication time is about ten weeks. More than twenty years of experience guarantee authors the best possible service. To reach the goal of rapid publication at a low price the technique of photographic reproduction from a camera-ready manuscript was chosen. This process shifts the main responsibility for the technical quality considerably from the publisher to the authors. We therefore urge all authors and editors of proceedings to observe very carefully the essentials for the preparation of camera-ready manuscripts, which we will supply on request. This applies especially to the quality of figures and halftones submitted for publication. In addition, it might be useful to look at some of the volumes already published. As a special service, we offer free of charge LaTeX and TeX macro packages to format the text according to Springer-Verlag's quality requirements. We strongly recommend that you make use of this offer, since the result will be a book of considerably improved technical quality. To avoid mistakes and time-consuming correspondence during the production period the conference editors should request special instructions from the publisher well before the beginning of the conference. Manuscripts not meeting the technical standard of the series will have to be returned for improvement.

For further information please contact Springer-Verlag, Physics Editorial Department II, Tiergartenstrasse 17, D-69121 Heidelberg, Germany

Pierre-Antoine Bois Emmanuel Dériat
Renée Gatignol Alain Rigolot (Eds.)

# Asymptotic Modelling in Fluid Mechanics

Proceedings of a Symposium in Honour of
Professor Jean-Pierre Guiraud
Held at the Université Pierre et Marie Curie,
Paris, France, 20-22 April 1994

Springer

**Editors**

Pierre-Antoine Bois
Bât M3, Cité Scientifique
Université de Lille I
F-59655 Villeneuve d'Ascq, France

Emmanuel Dériat
ONERA, Direction de l'Aérodynamique
29, ave. de la Division Leclerc, BP 72
F-92322 Châtillon Cedex, France

Renée Gatignol
Alain Rigolot
Laboratoire de Modélisation en Mécanique
Université Pierre et Marie Curie, Tour 66
4 Place Jussieu
F-75252 Paris Cedex 05, France

ISBN 978-3-662-13998-1 ISBN 978-3-540-49265-8 (eBook)
DOI 10.1007/978-3-540-49265-8

CIP data applied for


Originally published by Springer-Verlag Berlin Heidelberg New York in 1995
Softcover reprint of the hardcover 1st edition 1995

Typesetting: Camera-ready by the authors
SPIN: 10481038 55/3142-543210 - Printed on acid-free paper

## Preface

From the 20th to the 22nd of April 1994, an international conference was held at the University Pierre et Marie Curie (Paris VI) in honour of Professor Jean-Pierre Guiraud. This symposium, which was organised by some of his former students and his colleagues of the Laboratoire de Modélisation en Mécanique, was devoted to asymptotic modelling in fluid mechanics, a field in which he has worked all his life.

There were more than 100 participants and fourteen lectures and thirteen communications were given by scientists from six different countries.

This volume consists of contributions covering four fields addressed at the symposium: vorticity and turbulence, hydrodynamic instability, non-linear waves, and aerodynamics and rarefied gas flows. An introductory paper, *"Growing up with Asymptotics"* resumes the opening conference by Professor Milton Van Dyke. A concluding paper , *"Going on with Asymptotics"*, has been written by Jean-Pierre Guiraud for this book. This paper is aimed at giving didactic insights into other topics pertaining to asymptotics, explaining the role of asymptotic methods and asymptotic modelling in applied sciences, as well as their interest for the research of today.

We gratefully acknowledge the support from our sponsors which made this symposium possible: Ministère de l'Enseignement Supérieur et de la Recherche, Centre National de la Recherche Scientifique, Université Pierre et Marie Curie, U.F.R. de Mécanique de Paris VI, Centre Européen Luc Gauthier - CSAMI, Centre National d'Etudes Spatiales, Commissariat à l'Energie Atomique (D.A.M.), Délégation Générale à l'Armement (D.R.E.T.), Electricité de France, Gaz de France, G.I.A.T. Industries (especially C. François).

Finally, we express our gratitude to Professor Beiglböck and Springer-Verlag for publishing this volume.

P.A. Bois
E. Dériat
R. Gatignol
A. Rigolot

Paris,
December 1994

# DE LA MECANIQUE ET DES HOMMES

Paul GERMAIN

Académie des Sciences

Ce livre présente un ensemble de mises au point et de contributions originales autour du thème de la modélisation asymptotique, c'est-à-dire de "l'art de manier les techniques de perturbation, lorsqu'elles ne sont pas uniformément valables, pour poser les problèmes les plus simples possible sous forme d'un puzzle dont on sait assembler les morceaux". L'ouvrage consacré à l'un des outils les plus performants pour l'attaque des problèmes non linéaires est le fruit de la collaboration d'éminents spécialistes de mécanique des fluides venus à Paris pour confronter leurs vues et leurs résultats autour de Jean-Pierre Guiraud, l'auteur de la définition de la modélisation asymptotique rappelée plus haut, au moment où ce dernier décidait de cesser ses activités de Professeur. Un livre tel que celui-ci est le plus bel hommage et le plus beau cadeau que peuvent donner les scientifiques à l'un d'entre eux qui, par son oeuvre et par son rayonnement, a très efficacement contribué aux progrès de sa discipline. Hommage discret où la personne que l'on honore n'est pas célébrée par des discours mais par toute l'ouverture à la créativité des chercheurs dans le champ où elle a travaillé. Cadeau d'autant plus précieux que la synthèse ainsi offerte, parce qu'elle représente un très bon état de la question et qu'elle trace des perspectives prometteuses, recevra un accueil qui prouvera son utilité. Qu'il me soit permis d'exprimer ma conviction qu'il en sera bien ainsi avec cet ouvrage.

Le progrès des connaissances scientifiques est oeuvre collective et la personnalité des savants, même des plus grands, n'a en définitive qu'un intérêt anecdotique. Sans Jean-Sébastien Bach, nous n'aurions pas les Concertos Brandebourgeois. Mais le dynamisme scientifique aurait tôt ou tard découvert la gravitation universelle, sans Newton, et la relativité sans Einstein. Il n'en reste pas moins que les connaissances scientifiques sont élaborées par des hommes - j'entends naturellement ici des hommes et des femmes - et que, pour eux, l'anecdote c'est leur vie.

Le lecteur découvrira dans la remarquable post-face qui clôt cet ouvrage, la personnalité du chercheur qu'est Jean-Pierre Guiraud, sa soif d'apprendre, l'étendue extraordinaire de ses connaissances, sa passion de repenser par lui-même, son enthousiasme, ses vues sur l'avenir, bref, son étonnante et merveilleuse jeunesse intellectuelle. Notre ami Milton Van Dyke dans la conférence d'ouverture du colloque a bien mis en évidence comment Jean-Pierre Guiraud a très tôt accompagné, développé, enrichi ce domaine de "l'asymptotique" dont il est devenu un maître. Ces deux textes donnent une belle image du savant que nous voulions honorer. Pour répondre à l'aimable proposition des organisateurs du colloque m'invitant à écrire quelques mots en guise d'avant-propos, il m'a paru préférable de chercher à mettre en évidence quelques traits de son histoire personnelle et de son caractère plutôt que d'analyser la qualité de l'ensemble de son oeuvre et de ses découvertes, à faire entrevoir l'homme, sa profonde originalité, son parcours atypique, ses choix particuliers d'existence. Je le fais en pensant très spécialement aux jeunes qui s'interrogeraient sur l'éventualité d'entreprendre une carrière de chercheur. Il y a bien des voies pour s'y lancer, bien des manières de la conduire. Si la recherche est à bien des égards une entreprise collective, on peut la vivre dans le respect de ses choix d'existence et dans la fidélité à ses valeurs.

C'est vers la fin de l'année 1955 que j'ai vu Jean-Pierre Guiraud pour la première fois. J'étais collaborateur extérieur à l'ONERA et la direction avait bien voulu me confier l'animation d'une petite équipe de jeunes chercheurs qui jouissait d'une autonomie pratiquement totale dans la conduite de ses travaux au sein de la division d'aérodynamique théorique. Jean-Pierre était candidat pour occuper un poste d'ingénieur à l'Office. Il m'était adressé par le directeur du personnel qui l'avait accueilli et l'avait interrogé et il venait de rencontrer le directeur technique Robert Legendre. Mon équipe ne disposait d'aucun poste libre et je me demandais pourquoi il me fallait recevoir un candidat destiné à un autre service.

J'ai découvert un Ingénieur des Mines de Paris, qui à vrai dire, ne semblait pas avoir eu une activité professionnelle bien nette durant les quatre années qui s'étaient écoulées depuis sa sortie de l'école. Mais il avait suivi de nombreux cours à la Sorbonne. Il avait lu de bons traités - en particulier les tomes du cours d'analyse de Goursat. En parlant avec lui, en l'interrogeant sur ses lectures et sur les matières qu'il avait étudiées, j'ai été très vite impressionné par la qualité de ses connaissances et la profondeur de sa compréhension. Je lui racontais ce que nous cherchions à faire avec mes jeunes collaborateurs et dans quel esprit nous travaillions. J'ai senti qu'il était très intéressé. Dès son départ, j'ai fait savoir que je voulais absolument que Jean-Pierre Guiraud soit embauché et affecté dans ma petite équipe. J'ai demandé avec insistance à être reçu par le Directeur Général Maurice Roy. Ce dernier accepta. J'ai eu rapidement l'impression que j'aurais gain de cause, mais Maurice Roy prenait plaisir, me semble-t-il, à énoncer des objections pour stimuler mon plaidoyer. La réussite de l'entreprise est, j'en suis convaincu, la plus belle contribution que j'ai pu faire au développement de la mécanique des fluides française.

Comment ne pas évoquer avec émotion les six années où nous avons travaillé ensemble. Nous partagions le même bureau. J'y venais une fois ou deux par semaine. Nous échangions nos idées, nos petites découvertes, celles que nous avions faites dans nos lectures, celles qui résultaient de nos recherches. Jean-Pierre avait une capacité de travail extraordinaire. Je m'employais à tempérer son ardeur car j'avais vraiment peur qu'il ne tombe malade. Six années durant lesquelles nous avons appris ensemble, vibré ensemble et où s'est forgée notre commune vision de la discipline à laquelle nous étions résolus à vouer notre existence. Une vision nourrie de réflexions stimulées par des lectures en provenance de l'étranger qui était souvent en avance sur nous. Nous cherchions à repenser par nous-mêmes nos savoirs, à les restructurer pour essayer d'aller plus loin. Si j'étais au début le meneur de ces échanges, vu la différence d'âge, les rôles se sont rapidement inversés. Mais je suis fier d'avoir fait partager à Jean-Pierre, mon ambition d'entraîner d'autres chercheurs pour apporter au concert de la mécanique mondiale, une contribution de qualité tout imprégnée de la tradition culturelle française.

Cette période où nous vivions quotidiennement ensemble a été brusquement interrompue par un tournant dans ma carrière qui m'obligea à consacrer une partie notable de mes activités à d'autres objectifs que la mécanique. Nos destinées se sont séparées; les occasions de discuter sur des questions de mécanique sont devenues rares, trop rares. Mais à chaque fois qu'il nous a été donné de pouvoir le faire, nous avons pu dialoguer avec la même facilité, la même soif de comprendre, le même plaisir à partager que par le passé; comme si nous reprenions une conversation interrompue. C'est sans doute que, pour nous deux, la mécanique a été plus qu'un métier, plus qu'un jeu passionnant; une composante essentielle de notre personnalité et de notre manière de voir le monde. Je suis sûr de n'être pas le seul à avoir vécu une amitié de cette nature et de cette qualité avec Jean-Pierre Guiraud; une amitié qui lui vaut la reconnaissance affectueuse des collègues et élèves qui ont eu le privilège d'en être bénéficiaire. Je ne suis ici que leur porte parole étant parmi eux le plus ancien.

Quelques mots brefs et discrets, en terminant, sur l'homme et son histoire. Jean-Pierre en raison des déplacements de sa famille et surtout de la guerre a eu une scolarité très mouvementée. Il a échappé à la formation rigide et bien encadrée que donne aux élèves notre système éducatif. Il a pu ainsi donner libre cours à ses goûts personnels, littérature, théâtre, cinéma, musique et aussi mathématiques; pour le plaisir et non pour réussir. Certains rêveront peut-être à ce qu'il aurait pu devenir avec des études plus régulières et mieux maîtrisées. Pour moi, je pense que c'est cette histoire hors du commun largement ouverte sur les aspirations de sa jeunesse qui lui a valu de préserver et de développer sa faculté de créativité, faculté souvent muselée par un système scolaire qui force les élèves dans un moule unique. Jean-Pierre est un être libre et qui se dévoue pour les autres, ses élèves, ses collègues, son laboratoire, sa discipline. Mais c'est aussi un être fragile. Il a des convictions mais n'est pas décidé à se battre pour les promouvoir. Il n'aime ni les responsabilités ni la notoriété. Il ne souhaitait pas que sa candidature soit présentée à l'Académie. C'est un sage respecté et aimé comme tel. L'estime de ceux qui le connaissent, et ce livre en est un témoignage, est sans doute, ce qu'il apprécie le plus.

Il a eu la bonne fortune de rencontrer Denise qui travaillait dans notre petite équipe. Elle l'a bien aidé à surmonter sa fragilité. Si en dépit de toutes les difficultés auxquelles ils ont dû faire face, Jean-Pierre a pu tenir pour accomplir l'oeuvre à laquelle ce livre rend hommage, il le doit certes à ses dons personnels, mais aussi à la présence attentive et compréhensive de Denise. Qu'elle veuille donc bien partager avec lui notre amicale reconnaissance.

Paul Germain

# Contents

## Growing up with Asymptotics

## Part III Non-Linear Waves



## Part IV Aerodynamics and Rarefied Gas Flows



## Going on with Asymptotics

# Growing up with Asymptotics

# Growing up with Asymptotics

Milton Van Dyke

Departments of Mechanical Engineering and of Aeronautics & Astronautics,
Stanford University, Stanford, California 94305-4040, USA

## Introduction: A generation of theoretical fluid mechanicians

When I was invited to help celebrate Jean-Pierre's career, I proposed to "view his work against the development of asymptotic methods in fluid mechanics." But with his typical modesty, Jean-Pierre protested that this colloquium should not, to any extent, be devoted to his own work; and after discussion with him, Prof. Rigolot suggested that I treat my own experience with asymptotic methods in fluid mechanics.

So I decided to do both. I realized that there is a generation of fluid mechanicians, born in the 1920s and 1930s, who may be said to have grown up with asymptotics—who have spent a good part of their careers exploring the mysteries of regular and singular perturbations, across the spectrums of Reynolds number and Mach number. Jean-Pierre is an outstanding representative from France, I am a delegate from the United States, and in this audience are members of this fraternity from a number of European countries.

Thus I propose to survey the progress that we have made—Jean-Pierre and I, and the rest of this community—and to illustrate it with examples of his work and mine, as representative of the research that we have all carried out.

## Regular perturbations

We all started with regular perturbations—the simplest sort of asymptotic expansions, which remain uniformly valid throughout the region of interest. Guiraud, for example, in some of his first papers, published in *Comptes Rendus* , perturbed the simple solution for supersonic flow past a wedge. Thus he treated nearby shapes—first in the approximation of hypersonic small-disturbance theory (Guiraud 1957a) and later for any supersonic flow (Guiraud 1957b).

He used a more unusual perturbation to estimate the nature of the singularity at the vertex of a delta wing in subsonic flow. Germain had shown that the velocity potential is a homogeneous function of the space coordinates of degree $\nu$ (and the velocity and pressure homogeneous of degree $\nu$ -1); and in 1956 Robert Legendre calculated the degree of homogeneity asymptotically for slight sweep of the leading edge, while Guiraud did the same for the slender wing, of high sweep, obtaining

$$\nu \sim 1 - (1 - \cos\gamma)/2 + \ldots \quad = 1 - \gamma^2/4 + \ldots$$

where $\gamma$ is the semi-vertex angle (Guiraud 1956). The first form, which is that of Guiraud, happens to yield the correct value $\nu = 1/2$ in the limit of the straight trailing edge ($\gamma = \pi/2$), but that is only a coincidence. Thirteen years later Brown & Stewartson 1969) calculated the next term in the second form of this expansion, and found that it contains the fourth power multiplied by the logarithm of $\gamma$. This probably means that the series is *purely asymptotic* , with zero radius of convergence. Over the years we have learned that logarithms of the perturbation quantity often arise unexpectedly as in this example.

For my part, my Ph.D. thesis at Caltech was devoted to extending the rather well-developed linearized theory of supersonic flow to a second approximation (Van Dyke 1952). I was delighted to discover (late one night) that for axisymmetric as well as plane flow a particular integral of the nonhomogeneous iteration equation can be formed from the first approximation. Thus the second-order problem is reduced to an equivalent first-order one.

Guiraud, too, soon carried regular perturbations to higher than first order. In a paper presented at a meeting in Bristol (Guiraud 1959), he systematized the so-called newtonian theory of hypersonic flow past a body of arbitrary shape as a double expansion carried to second order in powers of both $\gamma - 1$ (where $\gamma$ is now the adiabatic exponent of the gas) and $1/M^2$ and—for the special case of the inclined circular cone—a triple expansion to second order also in the angle of attack.

After my thesis, I became infatuated with second approximations in powers of thickness, calculating subsonic as well as supersonic flow past airfoils and slender bodies of revolution, and supersonic flow past slowly oscillating airfoils (Van Dyke 1953, 1955, 1959).

## Singular perturbations

All these regular perturbations ignore viscosity; they involve Mach number but not Reynolds number. The reason is partly historical: the 1950s were a decade of rapidly increasing speed of human flight, following the breaking of the "sound barrier" in 1947. But asymptotics also played a part.

Prandtl taught us in 1904 that flow at high Reynolds number is a singular perturbation problem. But it was only in the late 1950s that Kaplun & Lagerstrom and Proudman & Pearson taught us that the other extreme of vanishing Reynolds number also leads to nonuniformities, and gave us general techniques for handling those and many other singular perturbations.

My beloved mentor, Paco Lagerstrom insisted that there is no such thing as a singular perturbation *problem* , there are only singular perturbation *solutions*.. [However, he violated this rule in the very first sentence of his book (Lagerstrom 1988).] For whether an asymptotic expansion is uniformly valid or not depends on how the variables are chosen. This is strikingly illustrated, in the most familiar example of singular perturbations, by Saul Kaplun's discovery, for Prandtl's boundary layer, of what he termed *optimal coordinates* (Kaplun 1954). A single regular expansion holds uniformly throughout both the viscous layer and the outer inviscid flow when expressed in terms of these coordinates. Unfortunately, except when they can be cleverly guessed (for example, as parabolic coordinates for the standard problem of the semi-infinite plate), optimal coordinates can be found only by first solving separately the boundary layer and the outer flow using any convenient (doubtless non-optimal) coordinates.

Can we predict a singular perturbation? We cannot in a purely mathematical problem; but dimensional reasoning can be invoked in a physical problem. Suppose the perturbation quantity is the ratio of two lengths (geometric or not), say $\varepsilon = s/L$. Then the conventional procedure of introducing dimensionless coordinates leads to a dilemma: for example, should we refer $x$ to $s$ or to $L$? The answer is that only exceptionally (as with optimal coordinates) will either choice alone suffice. In general we will need both scalings in adjacent regions, or perhaps in the same one. Thus we conclude with certainty that

> An asymptotic expansion will be regular if the perturbation quantity is not the ratio of two lengths;

and we can conclude with near certainty that

> An asymptotic expansion will probably be singular if the perturbation quantity is the ratio of two lengths.

(We could write, "the ratio of two lengths or of two times," but every physical problem has at least one space coordinate, whereas only an unsteady problem has a time coordinate.)

The lengths need not be geometric ones. Indeed, one of the most familiar of perturbation quantities, the Reynolds number $R = UL/\nu$, can be interpreted as the ratio of a geometric length $L$ to a viscous length $\nu/U$. At low Reynolds numbers, distances must be referred to a body dimension $L$ near the body and the viscous length $\nu/U$ far away, whereas at high Reynolds number they are referred to $L$ far away and to the combination $(L\,\nu/U)^{1/2}$ near the surface—the boundary-layer scale.

On the other hand, the other similarity parameter most familiar to fluid mechanicians, the Mach number, is the ratio of two speeds—and its square (in which form it often appears) is the ratio of two energies—but in no reasonable way can it be regarded as the ratio of two lengths. Thus an asymptotic expansion for large or small Mach number is assuredly a regular perturbation (the latter being the "Janzen-Rayleigh" or "$M^2$" expansion).

Unfortunately, however, most perturbation quantities are ratios of two lengths, so most asymptotic expansions in mechanics are singular perturbations. For example, Guiraud's analysis of the slender delta wing appears to be a regular perturbation; but the perturbation quantity, the semi-vertex angle $\gamma$, is clearly the ratio of two lengths; and in higher approximations Brown & Stewartson encountered a singular perturbation.

How do we deal with singular perturbations? As we grew up in the 1950s we were offered a confusing choice of techniques, each with a variety of names. But with time we have settled upon two principal methods, with accepted names. The choice between them depends upon the complexity of the problem.

In the simpler category of problems, one scaling of dimensionless coordinates or the other (e.g. $x/s$ or $x/L$) is dominant in each of two (or more) adjacent regions. Then the appropriate technique is the generalization of Prandtl's boundary-layer idea that was previously known by such names as "the method of inner and outer expansions," but is now generally called the *method of matched asymptotic expansions.*

In the more complicated category of problems, both scalings are operative everywhere. Then the appropriate technique is what was once even facetiously called "two-timing," but is now generally known as *the method of multiple scales.* Our description suggests that multiple scales is the more general technique—that any problem that can be solved by matched expansions can be solved by multiple scales, but not conversely—and this seems to be confirmed by experience.

## Matched asymptotic expansions

The boundary-layer idea—or the method of matched asymptotic expansions—has its roots in the nineteenth century, beginning apparently with Laplace. In 1805, in his *Mécanique Céleste* , he calculated the asymptotic shape of a large drop of mercury resting on a horizontal glass plate. Balancing the forces of gravity and surface tension yields a nonlinear ordinary differential equation. It involves a single dimensionless parameter that is the ratio of two lengths, the cube root of the volume of mercury and the square root of the ratio of surface tension to density times gravity. When that ratio is large, the drop is spread thin. The governing equation can therefore be linearized over most of the surface, where the slope is small, to give Bessel's equation. That approximation fails in a narrow "boundary layer" near the rim; but there the problem is approximately plane, and the solution is known in closed form. Laplace matched those two results to complete the first approximation. It is interesting to observe that Lord Rayleigh later obtained the same result independently, and upon learning that Laplace had anticipated him by 110 years, hastily carried out the second approximation to lend some novelty to his paper.

In the last third of the nineteenth century the same idea was applied to problems in elasticity, electrostatics, acoustics, and viscometry by such giants as Maxwell, Helmholtz, Kirchhoff, Rayleigh, and Hertz (Van Dyke 1994). However, it remained for Prandtl, at the beginning of our century, to systematically develop and popularize the idea as applied to the laminar boundary layer. And in the middle of our century profound study of the idea and its development as a general technique were carried out by Kaplun and Lagerstrom.

It was evidently from them that Guiraud learned of the method, for he cited a paper of Kaplun & Lagerstrom (1957) when, as early as 1958—before he spent six months in California—he first applied the method of matched asymptotic expansions to hypersonic flow past a blunt-nosed plate (Guiraud 1958). In contrast to all the expansions mentioned previously for small values of a parameter, this is a *coordinate expansion* , for distances downstream large compared with the thickness of the nose. An entropy layer next to the surface consists of fluid that has passed through the steep part of the detached bow wave. It is matched in subtle fashion to an outer layer that is governed by hypersonic small-disturbance theory. In a series of papers continuing to 1964, Guiraud refined his analysis and proceeded to the second approximation, and so clarified the roles of blast-wave theory and the newtonian approximation.

One of these papers, written with Guiraud's future wife, Denise (Vallée & Guiraud 1963) introduced the terms "proximal" and "distal," which I much prefer to the historical "inner" and "outer." The latter lead to the ridiculous situation that in Laplace's drop of mercury the "outer" region is encircled by the "inner" one! I have tentatively proposed translating "proximal" and "distal" somewhat imprecisely as "local" and "global"; but I am open to suggestion.

In a very different problem Guiraud (1965) applied matched expansions to the focusing, when it approaches a caustic, of the N-wave from a supersonic airplane. A novel feature is the need for three different scalings in three successively smaller regions. In a felicitous phrase, Guiraud describes his objective as to study the innermost region with the aid of a *mathematical microscope.* Three different scalings arise also in his study (Guiraud 1974) of separation at the trailing edge of a thin three-dimensional wing, where the more familiar layered structure bears Stewartson's name of the *triple deck*.

I, like Guiraud, was fascinated by the breakdown of asymptotic theories at leading and trailing edges. Before I learned the formalism of matched expansions, I followed its spirit in deducing simple rules for correcting edges (Van Dyke 1954). For example, the first-order thin-airfoil theory at a round edge is corrected by multiplying it by the ratio of the exact solution for the osculating parabola to its thin-airfoil expansion.

After I learned to match properly, I devoted perhaps more years than I should have to second-order boundary-layer theory (Van Dyke 1962). Another application of matched expansions that I enjoyed was to the lift of a low-speed wing of high aspect ratio (Van Dyke 1964). I was surprised to learn that Prandtl did not realize that his lifting-line theory, like his viscous boundary layer, is a singular perturbation.

## Multiple scales

The method of matched asymptotic expansions fails when several scalings are operative concurrently, and one must then turn to the method of multiple scales. This consists simply in introducing correspondingly disparate coordinates in the relevant direction (e.g. $x_1 = x$, $x_2 = \varepsilon x$), solving asymptotically for small $\varepsilon$ as if they were truly independent variables, and only at the end exploiting the freedom that they provide to suppress nonuniformity. This seems at first sight a retrograde procedure, since we usually strive to reduce, rather than increase, the number of independent variables (for example, by separation of variables); but in fact the longer (or slower) coordinate does not complicate the solution, because it appears only parametrically.

Despite its greater generality, the method of multiple scales is used less often in fluid mechanics than matched asymptotic expansions. I myself have never used it at all. It appears that most singular perturbations in fluid mechanics are of boundary-layer type, with disparate dominant scales in adjacent regions. The simultaneous dominance of two or more scales arises more often in problems of dynamics, for example with two different time scales in the slow damping of a rapid oscillator. In fluid mechanics, however, Guiraud, working with Zeytounian, has applied multiple scales to two quite different problems, where it is so clearly the appropriate method that it is mentioned in their titles. First, in the theory of axial-flow machinery we have (Guiraud & Zeytounian 1974):

> "Application du concept d'échelles multiples à l'écoulement dans une turbomachine axiale".

Clearly the long scale is that of the mean flow, and the short scale that of the individual blades, the small parameter being the reciprocal of the number of blades or of stages. Second (Guiraud & Zeytounian 1977) we have:

> "A double scale investigation of the asymptotic structure of rolled-up vortex sheets".

This applies to the rolled-up vortex sheets from the trailing edge of a high-aspect-ratio wing or from the leading edges of a delta wing, the small parameter being the number of turns.

## Strained coordinates

In the last forty years we have thus learned to choose between two standard methods for handling singular perturbations. The methods of matched asymptotic expansions and of multiple scales both seem to be reliable, in the sense that the process can be continued to yield a uniformly valid approximation to any order.

Most of the rival methods that were suggested during this period have been forgotten, but one lingers on. A few years before the Caltech group developed matching, Lighthill (1949) proposed what he called "a technique for rendering approximate solutions to physical problems uniformly valid." Because it was a generalization of an older technique due to Poincaré, and was applied (incompletely) to laminar flow past a finite flat plate by Kuo, Tsien (1956) called it the *Poincaré-Lighthill-Kuo* or *PLK method.* However, it is generally known today as the *method of strained coordinates.*

That name describes the essential idea, which is to expand the independent as well as the dependent variables, and take advantage of the additional freedom to suppress nonuniformity. The process can be simplified (Pritulo 1962) by expanding the solution rather than the problem. The method has not proved reliable, however. Lighthill (1961) himself advised against using it for elliptic partial differential equations; and Crocco (1972) found that even for hyperbolic equations the third approximation can be rendered uniform only by appealing also to the idea of multiple scales. Other objections have been raised, so that the method remains suspect.

Again I have never used strained coordinates; but Guiraud referred to the method twice during the early 1960s. In particular, in their detailed study of bluntness effects in hypersonic small-disturbance theory, Guiraud, Vallée & Zolver (1965) summarized its use by our Russian colleague Sychev. However, they express misgivings as to its generality, and regard the method of matched asymptotic expansions as superior.

## Asymptotics allied with numerics

Eight years ago, Guiraud and Zeytounian (1986) argued persuasively that although asymptotic techniques will in special circumstances continue to be used to derive closed-form solutions, their future role will be primarily as an adjunct to numerical simulation. They illustrated this view with a number of examples from fluid mechanics, concluding with several drawn from their own research.

From casual observation I suggest that the number of papers appearing in technical journals with the word "asymptotic" in their titles is actually increasing steadily. On the other hand, the cooperation between experts on asymptotics and numerics—which I agree is desirable and inevitable—seems not yet to have gained momentum.

Meanwhile, I have for some years been experimenting with a quite different alliance between asymptotics and the computer. This is the three-step process of first extending a regular perturbation series to high order by delegating the algebra to the computer, second analyzing the coefficients to unveil the structure of the solution in the complex plane of the perturbation quantity, and third on that basis improving the utility of the series (Van Dyke 1984). This method has had some undisputed successes; but in several problems we have been guilty of too rash extrapolation. Perhaps for this reason the method has not been wholeheartedly embraced by our colleagues. But we feel that time is on our side: as computers grow steadily more powerful, they will provide ample terms for reliable analysis.

## References


Brown, S.N. & Stewartson, K. (1969): "Flow near the apex of a plane delta wing", *J. Inst. Maths. Applics.* **5**, 206-216.

Crocco, L. (1972) : "Coordinate perturbation and multiple scale in gas dynamics", *Philos. Trans. Roy. Soc. London A* **272**,275-301.

Guiraud, J.-P. (1956): "Sur la nature de la singularité d'un écoulement de fluide compressible au voisinage de la pointe avant d'une aile delta en régime subsonique", *Comptes Rendus* **243**,2012-2014.

Guiraud, J.-P. (1957a): "Écoulements hypersoniques infiniment voisins de l'écoulement sur un dièdre", *Comptes Rendus* **244**,2281-2284.

Guiraud, J.-P. (1957b): "Écoulements supersoniques bidimensionnels derrière une onde de choc attachée", *Comptes Rendus* **245**,2474-2476.

Guiraud, J.-P. (1958): "Écoulement hypersonique d'un fluide parfait sur une plaque plane comportant un bord d'attaque d'épaisseur finie", *Comptes Rendus* **246**,2842-2844.

Guiraud, J.-P. (1959): "Newtonian flow over a surface—theory and application" in Proc. 11th Symp. Colston Res. Soc., (Bristol), 253-300.

Guiraud, J.-P. (1965): "Acoustique géométrique, bruit ballistique des avions supersoniques et focalisation", *J. de Mécanique* **4**,215-267.

Guiraud, J.-P. (1974): "Écoulement décollé au voisinage du bord de fuite d'une aile mince tridimensionnelle", *J. de Mécanique* **13**,409-432.

Guiraud, J.P., Vallée, D. & Zolver, R. (1965): "Bluntness effects in hypersonic small disturbance theory" in *Basic Developments in Fluid Dynamics,* ed. by M. Holt (Academic Press) 127-247.

Guiraud, J.-P. & Zeytounian, R. Kh. (1974): "Application du concept d'échelles multiples à l'écoulement dans une turbomachine axiale", *Int. J. Eng. Sci.* **12**,311-330.

Guiraud, J.P. & Zeytounian, R. Kh. (1977): "A double-scale investigation of the asymptotic structure of rolled-up vortex sheets", *J. Fluid Mech.* **79**,93-112.

Guiraud, J.-P. & Zeytounian, R. Kh. (1986): "Asymptotic modelling of fluid flows. General features", *J. Méc. Théor. Appl.* **Numéro spécial**, 1-24.

Kaplun, S. (1954): "The role of coordinates in boundary-layer theory", *Z. Angew. Math. Phys.* **5**,111-135.

Kaplun, S. & Lagerstrom, P.A. (1957): "Asymptotic expansions of Navier-Stokes solutions for small Reynolds numbers", *J. Math. Mech.* **6**,585-593.

Lagerstrom, P.A. (1988): *Matched Asymptotic Expansions* (Springer).

Lighthill, M.J. (1949): "A technique for rendering approximate solutions to physical problems uniformly valid", *Philos. Mag.* 7, **40**,1179-1201.

Lighthill, M.J. (1961): "A technique for rendering approximate solutions to physical problems uniformly valid", *Z. Flugwiss.* **9**,267-275.

Pritulo, M.F. (1962): "On the determination of uniformly accurate solutions of differential equations by the method of perturbation of coordinates", *J. Appl. Math. Mech.* **26**,661-667.

Tsien, H.S. (1956): "The Poincaré-Lighthill-Kuo method", *Advances in Applied Mechanics* **4**,281-349.
Vallée, D. & Guiraud, J.-P. (1963): "Écoulement hypersonique asymptotique d'un gaz parfait non visqueux non conducteur de la chaleur sur une plaque plane émoussée symétrique sans incidence", *Comptes Rendus* **256**,1214-1216.
Van Dyke, M.D. (1952): *A study of second-order supersonic flow theory* in NACA Tech. Rept. 1081.
Van Dyke, M.D. (1953): "On second-order supersonic flow past a slowly oscillating airfoil", *J. Aeronaut. Sci.* **20**,61.
Van Dyke, M.D. (1954): *Subsonic edges in thin-wing and slender-body theory* in NACA Tech. Note 3343.
Van Dyke, M.D. (1955): *Second-order subsonic airfoil-section theory and its practical application* in NACA Tech. Note 3390.
Van Dyke, M.D. (1959): *Second-order slender-body theory—axisymmetric flow* in NACA Report R-37.
Van Dyke, M.D. (1962): "Higher approximations in boundary-layer theory", *J. Fluid Mech.* **14**,161-177, 481-495.
Van Dyke, M.D. (1964): "Lifting-line theory as a singular perturbation problem", *Appl. Math. Mech.* **28**,90-101.
Van Dyke, M.D. (1984): "Computer-extended series" in Ann. Rev. Fluid Mech. **16**,287-309.
Van Dyke, M.D. (1994): "Nineteenth-century roots of the boundary-layer idea", *SIAM Rev.* **36**,415-424.

# Part I

# Vorticity and Turbulence

# Instabilities and Vortex Patterns in Circular Couette Flow with Axial Density Stratification

E. J. Hopfinger[1], B. M. Boubnov [1,2] and E. B. Gledzer [1,2]

[1] LEGI/IMG, CNRS–UJF–INPG,
B.P. 53, 38041 Grenoble Cedex 9, France
[2] Institute of Atmospheric Physics, 109017,Moscow,Russia

**Abstract:** The instability and transitions in flow regimes of circular Couette flow with axial density stratification, characterized by the buoyancy frequency $N$, has been investigated experimentally for different values of the non-dimensional gap width $\epsilon$ between the cylinders. It is shown that onset of instability and the structure of the resulting flow regimes, namely non-axisymmetric vortices, Taylor vortices and the related descrete layer formation strongly depend on $N$ in addition to Reynolds number and $\epsilon$. The linear stability analysis of this problem has been performed for the asymptotic case of $\epsilon \to 0$.

## 1 Introduction

One of the classical examples of hydrodynamic instability is the flow of fluid confined to the annulus between concentric, rotating cylinders. Since the pioneering works by Couette (1890) and Taylor (1923) a very large number of experimental and theoretical studies have considered different aspects of instability and transitions of this flow configuration (see for instance Di Prima & Swinney,1981; Andereck et al.,1986; Chossat & Iooss,1994).Various modifications of this problem have also received considerable attention and these include the influence of axial flow (Gravas & Martin,1978; Lueptov et al.,1992), unsteadyness of the rotation rate (Cooper et al.,1985) and the effect of radial temperature or density variations with and without an axial gravitational field (Snyder & Karlson,1964; Yao & Rogers,1989; Kubotani et al.,1989; Ali & Weidman,1990).

In the simplest case of Couette-Taylor flow, with only the inner cylinder rotating, a variety of different regimes is observed: Taylor vortices, wavy vortices, modulated wavy vortices, turbulent Taylor vortices. When additional effects are included (Coriolis force, velocity shear, radial density stratification etc.) significant changes in the stability occur and the flow states reveal a rich variety of phenomena. Andereck et al.,1985, for instance found more than fifteen principal flow regimes between independently rotating cylinders in homogeneous fluid. Surprisingly, the case of the interaction of centrifugal and buoyancy forces, namely, Couette flow with axial density stratification has not received much attention. The only previous theoretical consideration of this configuration is by Thorpe

(1966). He investigated, however, only the case of a small difference between the angular velocities of the two cylinders, necessary for the analogy of this problem with rotating Bénard convection (mentioned by Chandrasekhar (1961), see also Veronis (1970)).

The aim of the present study is to investigate experimentally and with the help of linear stability theory, the instabilities and transition regimes in Couette flow with axial, linear density statification for the case when only the inner cylinder rotates. It is shown that density stratification has a strong effect on the onset of instability and the resulting vortex structures, giving rise to layer formation. It is noteworthy to mention that descrete layers are formed.

The main non-dimensional parameters of the problem are the usual Reynolds number $Re = \Omega a(b-a)/\nu$ (see e.g. Andereck et al.,1985), the non-dimensional gap width $\epsilon = \frac{b-a}{a}$ ($a$ and $b$ are the radii of inner and outer cylinders respectively, $\Omega$ the rotation rate of the inner cylinder, and $\nu$ the viscosity which is taken constant, although, generally there exists a weak dependence of $\nu$ on density). The stratification, expressed by the Brünt-Väisälä frequency $N = \left(-\frac{g}{\rho}\frac{\partial\rho}{\partial z}\right)^{1/2}$, ($\rho$ is the density and $g$ the gravitational acceleration), introduces an additional non-dimensional parameter which is the Froude number $Fr = \Omega/N$. The stability analysis considers the asymptotic limite $\epsilon \to 0$ only, whereas experiments are performed for large values of $\epsilon$ aswell.

# 2 Experimental apparatus and procedure

The experimental installation for circular Couette flow in stratified fluid is similar to those which are used for studying Couette-Taylor flow in homogeneous fluid. It consists of long, coaxial transparent plexiglass cylinders with the outer cylinder, the bottom and upper surface being at rest and the inner cylinder rotating with constant angular velocity $\Omega$. The outer cylinder has an inner radius of $b = 51mm$, while three different inner cylinders with outer radii $a$=20, 30 and 40 mm were used, giving respectively non-dimensional gap widths $\epsilon$=1.55, 0.7 and 0.275. The last case of $\epsilon = 0.275$ permits to interpret the results within the small gap approximation theory and is still suitable for the study of the vortex and layer structures. The length of the cylinders $L$ was taken large enough: $L = 573mm$, giving $\Gamma = L/(b-a) = 52, 27$ and 18 for $\epsilon = 0.275, 0.7$ and 1.55 respectively. In most experiments the upper surface of the cylinder was rigid, and only in experiments with large density gradient, when $N > 1.5\,rad/s$, the upper boundary was free, because, in order to reach a large value of $N$ it was necessary to fill the space between the cylinders to a height less than $L$. In our experiments, stable density stratification protects against propagation of disturbances from the top and bottom surfaces to the fluid interior and there is practically no influence of end boundary conditions on the flow regimes (which is not the case in non-stratified fluid, where end boundary conditions can be very important in the onset of instability).

The inner cylinder rotation for a given $\Omega$ was maintained constant within 1% and the value of $\Omega$ could be changed continiously from 0.1 to $3.7 rad/s$. Constant cylinder rotation was reached in less than $2s$ and when $\Omega < \Omega_{cr}$ a stable, steady state flow was established in a time of about 1 minute. The time for onset of instability of the zonal axisymmetric flow depends how close the rotation rate is to critical conditions.

Generally, experiments were made by starting with the fluid rest and by setting the rotation of the inner cylinder at the desired value. Changing the rotation rate of this cylinder from one value to another was used only near onset of instability of the azimutal flow and also in special experiments devoted to studying the variations of the formed layer structures and transitions between different regimes.

The linear stratification was accomplished by using a salt solution and the standard "double-bucket" filling method. The time of filling was of the order of two hours. Normally, stratification measurements were made at four heights, but in some controlled experiments these measurement points were closer spaced. The linearity of the density profile was also verified with the help of shadowgraph, in which the outer surface of the inner cylinder appears as a line inclined at some angle to the vertical, proportional to the Brünt-Väisälä freguency $N$. The deviations of the density profiles from linearity in the main part of the fluid column ($0.1 < z/L < 0.9$, where $z$ is the vertical coordinate) were less then 5%. The value of $N$ was changed in the limits $0.35 < N < 1.78 rad/s$. After the experiment with one rotation rate $\Omega$ was finished and the formed layers were allowed to diffuse , the linearity of the main stratification was checked and a new experiment could be started.

In order to reduce optical distorsions due to the curvature of the cylinders, the apparatus was placed in a large square box filled with water (also there is water in the inner cylinder). For the flow visualizations two different and complementary techniques were used: a shadograph technique, sensitive to the second derivative of density which allows to see the horizontally everaged density structures, and particle streak line methods to visualize the vortex motion. Aluminium powder and kalliroscope particles were used in this case. The aluminium particle are almost isotropic and move with the fluid and show the motions in the vortices, while plane kalliroskope flakes outline the total structure of the flow.

# 3 Experimental Results

## 3.1 Narrow gap ($\epsilon$=0.275)

### 3.1.1 Flow regimes

The diagram of observed flow regimes for narrow gap in the ($\Omega$,$N$) parameter space is shown in Fig.1.

Here, we briefly describe the main regimes and transitions between them with more quantitative details being given later. For each value of $N$, the rotation rate $\Omega$ was increased by small increments until the critical value of $\Omega_c(N)$ or $Re_c(N) = \Omega_c(N)a(b-a)/\nu$ for onset of instability was reached. For $Re < Re_c(N)$ the flow has only an azimuthal component of velocity (circular Couette flow) with no changes in the density field along the vertical axis (regime A, Fig.1). This steady flow arises shortly after the beginning of rotation of the inner cylinder. For $\Omega \geq \Omega_c(N)$ on the other hand, the shadowgraph images indicate vertical variations in density with the appearance of well identifiable layers of nearly equal height. By this means it is possible to determine $\Omega_c$ as a function of $N$. In the neutral case of $N = 0$ the critical value of $\Omega$ (needed for comparison with the results of other studies and to test the theoretical value) was determined from the onset of the motion of particles suspended in the fluid. For the present experimental conditions

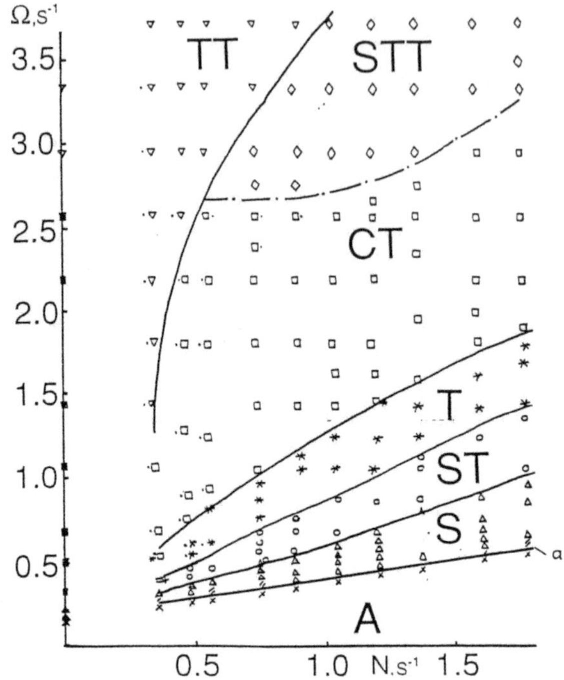


Figure 1: Main instability and flow regimes for stratified circular Couette flow ($\epsilon = 0.275$): ×, Azimuthal flow; ▲, Taylor vortices at $N = 0$; ■, wavy Taylor vortices at $N = 0$; △, $n_l = 1/2$, S regime, $N \neq 0$; /, no layers, $n = 1$; ∗, $n_l = 1$, T regime; ∘, transition between $n_l = 1/2$ and $n_l = 1$, ST regime; □, $n_l = 2$, CT regime; ◇, turbulent motion with $n_l = 2$, STT regime; ▽, turbulent mixing, TT regime.

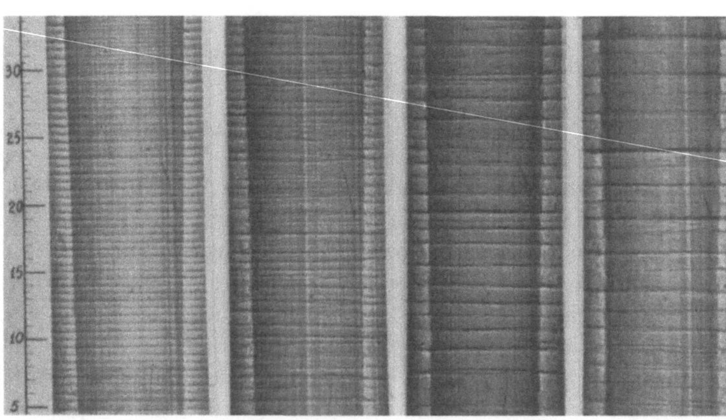


(a) $\Omega = 0.6\,\mathrm{s}^{-1}$ (b) $\Omega = 0.9\mathrm{s}^{-1}$ (c) $\Omega = 1.2\mathrm{s}^{-1}$ (d) $\Omega = 2.4\mathrm{s}^{-1}$

Figure 2: Shadowgraph visualizations of the change in layer height for different angular velocities $\Omega$ of the inner cylinder for $\varepsilon = 0.275, N = 1.2\mathrm{s}^{-1}$. (a), $h \approx (b-a)/2$; (c), $h \approx (b-a)$; (d), $h \approx 2(b-a)$; (b), transition from $h \approx (b-a)/2$ to $h \approx (b-a)$.

$\Omega_c(0) \simeq 0.2$, giving a Reynolds number $Re_c(0) = \Omega_c a(b-a)/\nu \simeq 90$, in agreement with the expected value (Di Prima & Swinney,1981). For $N = 0$ the instability of circular Couette flow gives rise to axisymmetric Taylor vortices of vertical size equal to the width of the gap.

For stratified Couette flow when $N$ is large enough ($N > 0.4$) and $\Omega$ just above $\Omega_c(N)$, the first unstable structures are of height approximately equal to one half of the gap width (region S in Fig.1), giving $n_l = \frac{h}{b-a} \approx \frac{1}{2}$, where $n_l$ is the non-dimensional layer height or layer height factor. The boundary of transition between regimes A and S, i.e. the neutral stability curve is curve $a$ in Fig.1. With increasing $\Omega$, the usual Taylor vorticies are also appearing. Steady Taylor vortices with $n_l = 1$ are observed for values of $\Omega$ and $N$ lying in region T, Fig.1. Between the flowstates S and T there exists a transition region ST in which both regimes with $n_l \approx 1/2$ and $n_l = 1$ are observed. When we move to larger $\Omega$ (for the same $N$) the Taylor vortices become more and more unsteady and neighbouring vortices begin to interact with each other. As a result, instead of wavy vortex regimes, observed in homogeneous fluid for large $\Omega$ (see Andereck et al.,1985 and square black symbols in Fig.1), layers of $n_l \approx 2$ arise with two compact Taylor vortices inside; this is region CT in Fig.1. From Fig.1 one can see that onset of instability and the boundaries between the main regimes are characterized by a Froude number $\Omega/N$ rather than by Reynolds number.

By further increasing $\Omega$, more complicated interactions between the vorticies may be observed (region STT in Fig.1). For small $N$ and not too large $\Omega$, transition to strong turbulent mixing occurs (regime TT) where it is very difficult to identify regular layers. It should also be remarked that for all regular regimes the formed layer structure is conserved during a long time after stopping the rotation of the cylinder (for example in regime T layers remained visible up to half an hour, corresponding to about 180 rotation periods). Shadowgraph images of the different layers corresponding to regimes S, ST, T and CT are shown in Fig.2 for $N = 1.2 rad/s^{-1}$.

Usually, instability of stratified, circular Couette flow is manifest (for these geometrical parameters) by the appearance of density layers on the shadograph images. The height $h$ of these layers strongly depends on the buoyancy frequency $N$ or, more precisely, on the Froude number $\Omega/N$. The value of $h$ varied between 5.5mm to 8.5mm (for a gap width $d = b - a = 11mm$). In Fig.3 the dependence of $h/d$ (defined near the onset of instability) is presented as a function $\Omega/N$. From this Figure it is clearly seen that when $\Omega/N \leq 0.5$ there is a well pronounced dependency of the layer height on Froude number and this dependency may be expressed by the empirical formula

$$n_l = \frac{h}{d} = C_1 \frac{\Omega}{N} = C_1 Fr, \quad C_1 \approx 1.38 \tag{1}$$

The vortices responsable for the layer formation of height $h$ have a vertical wave length $\lambda = 2h$ giving a wave length factor defined as $n = \lambda/2d$ equal to $h/d$. The vortex filament at onset of instability is, however, not of axisymmetric form. Such a non-axisymmetric motion at onset of instability of the azimuthal flow can be explained by considering that when stratification is important the vortices are of size less than the gap width and can, therefore, freely move in the horizontal plane. Since adjacent vortices are counter-rotating, vortex pairs tend to move at one point away from the inner cylinder and one vortex size away toward it. Such a row of vortices will not be stable and may give rise to oscillatians

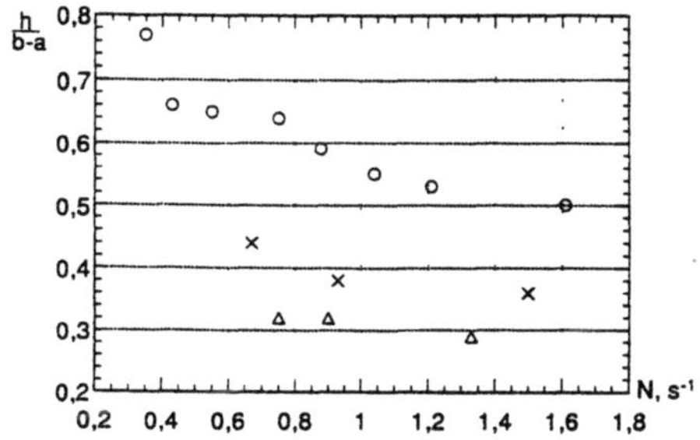


Figure 3: Non-dimensional minimum layer height as a function of $N$ for $\epsilon = 0.275(\circ)$, $0.7(\times)$, $1.55(\triangle)$

and be non-axisymmetric (a stable axisymmetric state would require vortices which are elongated in horizontal direction to fill the gap). As a result we can have wave motions in azimuthal direction with wave number $m$; in our case $m$=2.

The time it takes from the start of cylinder rotation until layers become visible depends on how close $\Omega$ is to the critical rotation rate. Very close to critical conditions ( $\Omega = 1.1\Omega_c$) a layered density structure becomes visible after about 20 rotation periods. Under these conditions the layers are, however, not very steady. For conditions shown in Fig.2a, corresponding to about 1.4 $\Omega_c$, there is more intensive mixing and very stable layers are established in about 30 rotation periods. On shadograph images such as shown in Fig.2a, the density interfaces (dark lines) are clearly at fixed position. The vortex structure in this regime is presented in Fig.4a.

The Froude number $Fr = \Omega/N$ is a measure of the relative importance of centrifugal to buoyancy forces. For the case when $Fr >> 1$ centrifugal force dominates the motion and the influence of stratification is a small additional effect. This fact is reflected in the flow in regimes TT and STT where the influence of stratification is negligible. The opposite is the case in regime S where the Froude number is less then 1. In this case stratification controls the size of vortex structure.

For some Froude numbers onset of instability occurs via weak vortices of size $h \approx d$ which cause no layer formation. The existence of these vortices, marked by inclined dashes (/) on Fig.1, may be explained as a result of some resonant interactions between buoyancy $N$ and rotation $\Omega$ frequencies: all of these points are near the points of intersection of stability curve ($a$) and lines $Fr = 1; 1/2$ and $1/3$.

When the Froude number is near unity or larger, the centrifugal effects dominate the instability mode and Taylor vortices appear. These vortices are toroidal and are of equal vertical and horizontal size and are equal to the gap between the cylinders. The corresponding vertical wave length is $\lambda = 2d$ giving $n = 1$. Taylor vortices exist in regime T (Fig.1) and the corresponding layer height factor is $n_l = 1$ (Fig. 2c). The time required to reach a steady state is about 30 rotation periods after beginning of inner cylinder rotation. Although vortices are formed in less then 5 rotation periods the system of vortices reaches a steady state only after some ajustment. Examples of

Figure 4: Kalliroscope particle visualizations for different regims ($N = 0.92\mathrm{s}^{-1}$): (a), $\Omega = 0.43s^{-1}$; (b), $\Omega = 0.92s^{-1}$; (c), $\Omega = 1.82s^{-1}$.

Taylor vortices, visualized by kalliroscopic particles and aluminium powder, are presented in Fig.4b, showing well developed vortices with very weak interaction between them. Neighbouring vortices have opposite rotation.

The main difference between the regimes with $n_l = 1$ and $n_l = 2$ (regime CT on Fig.1) is that the interfaces separating layers (dark lines on shadograph image) are in the T regime with $n_l = 1$ of approximately equal intensity (Fig.2c); for $n_l = 2$ on the other hand the intensity of every second interface is weaker (Fig.2d). This effect is even better displayed when the rotation of the inner cylinder is stopped and after some time only the stronger interfaces, separating layers with $n_l = 2$, remain. At later times these diffuse also. In the CT regime, Taylor vortices are also the main structures, but there are additional vortex interactions by pairs and a weaker mode $n = 2$ is present. The interaction is such that the centers of neighbouring vortices sometimes disappear and re-emerged, and it is also possible that one of the interacting vortices may even change the sense of rotation. In Fig.4c the interaction by pairs and some presence of mode $n = 2$ is indicated by the absence of kalliroscope particles between compact vortex pairs with mixing inside pairs. The particle streak pictures show, except for two compact Taylor vortices an apparent flow over a height equal to $2d$. These results may be interpreted as interactions between the main mode with $n = 1$ and a mode $n = 2$.

The transition and turbulent regimes are characterized by the absence of steady layers. Transition regimes are observed in some cases near the critical curve (a) at onset of instability, in the region between regimes with different $n_l$, in particular between $n_l \approx 1/2$ and $n_l = 1$ and also for large values of $\Omega$. The transition regimes between steady states exhibit complex wave motions near the interfaces (internal waves). In the turbulent regime it is thought that stratification has only a very weak effect on the dynamics with the internal waves taking only a small fraction of the energy.

## 3.2 Linear stability analysis

In cylindrical coordinates the equations of incompressible stratified viscous flow are:

$$\begin{cases} \dfrac{\partial u_r}{\partial t} + (\mathbf{u} \cdot \nabla) u_r - \dfrac{u_\varphi^2}{r} = -\dfrac{1}{\rho_0} \dfrac{\partial p'}{\partial r} + \nu \left( \Delta u_r - \dfrac{u_r}{r^2} - \dfrac{2}{r^2} \dfrac{\partial u_\varphi}{\partial \varphi} \right), \\ \dfrac{\partial u_\varphi}{\partial t} + (\mathbf{u} \cdot \nabla) u_\varphi + \dfrac{u_r u_\varphi}{r} = -\dfrac{1}{\rho_0 r} \dfrac{\partial p'}{\partial \varphi} + \nu \left( \Delta u_\varphi - \dfrac{u_\varphi}{r^2} + \dfrac{2}{r^2} \dfrac{\partial u_\varphi}{\partial \varphi} \right), \\ \dfrac{\partial w}{\partial t} + (\mathbf{u} \cdot \nabla) w = -\dfrac{1}{\rho_0} \dfrac{\partial p'}{\partial z} - g \dfrac{\rho'}{\rho_0} + \nu \Delta w, \\ div \mathbf{u} = 0, \\ \dfrac{\partial \rho'}{\partial t} + w \dfrac{d\rho_0}{dz} = 0, \end{cases} \tag{2}$$

where $(u_r, u_\varphi, w)$ are the components of velocity respectively in directions $(r, \varphi, z)$. In (2) the Boussinesq approximation has been used with

$$\rho(\mathbf{x}) = \rho_0(z) + \rho'. \tag{3}$$

In order to reduce the complexity, the density diffusion term in the mass conservation has been neglected. This is justified on physical grounds, because in the experiments the Schmidt number is large.

The steady axisymmetric flow solution of (2) is:

$$u_\varphi^0 = Ar + \frac{B}{r}, \; u_r^0 = 0, \; w^0 = 0, \tag{4}$$

where

$$A = -\Omega \frac{a^2}{b^2 - a^2} = -\Omega \frac{1}{\mu^2 - 1}, \; B = -Ab^2 = \Omega a^2 \frac{\mu^2}{\mu^2 - 1}, \; \mu = \frac{b}{a},$$

and $\Omega$ is the rotation rate of the inner cylinder; $u_\varphi^0|_{r=a} = \Omega a$ and the outer cylinder is at rest, $u_\varphi^0|_{r=b} = 0$.

For linear density stratification, $\frac{d\rho_0}{dz} = \alpha = const < 0$, the linearized equations (2) for axisymmetric $(\partial/\partial\varphi = 0)$ disturbances, periodic in axial direction,

$$(u_r, u_\varphi, w)(r, z, t) = (\tilde{u}_r, \tilde{u}_\varphi, \tilde{w})(r) e^{i(\omega t + kz)} \tag{5}$$

are:

$$\begin{cases} i\omega \tilde{u}_r - 2\left(A + \dfrac{B}{r^2}\right) \tilde{u}_\varphi = -\dfrac{\partial}{\partial r} \dfrac{p}{\rho_0} + \nu \hat{D} \tilde{u}_r, \\ 2A\tilde{u}_r + i\omega \tilde{u}_\varphi = \nu \hat{D} \tilde{u}_\varphi \\ \hat{D} = \Delta - \dfrac{1}{r^2}, \; \Delta = \dfrac{1}{r} \dfrac{\partial}{\partial r} r \dfrac{\partial}{\partial r} - k^2, \\ \dfrac{p}{\rho_0} = \dfrac{1}{k\omega} \left( N^2 - \omega^2 - i\omega\nu\Delta \right) \tilde{w}, \\ \tilde{w} = \dfrac{i}{kr} \dfrac{\partial}{\partial r} r \tilde{u}_r, \end{cases} \tag{6}$$

where the Brünt-Väisälä frequency $N^2 = -\frac{g}{\rho_0}\frac{d\rho_0}{dz} = const$, and density disturbances were excluded with the help of the last equation in (2). The boundary conditions for the system (6) for periodic disturbances with axial wave-length $\lambda$ are the following:

$$\tilde{u}_\varphi|_{r=a,b} = 0,\ \tilde{u}_r|_{r=a,b} = 0,\ \tilde{w}|_{r=a,b} = 0,\ \tilde{w}|_{z=0,\lambda/2} = 0 \tag{7}$$

The last condition in (7) gives, according to (5)

$$k = 2\pi/\lambda = \frac{\pi}{nd}, \tag{8}$$

where $n$ is the wave length factor introduced before and $1/n$ corresponds to the number of vorticies on a height equal to one gap width ($d = b - a$).

Excluding from the first equation in (6) $\frac{p}{\rho_0}, u_r$ and $\tilde{w}$, we obtain the equation (from hereon we drop the overbar $\tilde{}$ )

$$\left[(\nu\hat{D} - i\omega)\hat{D} + i\frac{N^2}{\omega}(\hat{D} + k^2)\right](\nu\hat{D} - i\omega)u_\varphi = 4k^2 A\left(A + \frac{B}{r^2}\right)u_\varphi \tag{9}$$

with the boundary conditions that follow from (7) and the second equation in (6):

$$u_\varphi|_{r=a,b} = 0,\ (\nu\hat{D} - i\omega)u_\varphi|_{r=a,b} = 0,\ \frac{\partial}{\partial r}(\nu\hat{D} - i\omega)u_\varphi|_{r=a,b} = 0. \tag{10}$$

Since $(\nu\hat{D} - i\omega)u_\varphi = 2Au_r$ it is easy to see that for $N = 0$ this equation transforms to the known system (see Di Prima & Swinney, 1981) for Couette-Taylor instability.

Introducing the new variables

$$R = (kr)^2,\ u_\varphi = R^{-\frac{1}{2}}G, \tag{11}$$

we obtain the following equation for $G$:

$$\mathcal{K}_3\left(R\frac{d^2}{dR^2}\right)^3 G + \mathcal{K}_2\left(R\frac{d^2}{dR^2}\right)^2 G + \mathcal{K}_1\left(R\frac{d^2}{dR^2}\right)G = \mathcal{K}_0 G, \tag{12}$$

where

$$\begin{aligned}
\mathcal{K}_3 &= \left(4\frac{\nu k^2}{\Omega}\right)^2,\ \mathcal{K}_2 = -\left(4\frac{\nu k^2}{\Omega}\right)\left(3\frac{\nu k^2}{\Omega} + 2i\frac{\omega}{\Omega} - i\frac{N^2}{\omega\Omega}\right),\\
\mathcal{K}_1 &= \left(\frac{\nu k^2}{\Omega} + i\frac{\omega}{\Omega}\right)\left(3\frac{\nu k^2}{\Omega} + i\frac{\omega}{\Omega} - i\frac{N^2}{\omega\Omega}\right),\\
\mathcal{K}_0 &= \frac{1}{4(\mu^2 - 1)^2}\left[4\left(1 - \frac{(ka\mu)^2}{R}\right) + (\mu^2 - 1)^2\left(\frac{\nu k^2}{\Omega} + i\frac{\omega}{\Omega}\right)^2\right],
\end{aligned}$$

with the boundary conditions that follow from (10)

$$G|_{R=R_0,R_1} = 0,\ \frac{d^2}{dR^2}G|_{R=R_0,R_1} = 0,\ R_0 = (ka)^2,\ \ R_1 = (kb)^2 = R_0\mu^2, \tag{13}$$

$$\frac{d^3}{dR^3}G|_{R=R_0,R_1} = \frac{1}{4}\left(1+i\frac{\omega}{\nu k^2}\right)\frac{d}{RdR}G|_{R=R_0,R_1} = 0$$

It is of interest to consider first the inviscid solutions which give lower bounds and the dependence on $N$. For $\nu = 0$ in terms of the new variable

$$\xi = \frac{b^2 - r^2}{a^2} = \mu^2 - \frac{R}{R_0},\ 0 \le \xi \le \mu^2 - 1,\ R_0 = (ka)^2 = \left(\frac{\pi a}{n(b-a)}\right)^2, \tag{14}$$

and in the limit of small gap

$$\epsilon = \mu - 1 = \frac{b}{a} - 1 \to 0, \tag{15}$$

equation (12) reduces to

$$\frac{d^2}{d\xi^2}G = -\left(\frac{\pi}{n}\right)^2 \frac{\Omega^2}{N^2-\omega^2}\frac{\xi}{4\epsilon^4}G, \quad G|_{\xi=0;2\epsilon} = 0. \tag{16}$$

The solution of (16) is

$$G = \xi^{\frac{1}{2}} J_{\frac{1}{3}}(\eta), \quad \eta = \frac{2}{3}\left(\frac{p\xi^3}{4\epsilon^4}\right)^{\frac{1}{2}}, \quad p = \left(\frac{\pi}{n}\right)^2 \frac{\Omega^2}{N^2-\omega^2}, \tag{17}$$

where $J_{\frac{1}{3}}$ is the Bessel function. From boundary condition at $\xi = 2\epsilon$ it follows that

$$J_{\frac{1}{3}}(\eta_j) = 0, \quad \eta_j = \frac{2}{3}\left(\frac{2p}{\epsilon}\right)^{\frac{1}{2}}. \tag{18}$$

For the roots of equation (18) we have

$$\eta_1 = 2.9026, \quad \eta_2 = 6.0327. \tag{19}$$

From (17) and (18) we obtain

$$p = \frac{\epsilon}{2}\left(\frac{3\eta_j}{2}\right)^2, \quad \frac{\Omega}{N} = \sqrt{\frac{\epsilon}{2}}\frac{3\eta_j n}{2\pi}\frac{1}{\sqrt{1+\frac{\epsilon}{2}\left(\frac{3\eta_j n}{2\pi}\right)^2\frac{\omega^2}{\Omega^2}}}$$

and instability for given $n$ with $\omega = -is_r\Omega$ ($s_r$ is real), is possible for

$$\frac{\Omega}{N} = \sqrt{\frac{\epsilon}{2}}\frac{3\eta_j n}{2\pi}\frac{1}{\sqrt{1-\frac{\epsilon}{2}\left(\frac{3\eta_j n}{2\pi}\right)^2 s_r^2}} \ge \sqrt{\frac{\epsilon}{2}}\frac{3\eta_j n}{2\pi} \tag{20}$$

The bounds corresponding to the right hand side of (20) for $n = \frac{1}{2}$,1, and 2 (approximating the experimentally observed cell sizes) are shown in Fig.5 by dash-dotted lines. So, above these lines axisymmetric inviscid disturbances with appropriate $n$ are amplified.

For viscous solution $\nu \ne 0$ in terms of the variable

$$\xi = \epsilon x, \quad 0 \le x \le \mu + 1 = 2 + \epsilon, \tag{21}$$

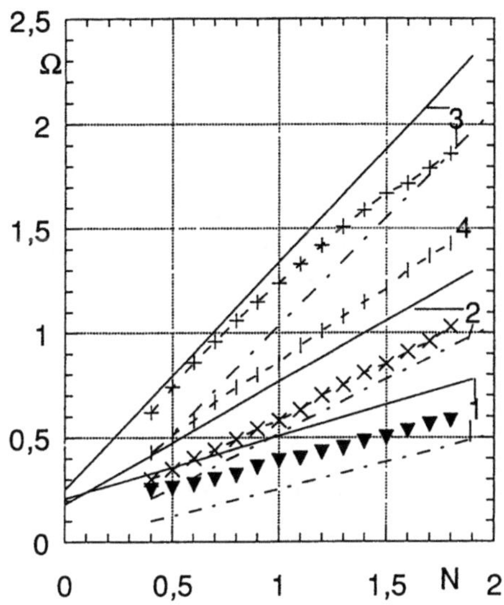


Figure 5: Neutral stability bounds for the existence of vortices with wave length factors $n = \frac{1}{2}$ (1), $n = 1$ (2) and $n = 2$ (3); ▼, ×, |, + , experiments; $- \cdot - \cdot - \cdot -$,linear, inviscid theory; —, linear, monotonic viscous theory; (4) marks the onset of steady Taylor vorticies.

instead of (14), equation (12) and the boundary conditions (13), in the limit of $\epsilon \to 0$ take the form

$$K_3 \frac{d^6}{dx^6} G + K_2 \frac{d^4}{dx^4} G + K_1 \frac{d^2}{dx^2} G + K_0 G = 0, \tag{22}$$

where

$$\begin{aligned}
K_3 &= 64\epsilon(\frac{\delta}{\Omega a^2})^2, \ \delta = \frac{\nu}{\epsilon^2}, \ K_0 = x\left(\frac{\pi}{n}\right)^2, \\
K_2 &= 16i\epsilon\frac{\delta}{\Omega a^2}\left[\frac{N^2 - 2\omega^2}{\Omega\omega} + 3i\left(\frac{\pi}{n}\right)^2 \frac{\delta}{\Omega a^2}\right], \\
K_1 &= 4\epsilon\left[\frac{\omega}{\Omega} - i\left(\frac{\pi}{n}\right)^2 \frac{\delta}{\Omega a^2}\right]\left[\frac{N^2 - \omega^2}{\Omega\omega} + 3i\left(\frac{\pi}{n}\right)^2 \frac{\delta}{\Omega a^2}\right],
\end{aligned}$$

$$G|_{x=0,2} = \frac{d^2}{dx^2} G|_{x=0,2} = 0, \ \frac{d^3}{dx^3} G|_{x=0,2} = \left(\frac{\pi}{2n}\right)^2 \left(1 + i\frac{\omega a^2 n^2}{\delta(\pi)^2}\right) \frac{d}{dx} G|_{x=0,2} = 0, \tag{23}$$

The simplest method to solve the boundary value problem (22), (23) is to use the middle layer approximation

$$x \equiv x_0 \approx 1. \tag{24}$$

For constant $x$ (approximation (24)) and

$$\frac{\omega}{\Omega} = -is_r, \tag{25}$$

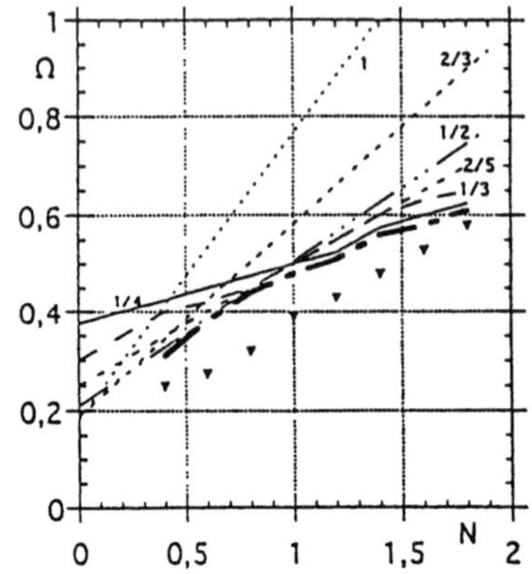


Figure 6: Neutral stability bounds for vortices with wave length factors $n = 1, \frac{2}{3}, \frac{1}{2}, \frac{2}{5}, \frac{1}{3}, \frac{1}{4}$; $\nabla$, experimental bound of instability onset ; —— · —— · —— · ——, envelope curve for monotonic instability with the most unstable wavenumbers.

where $s_r$ is real (and $s_r\Omega > 0$ for instability which is of monotonic type), equation (22) and the boundary conditions in (23) are real. Solutions of (22) can be obtained in the form

$$G = \sum_{i=1}^{6} C_i e^{k_i x},$$

where $k_i, i = 1, ..., 6$ are the roots (real or complex) of the equation that follows from (22). The six boundary conditions (23) for $x = 0$ and $x = 2$ give the determinant. By setting the determinant equal to zero we obtain the eigenvalue problem which contains $s_r, \delta, \frac{\pi}{n}, \Omega a^2, \frac{N^2}{\Omega^2}, \epsilon$ as parameters.

On Fig.5 the solid lines are the neutral stability curves for wave number factors $n = \frac{1}{2}, 1, 2$. In the neutral case $N = 0$ the disturbances with $n = 1$ are first amplified if $\Omega > 0.18$. This value corresponds to the expected critical value (for $\frac{a}{b} = 0.8, \Omega_c \approx 0.2$, see Andereck et al.(1986)). For $N > 0.4$ it is seen from Fig.5 that the experimental values for onset of instability for $n = \frac{1}{2}, 1, 2$ fall between the corresponding theoretical bounds obtained from monotonic viscous and inviscid linear stability solutions.

On Fig.6 the neutral, viscous stability curves are plotted for $n = 1, 2/3, 1/2, 2/5, 1/3, 1/4$ showing the lower bound of viscous monotonic axisymmetric instability indicated by the thick dash-dotted envelope curve. This curve is not a linear function of $N$ and exhibits even a minimum around $N = 1.2$. The difference between this bound of instability and experiments suggests the existence of an oscillatory or non-axisymmetric components of instability.

# 4 Wide gaps

In our experimental study we used also two large gaps $\epsilon = 0.7$ and $1.55$ which do not satisfy the small gap approximation. Nevertheless, as in the small gap experiments, the critical value of $\Omega$ for onset of instability is also a nearly linear function of $N$. The experimental points collapse reasonably well onto one curve when plotted in a way shown in Fig.7, where $\Omega_c(0)$ is the critical value of $\Omega$ for homogeneous fluid, according to Di Prima & Swinney (1987).

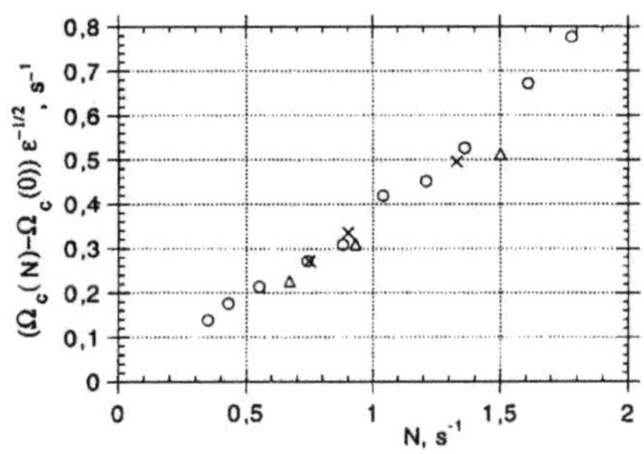


Figure 7: Critical curve for onset of instability for different nondimensional gap widths $\epsilon$. $\epsilon = 0.275(\circ)$, $0.7(\times)$, $1.55(\triangle)$. $\Omega_c(0)$ is the critical value of $\Omega$ for homogeneous fluid taking from our experiments for $\epsilon = 0.275$ and from Di Prima & Swinney (1987) for the large gaps.

The main reason for using wide gaps was to give some answer to the question whether or not the minimum layer height depends on the gap width. The minimum layer heights normalized by the gap width $b-a$ are presented in Fig.3. There is a definite dependence on the gap width with $h_{min}$ being about $(b-a)/2$ for $\epsilon = 0.275$ and decreasing to about $(b-a)/3$ for $\epsilon = 1.55$. It is not clear at this stage whether $h_{min}$ depends on $\epsilon$ or $(b-a)$ keeping for instance $\epsilon$ the same (dependence on Reynolds number).

The vortex structures and resulting layers in the wide gap $\epsilon = 1.55$ are asymmetric (Fig.8b) and have the appearance of a double helix as is shown in Fig.8a. A pair of vortices originates at the inner cylinder and propagate toward the outer boundary, mixing fluid in between them. On the opposite side a density interface forms in the central plane of the vortex pair. The vortex pairs on diametrally opposite sides are shifted vertically by one vortex size or layer height. How these join is not clear; a local spiral structure is one possibility. The whole pattern rotates with a constant velocity less than $\Omega$. For the large angular velocities the flow regimes in the large gap case are again similar to the Taylor vortices regimes ($n = 1$). Taylor vortices of size equal to the gap width are present (or interacting Taylor vortices). The main difference here, compared with the small gap, is stronger turbulent motions and hence more intensive mixing in the layers.

# 5 Discussions and conclusions

The main result of the experiments is the formation of layers in the stratified fluid with their height taking on discrete values instead of changing in a continuous fashion with $\Omega$. For the configuration considered, the smallest layer height is less then the gap width between the cylinders and is about half the gap width at small $N$ decreasing with increasing $N$. The origin for this descrete layer formation is the existence of descrete vortex structures resulting from the most amplified wave length at different values of $(\Omega, N)$. Near onset of instability the most amplified mode has a wave length factor of $n \approx 1/2$ and the instability is non-axisymmetric with azimuthal mode $m = 2$. Linear stability theory indicates that the most unstable mode is of oscillatory type. At larger values of

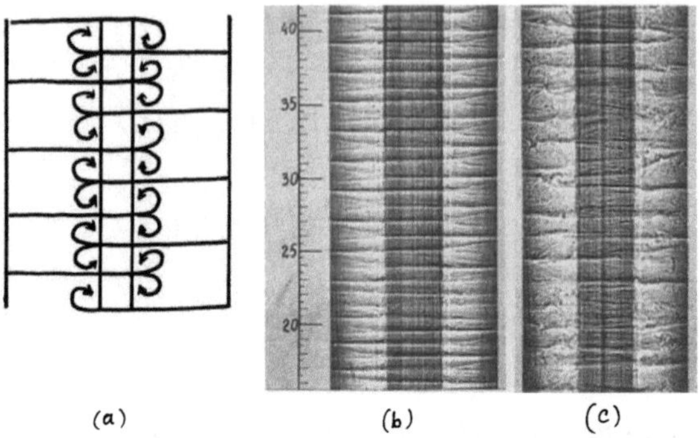


Figure 8: Sketch of vortex motions (a) and corresponding shadograph pictures (b, $\Omega = 1.25s^{-1}$) for the case of $\epsilon = 1.55$. Case (c, $\Omega = 2.57s^{-1}$) is a nearly turbulent state.

$\Omega$ the instability is axisymmetric and modes $n = 1$ and $n = 2$ are successively amplified. The azimuthal wave instability observed in homogeneous fluid is suppressed by stratification. Wave motions exist, however, in the form of internal waves. Their influence on the dynamics of the flow is secondary because, generally, the energy contained in these waves is a small fraction of the energy in the coherent vortex structures. In the viscous theory internal wave motion is one branch solution which has not been looked for.

The experiments are conducted with a fluid with large Schmidt number (about 2000) and it is, therefore, justified to neglect density diffusion in the mass conservation equation. In the ocean the Prandtl number is about 10 which can still be considered as large. For lower Prandtl number (order 1) the descrete mode selection still exists (Orlandi, 1994) but stable density layers could, probably, only form if the Peclet number is large, $Pe = \frac{uh}{\kappa} \gg 1$, where $u$ is a characteristic velocity of the vortices and $\kappa$ the diffusivity.

The experiments have been conducted with the idea of the possible relevance of the results to equatorial jet formation. Measurements by Firing (1987) show along longitudes isovelocities which give the image of flattened cells of aspect ratio (width to depth) of about 300. The largest aspect ratio in the present experiments is 2 to 3 only with an increase in aspect ratio when the gap width increases. The question is whether a large gap and large Reynolds number could give an aspect ratio as large as 300.

The authors wish to thank Back Lien Hua who was at the origin of this work. Many discussions with her during the cause of the research were also most valuable. This research has been financially supported by grant IFREMER No 93-2-424151 LPO. E.Gledzer was supported by DRET (contract 93-1100).

# References

[1] Ali, M. & Weidman, P. D. "On the stability of circular Couette flow with radial heating," J.Fluid Mech. **220** (1990) 53.

[2] Andereck, C. N., Liu, S. S. & Swinney, H. L. “Flow regimes in a circular Couette system with independently rotating cylindes,” J.Fluid Mech.**164** (1986) 155.

[3] Chandrasekhar, S. “Hydrodynamics and Hydromagnetic Stability,” Oxford University Press., 1961.

[4] Chossat, P. & Iooss, G. “The Couette-Taylor Problem”, Applied Math. Sci., **102**, Springer Verlag, 1994.

[5] Cooper, E. R., Jankowski, D. F., Neitzel, G. P. & Squire, T. H. “Experiments on the onset of instability in unsteady Couette flow,” J.Fluid Mech **161** (1985) 97.

[6] Couette, M. “Études sur le frottement des liquides,” Ann. Chim. Phys. **21** (1890) 433.

[7] Firing, E. J. “‘Deep zonal currents in the central equatorial Pacific,” J. Mar. Res. **45** (1987) 791.

[8] Gravas, N. & Martin, B. W. “Instability of viscous axial flow in annuli having a rotating inner cylinder,” J.Fluid Mech. **86**, (1978) 385.

[9] Kubotani, H., Miyama, S. M., Sekiya, M. & Kojima, Y. “The Surface Wave Instability of Stratified Incompressible Cylinders with Differential Rotation,” Progr.Theor.Phys. **82** (1989) 523.

[10] Lueptov, R. M., Docter, A. & Min, K. “Stability of axial flow in an annulus with a rotating inner cylinder,” Phys.Fluids A **4** (1992) 2446.

[11] Orlandi, P. “Stratified Taylor-Couette flow: numerical simulation,” 4-th Int.Symp.on Stratified Flows (1994).

[12] Di Prima, R. C. & Swinney, H. L. “Instabilities and transition in flow between concentric rotating cylinders,” in *Topics in Applied Physics, Hydrodynamic Instability and the Transition to Turbulence*, edited by H. L. Swinney & J. P. Collub ( Springer-Verlag,New York, 1981) **45** 139.

[13] Snyder, H. A. & Karlsson, S. K. “Experiments on the stability of Couette motion with a radial thermal gradient,” Phys.Fluids **7** (1964) 1696.

[14] Taylor, G. I. “Stability of a viscous fluid contained between two rotating cylinders,” Philos.Trans.R.Soc.London Ser.A **233** (1923) 289.

[15] Thorpe, S. A. “The stability of Stratified Couette Flow,” Notes on 1966 Summer Geophys.Fluid Dynam.,Woods Hole Oceanographic Inst. (1966) 80.

[16] Veronis, G. “The analogy between rotating and stratified fluids,” Annual Review of Fluid Mechanics (1970) 37.

[17] Yao, L. S. & Rogers, B. B. “The linear stability of mixed convection in a vertical annulus,” J.Fluid Mech. **201** (1989) 279.

# Vortices subjected to non-axisymmetric strain – unsteady asymptotic evolution


## H.K. Moffatt

Department of Applied Mathematics and Theoretical Physics,
Silver Street, Cambridge, CB3 9EW, UK


### 1. Introduction

Much interest has been recently re-focused on the problem of concentrated vortices and their interactions. Much of this interest has been stimulated by the discovery of such structures in direct numerical simulations of turbulence (see, for example, Jimenez et al 1993, and references therein). The relevance of stretched vortices to the problem of turbulence was first pointed out by Burgers (1948), and has recently been advocated by Moffatt, Kida & Ohkitani 1994 (hereafter MKO'94) who describe the situation in the following terms:

> "Vortices have often been described as the 'sinews' of fluid motion. Just as sinews serve to connect a muscle with a bone or other structure, so the concentrated vortices of turbulence serve to connect large eddies of much weaker vorticity; and just as sinews can take the stress and strain of muscular effort, so the concentrated vortices can accommodate the stress associated with the low pressure in their cores and the strain imposed by relative motion of the eddies into which they must merge at their ends."

The particular problem addressed by MKO'94 was that of a vortex subjected to non-axisymmetric uniform strain

$$\mathbf{U} = (\alpha x\,,\ \beta y\,,\ \gamma z) \tag{1.1}$$

with $\alpha + \beta + \gamma = 0\ ,\ \alpha < \beta < 0 < \gamma$, for which a steady high-Reynolds-number asymptotic solution may be obtained in which the leading order vorticity distribution is simply

$$\boldsymbol{\omega} = (0, 0, \omega_0(r)) \tag{1.2}$$

with $r^2 = x^2 + y^2$. The form of $\omega_0(r)$ is determined by a solvability condition at order $R_\Gamma^{-1}$, where

$$R_\Gamma = \Gamma/\nu \tag{1.3}$$

and $\Gamma$ is the total circulation of the vortex; this solvability condition in effect selects from the infinite family of 'inviscid' solutions of the form (1.2) the particular (Burgers) solution

$$\omega_0(r) = \frac{\gamma\Gamma}{4\pi\nu} \exp\left[-\frac{\gamma r^2}{4\nu}\right] \quad . \tag{1.4}$$

This solution is selected as a result of the persistent (long-term) effect of viscosity; the vortex experiences (at leading-order) the average strain field (averaged around the axis $Oz$ of positive strain rate) which favours an axisymmetric solution. Departures from axisymmetry appear at order $R_\Gamma^{-1}$, the vorticity distribution then having the form

$$\omega(r,\theta) = \omega_0(r) + \lambda R_\Gamma^{-1}\Omega(r)\sin 2\theta \quad , \tag{1.5}$$

where $\lambda = (\alpha - \beta)/(\alpha + \beta)$, length is now scaled on the Burgers scale

$$\delta = (\nu/\gamma)^{\frac{1}{2}} \tag{1.6}$$

and vorticity on $\gamma\Gamma/\nu$; $\Omega(r)$ is then a universal function determined by numerical solution of a second-order ordinary differential equation with boundary conditions that $\Omega(r)/r^2$ is finite as $r \to 0$, and $\Omega(r)$ is exponentially small as $r \to \infty$.

The asymptotic solution (1.5) may be extended to higher order in the small parameter

$$\epsilon = R_\Gamma^{-1} = \nu/\Gamma \quad , \tag{1.7}$$

the solution at each level $O(\epsilon^n)$ being completed by imposition of a solvability condition at $O(\epsilon^{n+1})$. At the level $n = 2$, a uniformly valid approximation to the field of dissipation of energy

$$\Phi = 2\nu s_{ij}\, s_{ij} \quad , \quad s_{ij} = \tfrac{1}{2}\left(\frac{\partial u_i}{\partial x_j} + \frac{\partial u_j}{\partial x_i}\right) \tag{1.8}$$

may be obtained. The contours of constant $\Phi$ obtained by Moffatt et al (1994) agree well with previously computed contours, using a finite-difference numerical scheme for solution of the governing nonlinear partial differential equations (obtained from the exact Navier-Stokes equations), and provide evidence that the asymptotic solution is accurate for $\lambda\epsilon \lesssim 0{\cdot}005$, and becomes increasingly more accurate with increasing Reynolds number.

Since a steady-state solution may take a long time to establish (starting from an arbitrary initial condition), and since moreover a steady solution may be subject to modification due to perturbation vorticity advected inwards 'from infinity' by the strain field (1.1), or due to interaction with other vortices (by the processes of merger and/or cancellation considered in a numerical study by Buntine & Pullin 1989), it is natural to investigate whether the asymptotic procedure described above may be extended to deal with unsteady evolution. A clue is provided by the work of Neu (1984) who showed that such an approach is indeed possible for two-dimensional strain fields ($\beta = 0$, $\alpha = -\gamma$ in the notation of the present paper). Here, we adapt this approach to accommodate the general strain field (1.1) and discuss the interesting perspectives opened up by this type of analysis.

## 2. Long-term quasi-axisymmetric evolution

We consider a vorticity field

$$\boldsymbol{\omega} = (0, 0, \omega(x, y, t)) \tag{2.1}$$

subjected to the action of the uniform irrotational strain (1.1). We may suppose that $\omega$ is localised in the sense that $|\omega|$ is exponentially small for large $r = (x^2 + y^2)^{\frac{1}{2}}$, and that the total circulation

$$\Gamma = \int\int \omega(x, y, t)\, dx\, dy \tag{2.2}$$

is finite and non-zero. We assume that the Reynolds number $R_\Gamma = \Gamma/\nu$ is large. The evolution of $\omega$ is described by the nonlinear vorticity equation. In polar coordinates $(r, \theta)$ with $x = r\cos\theta$, $y = r\sin\theta$, this equation takes the form

$$\begin{aligned}\frac{\partial\omega}{\partial t} - \frac{1}{r}\frac{\partial(\psi,\omega)}{\partial(r,\theta)} =& \gamma\left(\omega + \tfrac{1}{2}r\frac{\partial\omega}{\partial r}\right) + \nu\nabla^2\omega \\ &+ \tfrac{1}{2}(\beta - \alpha)\left(\cos 2\theta\, r\frac{\partial}{\partial r} - \sin 2\theta\,\frac{\partial}{\partial\theta}\right)\omega \quad ,\end{aligned} \tag{2.3}$$

where $\psi$ is the streamfunction of the rotational ingredient of the flow, i.e.

$$\nabla^2\psi = -\omega \quad . \tag{2.4}$$

Note that when $\beta = \alpha$ (the axisymmetric case), equation (2.3) admits the steady solution (1.4).

We now non-dimensionalise (2.3), (2.4) with the substitutions

$$r \to (\nu/\gamma)^{\frac{1}{2}} r\ , \quad t \to \gamma^{-1} t\ , \quad \psi \to \Gamma\psi \quad . \tag{2.5}$$

Equation (2.3) then becomes

$$-\frac{1}{r}\frac{\partial(\psi,\omega)}{\partial(r,\theta)} = -\epsilon L_0\omega - \epsilon\lambda L_1\omega \quad , \tag{2.6}$$

where $\epsilon = R_\Gamma^{-1}$ , $\lambda = (\beta - \alpha)/\gamma$, and

$$L_0 = 1 + \tfrac{1}{2}r\frac{\partial}{\partial r} + \nabla^2 - \partial/\partial t \quad , \tag{2.7}$$

$$L_1 = \tfrac{1}{2}\left(\cos 2\theta\, r\frac{\partial}{\partial r} - \sin 2\theta\frac{\partial}{\partial\theta}\right) \tag{2.8}$$

Equation (2.6) is in the form obtained by MKO'94, but the operator $L_0$ is modified by the inclusion of the term $-\partial/\partial t$. Note that by scaling time on $\gamma^{-1}$, we focus on evolution on a long slow time-scale (relative to the fast time-scale $(\nu/\Gamma)\gamma^{-1}$). Equation (2.6) cannot therefore describe the details of vortex merger, for example, which occurs on a time scale short compared with $\gamma^{-1}$ (see §4 below); but it *can* describe the slow evolution consequent upon such merger.

We now seek an asymptotic solution of (2.6) in the form

$$\psi(r, \theta, t) = \psi_0(r, \theta, t) + \epsilon\psi_1(r, \theta, t) + \epsilon^2\psi_2(r, \theta, t) + \ldots \tag{2.9}$$

with a corresponding expansion for $\omega$. At order $\epsilon^0$, we obtain the steady Euler equation

$$\frac{\partial(\psi_0, \omega_0)}{\partial(r, \theta)} = 0 \quad . \tag{2.10}$$

Here we focus on solutions of the form

$$\psi_0 = \psi_0(r, t) \quad , \quad \omega_0 = -\nabla^2 \psi_0 = \omega_0(r, t) \quad . \tag{2.11}$$

Thus the streamlines at leading order are circles $r =$ cst., and $t$ appears merely as a parameter. The dependence of $\psi_0$ on $r$ and $t$ can only be determined through application of a solvability condition at order $\epsilon^1$.

At order $\epsilon^1$, (2.6) gives

$$\frac{1}{r}\frac{\partial}{\partial \theta}\left(\frac{\partial \omega_0}{\partial r}\psi_1 - \frac{\partial \psi_0}{\partial r}\omega_1\right) = L_0\omega_0 + \lambda L_1 \omega_0 \quad . \tag{2.12}$$

The solvability condition is obtained by integrating over $\theta$, and is evidently

$$L_0\omega_0 \equiv \left(1 + \tfrac{1}{2}\, r\frac{\partial}{\partial r} + \nabla^2 - \frac{\partial}{\partial t}\right)\omega_0 = 0 \quad . \tag{2.13}$$

This equation determines a unique evolution of $\omega_0(r,t)$ from arbitrary initial condition

$$\omega_0(r, 0) = \hat{\omega}_0(r) \tag{2.14}$$

towards the asymptotic steady state

$$\omega_0(r,t) \sim \frac{1}{4\pi}\, e^{-r^2/4} \quad (t \to \infty) \quad , \tag{2.15}$$

i.e. the Burger's vortex. Equation (2.13) describes a 'quasi-axisymmetric' evolution, the rapid rotation within the vortex being such as to maintain axisymmetry at each instant $t$, despite the perturbing effect of the non-axisymmetric strain.

A particular solution of (2.13) which indicates the manner in which the asymptotic solution (2.15) is approached is given by

$$\omega_0(r,t) = \frac{1}{4\pi\delta^2}\, e^{-r^2/4\delta^2} \quad , \tag{2.16}$$

where

$$\delta^2 = 1 + (\delta_0^2 - 1)\, e^{-t} \quad . \tag{2.17}$$

A solution of precisely this kind was obtained by Neu (1984) for the case of a vortex subjected to plane strain ($\beta = 0$ in the present notation).

For $t \gg 1$, (2.16) may be expanded in the form

$$\omega_0(r,t) \sim \frac{1}{4\pi}\left(1 + (\frac{r^2}{4} - 1)(\delta_0^2 - 1)\, e^{-t} + O(e^{-2t})\right) e^{-\frac{1}{4}r^2} \quad . \tag{2.18}$$

Here the leading term carries the total circulation (normalised to unity) and the subsequent terms of the series have zero total circulation. The general solution of (2.13) (depending only on $r$ and $t$) with unit total circulation in fact has the form

$$\omega_0(r,t) = \frac{1}{4\pi}\left(1 + A\left(\frac{r^2}{4} - 1\right)e^{-t} + O\left(e^{-2t}\right)\right) e^{-\frac{1}{4}r^2} \quad , \tag{2.19}$$

where the constant $A$ depends on the initial conditions.

## 3. The perturbation from axisymmetry

Having satisfied the solvability condition (2.13), (2.12) may be integrated with respect to $\theta$ to give

$$\frac{1}{r}\left(\frac{\partial\omega_0}{\partial r}\psi_1 - \frac{\partial\psi_0}{\partial r}\omega_1\right) = \tfrac{1}{4}\lambda \sin 2\theta \, r\frac{\partial\omega_0}{\partial r} - q(r,t) \quad . \tag{3.1}$$

The solution for $\psi_1(r,\theta,t)$ has the form

$$\psi_1 = \lambda f(r,t)\sin 2\theta + g(r,t) \quad , \tag{3.2}$$

where

$$\frac{\partial^2 f}{\partial r^2} + \frac{1}{r}\frac{\partial f}{\partial r} - \frac{4}{r^2}f = (f - \tfrac{1}{4}r^2)\eta(r,t) \quad , \tag{3.3}$$

where

$$\eta(r,t) = -\frac{\partial\omega_0/\partial r}{\partial\psi_0/\partial r} \quad . \tag{3.4}$$

If we use the similarity solution (2.16), this gives

$$\eta(r,t) = -\frac{r^2}{4\delta^4(e^{r^2/4\delta^2} - 1)} \quad . \tag{3.5}$$

By the re-scaling

$$\hat{r} = r/\delta \quad , \quad \hat{f} = f/\delta^2 \tag{3.6}$$

equation (3.3) becomes

$$\frac{\partial^2 \hat{f}}{\partial \hat{r}^2} + \frac{1}{\hat{r}}\frac{\partial \hat{f}}{\partial \hat{r}} - \frac{4}{\hat{r}^2}\hat{f} = \left(\hat{f} - \tfrac{1}{4}\hat{r}^2\right)\hat{\eta}(\hat{r}) \quad , \tag{3.7}$$

where

$$\hat{\eta}(\hat{r}) = -\frac{\hat{r}^2}{4(e^{\hat{r}^2} - 1)} \quad . \tag{3.8}$$

The problem for $\hat{f}$ is thus reduced to precisely the form studied for the steady situation ($\delta = 1$) by MKO'94. All of the results concerning the elliptic distortion of the vortex may therefore be carried over directly to the unsteady situation considered here. The isovorticity contours and the dissipation contours are thus geometrically similar for the steady and unsteady problems, the only modification that is required being the introduction of the scaling factor $\delta(t)$.

We should note here however that this is not the only possibility. If a different zero-order solution of (2.13) is used, then $\eta(r,t)$ and $f(r,t)$ will be correspondingly modified, and more complex structures may be expected.

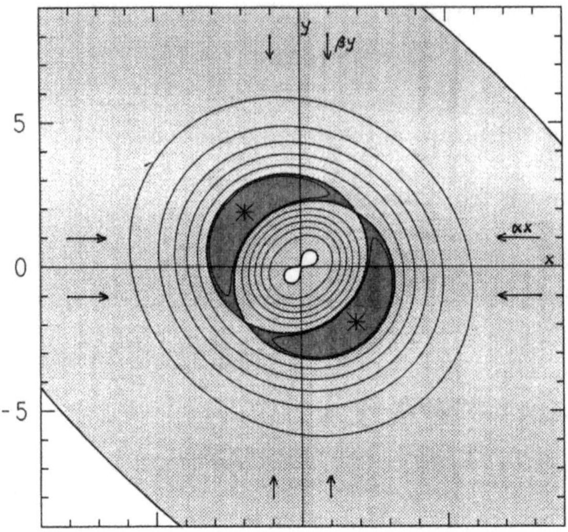


**Figure 1** [adapted from MKO'94]. Dissipation contours for a vortex subjected to non-axisymmetric strain. The maxima are in the dark shaded crescent-shaped regions and are indicated by *.

In general terms however the structure of the perturbed dissipation field may be anticipated from the following argument. For the gaussian structure (2.16), the dissipation field $\Phi$ is axisymmetric, with a minimum at $r = 0$, and a maximum on the circle $r = 2{\cdot}67\delta$ (where the radial shear is greatest). From a topological point of view, this distribution is degenerate; a non-axisymmetric perturbation proportional to $\cos 2\theta$ (equations (2.8) and (2.12) will replace the circle of maxima by just two isolated maxima at $r = 2{\cdot}67\delta\,,\, \theta = 3\pi/4\,,\, 7\pi/4$ (figure 1). These maxima are surrounded by two crescent-shaped regions determined by the separatrices of the field $\Phi(r,\theta)$. The minimum at $r = 0$ may also suffer a topological transformation: the calculation of MKO'94 (to order $\epsilon^2$) reveals that this minimum is split into two minima on the other diagonals $\theta = \pi/4$, $5\pi/4$, with a saddle point at the origin, the separatrix through the origin being a small figure-of-eight (figure 1).

For the time-dependent situation described by (2.16), the above dissipation field is quasi-static, time-dependence entering only through the radial scale $\delta(t)$. For large $t$, $\delta(t) \to 1$, and the behaviour settles down to the steady state treated by MKO'94.

## 4. Merging of vortices

If two vortices of like circulations (of order $\Gamma$) are driven towards each other by the strain field (1.1), then they will ultimately merge and form a single vortex in the manner described by Buntine & Pullin (1989). This process involves the engulfment of each vortex by the other, a double spiral pattern of vorticity contours being formed during the initial stage of merger before viscosity eliminates strong radial variations of vorticity.

The time-scale of merger when the Reynolds number (based on either vortex) is large may be estimated as follows. The length $\ell(t)$ of the spiral 'tongues' increases linearly with $t$ and the radial 'striation' scale therefore decreases like $t^{-1}$ (the area of the tongues being conserved during the nearly inviscid evolution. The local wave-number in the radial direction $k(t)$ therefore increases as $t$, and the effect of viscous interdiffusion of tongues from the two vortices may be represented by a factor

$$\exp -\nu \int_0^t k^2 dt = \exp\left\{ -C\nu \left(\frac{\Gamma}{\delta^3}\right)^2 t^3 \right\} \quad , \tag{4.1}$$

where $C$ is a constant of order unity. Hence the time-scale for the merging process is

$$t_M \sim t_0 R_\Gamma^{\frac{1}{3}} \sim t_V \, R_\Gamma^{-\frac{2}{3}} \quad , \tag{4.2}$$

where $t_0 = \delta^2/\Gamma$, $t_V = R_\Gamma t_0$. This is the time-scale conjectured by MKO'94 (p. 246), and it is not incompatible with the trend indicated by the computations of Buntine & Pullin (1989) for axisymmetric strain at Reynolds numbers $R_\Gamma/2\pi = 40$, 160, 640 and 1280.

We thus expect that when two like-signed vortices are swept together, the initial merging process occurs through the above spiral-tongue mechanism on a time-scale $t_0 R_\Gamma^{\frac{1}{3}}$; after this initial stage, the vorticity field is approximately axisymmetric and subsequent relaxation to the ultimate steady form (2.15) takes place on the viscous time-scale $t_V = t_0 R_\Gamma$, by the process described in §2 above. The mechanism is similar to that described by Rhines & Young (1982), and in the closely related problem of magnetic flux expulsion by Moffatt & Kamkar (1983).

## REFERENCES


Buntine, J.D. & Pullin, D.I. 1989 Merger and cancellation of strained vortices. *J. Fluid Mech.* **205**, 263-295

Burgers, J.M. 1948 A mathematical model illustrating the theory of turbulence. *Adv. Appl. Mech.* **1**, 171-199

Jimenez, J., Wray, A.A., Saffman, P.G. & Rogallo, R.S. 1993 The structure of intense vorticity in homogeneous isotropic turbulence. *J. Fluid Mech.* **255**, 65-90

Moffatt, H.K. & Kamkar, H. 1983 The time-scale associated with flux expulsion. *In* Stellar and Planetary Magnetism, Ed. A.M. Soward, Gordon & Breach, pp. 91-98

Moffatt H.K., Kida S. & Ohkitani, K. 1994 [MKO'94] Stretched vortices – the sinews of turbulence; large Reynolds number asymptotics. *J. Fluid Mech.* **259**, 241-264

Neu, J.C. 1984 The dynamics of stretched vortices. *J. Fluid Mech.* **143**, 253-276

Rhines, P.B. & Young, W.R. 1982 Homogenization of potential vorticity in planetary gyres. *J. Fluid Mech.* **122**, 347-367

# Some Contributions to Asymptotic Theory for Turbulent Shear Flows

Klaus Gersten

Institut für Thermo– und Fluiddynamik,
Ruhr–Universität Bochum
44780 Bochum, Fed. Rep. of Germany

**Abstract:** Contrary to laminar boundary layers which are correct asymptotic solutions of the full Navier–Stokes solutions for high Reynolds numbers turbulent boundary layers are usually not treated as asymptotically correct solutions. Hence, for commonly used prediction methods the turbulent boundary layers have to be determined for each Reynolds number separately.

It is shown for attached turbulent flows that the consequent application of asymptotic methods leads to a turbulent boundary–layer prediction which is independent of the Reynolds number.

Various aspects of turbulent shear–flow prediction are discussed from the viewpoint of asymptotic theory such as: equilibrium boundary layers, viscous superlayer, curvature, separation, low–Reynolds number modelling, and suction or blowing.

## 1 Introduction

Singular perturbation methods have been very successfully applied to investigate laminar flows at high Reynolds numbers. Prandtl's boundary layer theory and its extensions, e.g. double–deck and triple–deck theory, are well–known examples for powerful asymptotic methods.

But asymptotic methods can also be applied to solve the Reynolds–averaged flow equations for turbulent shear flows at high Reynolds numbers.

Although the Reynolds–averaged flow equations without additional turbulence–model equations are not a closed system of equations, singular perturbation methods can formally be applied to derive the flow equations and characteristics of their solutions in the limit of high Reynolds numbers. These solutions exhibit a multi–layer structure according to the asymptotic theory. Furthermore, this theory leads to a framework of equations and boundary conditions for the various layers which have to be satisfied by turbulence models. Hence, asymptotic theory can furnish the turbulence modellers with guidelines for their modelling.

The first papers on asymptotic theory for turbulent shear flows appeared in the early seventies, for example by Yajnik (1970), Mellor (1972), Bush, Fendell (1972), and Fendell (1972). Survey papers on asymptotic theory applied to turbulent shear flows have been given by Deriat, Guiraud (1986), Gersten (1987), Kluwick (1989), and Schneider (1991). In the text–book by Gersten, Herwig (1992) the turbulent shear flows have been treated consistently from the viewpoint of asymptotic theory.

In the following a few examples will be given to demonstrate that the application of asymptotic methods can be very useful to investigate turbulent shear flows at high Reynolds numbers.

## 2 Analytic Structure of Solutions for Attached Turbulent Shear Flows

The Reynolds–averaged Navier–Stokes equations for steady two–dimensional turbulent flows with constant fluid properties are:

$$\frac{\partial U}{\partial X} + \frac{\partial V}{\partial Y} = 0 \tag{1}$$

$$U\frac{\partial U}{\partial X} + V\frac{\partial U}{\partial Y} = -\frac{\partial P}{\partial X} - \epsilon^2\left[\frac{\partial \overline{U'^2}}{\partial X} + \frac{\partial \overline{U'V'}}{\partial Y}\right] + \frac{1}{Re}\left[\frac{\partial^2 U}{\partial X^2} + \frac{\partial^2 U}{\partial Y^2}\right] \tag{2}$$

$$U\frac{\partial V}{\partial X} + V\frac{\partial V}{\partial Y} = -\frac{\partial P}{\partial Y} - \epsilon^2\left[\frac{\partial \overline{U'V'}}{\partial X} + \frac{\partial \overline{V'^2}}{\partial Y}\right] + \frac{1}{Re}\left[\frac{\partial^2 V}{\partial X^2} + \frac{\partial^2 V}{\partial Y^2}\right]. \tag{3}$$

In this dimensionless form of the basic equations the lengths are based on a reference length $L_R^*$, the velocities on $U_R^*$, the pressure and the wall shear stress on $\rho^* U_R^{*2}$, but the fluctuation velocities on a reference friction velocity

$$u_{\tau R}^* = \sqrt{\tau_{WR}^*/\rho^*} = \epsilon\, U_R^* \quad , \tag{4}$$

which is supposed to be small compared to the reference velocity $U_R^*$. Dimensional variables are marked by an asterisk.

The system (1) to (3) contains the two small parameters $\epsilon$ and $1/\mathrm{Re}$ , where $\mathrm{Re} = U_R^* L_R^*/\nu^*$. For high Reynolds numbers these two values can be considered as perturbation parameters, which, however, are coupled by

$$\epsilon = \frac{\kappa}{\ln \mathrm{Re}} \quad , \tag{5}$$

as will be shown a posteriori. The Karman constant $\kappa = 0.41$ is universal. Although the system (1) to (3) is not closed without additional equations due to a turbulence model, the singular–perturbation method of matched asymptotic expansions can be formally applied to this system, see Mellor (1972). The main results for attached turbulent flows (non–zero wall shear stress) are:

1. The flow field consists of three regions: the inviscid non–turbulent outer flow (potential flow), the inviscid turbulent defect layer, and the viscous turbulent wall layer (viscous sublayer), see Fig. 1.
2. The wall layer is in the leading term universal and can be considered known, when the local wall shear stress is given.

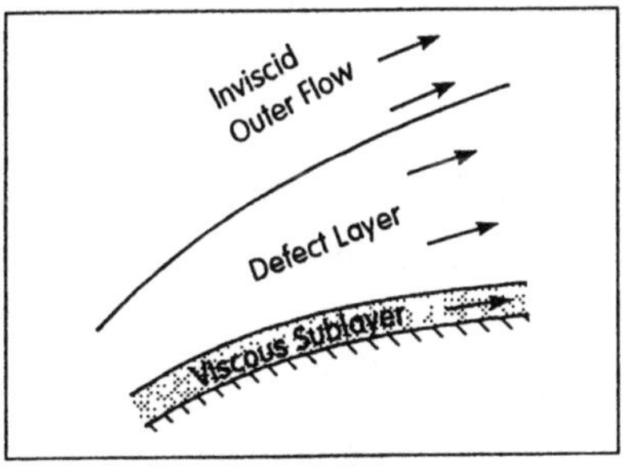


**Fig.1.** Multi–layer structure of turbulent shear flows at high Reynolds numbers

3. For a given pressure distribution resulting from the inviscid outer–flow solution, only the defect layer has to be determined.
In this layer the asymptotic expansions for the solution are:

$$\begin{aligned} U(X,Y,\epsilon) &= u_1(X,y) + \epsilon\, u_2(X,y) + \ldots \\ V(X,Y,\epsilon) &= \epsilon v_1(X,y) + \epsilon^2 v_2(X,y) + \ldots \\ P(X,Y,\epsilon) &= p_1(X,y) + \epsilon\, p_2(X,y) + \ldots \\ -\overline{U'V'} &= t_1(X,y) + \epsilon\, t_2(X,y) + \ldots \end{aligned} \tag{6}$$

where
$$y = \frac{Y}{\epsilon} \tag{7}$$

is the characteristic length scale for this defect layer.

The first–order solution is given by:

$$u_1 = U_{10}(X), \qquad v_1 = -y\, dU_{10}/dX, \qquad p_1 = P_{10}(X) = (1 - U_{10}^2)/2, \tag{8}$$

where $U_{10}(X)$ results from the inviscid outer solution.

The second–order solution has to satisfy the following system of equations ($p_2 = 0$):

$$\frac{\partial u_2}{\partial X} + \frac{\partial v_2}{\partial y} = 0 \tag{9}$$

$$U_{10}\frac{\partial u_2}{\partial X} + \frac{dU_{10}}{dX}\left[u_2 - y\frac{du_2}{dy}\right] = \frac{\partial t_1}{\partial y} \quad . \tag{10}$$

Since $u_2$ and $v_2$ are the defects of the velocities compared to the outer–flow velocities, the layer is called "defect layer". It is worth mentioning that (10) is a linear equation and does not contain $v_2(X,y)$, which for a known function $u_2(X,y)$ can be determined by (9).

The boundary conditions follow from the matching with the outer flow (at $y = \delta(X)$) as well as with the universal solution of the wall layer:

$$\begin{aligned} y \to 0: &\qquad \frac{\partial u_2}{\partial y} = \frac{\gamma(X)}{\kappa y}, \qquad t_1 = \gamma^2(X) \\ y = \delta: &\qquad u_2 = 0\,, \qquad\qquad t_1 = 0\quad , \end{aligned} \tag{11}$$

where

$$\gamma(X) = \frac{u_\tau(X)}{u_{\tau R}} = \frac{u_\tau(X)}{\epsilon} = O(1) \tag{12}$$

is the distribution of the local friction velocity.

By applying the transformation

$$\hat{y} = \frac{y}{\gamma(X)} \quad , \quad \hat{u}_2 = \frac{u_2}{\gamma(X)} \quad , \quad \hat{t}_1 = \frac{t_1}{\gamma^2(X)} \tag{13}$$

eqn.(10) leads to

$$U_{10} \frac{\partial \hat{u}_2}{\partial X} + \left[ \frac{dU_{10}}{dX} + \frac{U_{10}}{\gamma} \frac{d\gamma}{dX} \right] \left[ \hat{u}_2 - \hat{y} \frac{\partial \hat{u}_2}{\partial \hat{y}} \right] = \frac{\partial \hat{t}_1}{\partial \hat{y}} . \tag{14}$$

It will be shown a posteriori that one gets

$$\frac{U_{10}}{\gamma} \frac{d\gamma}{dX} = \frac{dU_{10}}{dX} + O(\epsilon) \quad . \tag{15}$$

Hence, the final form of the flow equation is:

$$U_{10} \frac{\partial \hat{u}_2}{\partial X} + 2 \frac{dU_{10}}{dX} \left[ \hat{u}_2 - \hat{y} \frac{\partial \hat{u}_2}{\partial \hat{y}} \right] = \frac{\partial \hat{t}_1}{\partial \hat{y}} \tag{16}$$

with the boundary conditions

$$\begin{aligned} \hat{y} \to 0 : &\qquad \frac{\partial \hat{u}_2}{\partial \hat{y}} = \frac{1}{\kappa\, \hat{y}} \quad , &\qquad \hat{t}_1 = 1 \\ \hat{y} = \hat{\delta} : &\qquad \hat{u}_2 = 0 \quad , &\qquad \hat{t}_1 = 0 \quad . \end{aligned} \tag{17}$$

With additional equations for $\hat{t}_1$ resulting from a turbulence model (16) and (17) can be solved independently of the Reynolds number. The boundary layer conditions are sufficient to determine $\hat{u}_2(X,\hat{y})$ and the distribution of the boundary–layer thickness $\hat{\delta}(X)$ for a given outer flow velocity distribution $U_{10}(X)$.

From the solution $\hat{u}_2(X,\hat{y})$ the following function can be found:

$$\hat{C}(X) = \lim_{\hat{y}\to 0} \left[\frac{1}{\kappa} \ln \hat{y} - \hat{u}_2(X,\hat{y})\right] \quad . \tag{18}$$

So far, the viscosity, i.e. the Reynolds number, was not involved.

A typical result of such a boundary layer calculation is shown in Figure 2.*

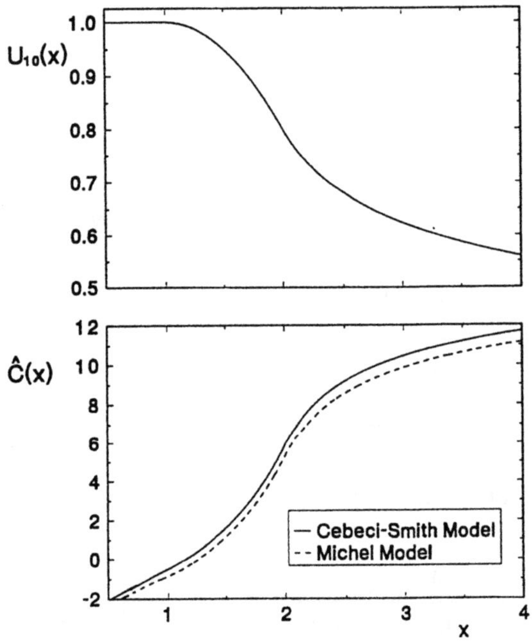


**Fig.2.** Main result of a boundary–layer calculation for turbulent flows at high Reynolds numbers. The function $\hat{C}(X)$ leads to the resistance law (21)

The law of resistance, i.e. the distribution of the local friction velocity and local wall shear stress, respectively, can be found by matching the velocity in the defect layer and the wall layer. This leads to

$$\frac{U_{10}(X)}{u_\tau(X)} = \frac{1}{\kappa} \ln \left[\mathrm{Re}\ u_\tau^2(X)\right] + C^+ + \hat{C}(X) \quad . \tag{19}$$

---

* I am obliged to Mr. Detlef Vieth, who carried out the analysis and the boundary–layer calculation.

This is an implicit formula for the local friction velocity distribution $u_\tau(X)$. For smooth walls the constant $C^+$ has the value $C^+ = 5.0$. For rough walls $C^+$ is a function of the wall roughness, see Gersten, Herwig (1992). Derivation of (19) with respect to X leads to (15). The friction coefficient based on the local outer–flow velocity

$$c_f(X) = \frac{2\ \tau_w^*(X)}{\rho^*\ U_{10}^{*2}(X)} \tag{20}$$

satisfies the equation

$$\sqrt{\frac{2}{c_f}} = \frac{1}{\kappa} \ln \left[\mathrm{Re}\ U_{10}^2(X)\ \frac{c_f}{2}\right] + C^+ + \hat{C}(X) \quad . \tag{21}$$

This is the universal resistance law for <u>all</u> attached turbulent flows at high Reynolds numbers.

In order to get an explicit formula for $c_f(X)$, Gersten and Herwig (1992) defined a new function $G(\Lambda;D)$ by

$$\frac{\Lambda}{G} + 2 \ln \frac{\Lambda}{G} - D = \Lambda \tag{22}$$

with the limit

$$\lim_{\Lambda \to \infty} G(\Lambda;D) = 1 \quad . \tag{23}$$

The explicit universal resistance law is now

$$c_f = 2 \left[\frac{\kappa}{\ln \mathrm{Re}}\ G(\Lambda;D)\right]^2 \tag{24}$$

where

$$\Lambda = \ln \mathrm{Re} \tag{25}$$

$$D = 2 \ln \kappa + \ln U_{10}^2(X) + \kappa \left[C^+ + \hat{C}(X)\right] \quad . \tag{26}$$

Eqn. (24) is also the a–posteriori justification for (5).

The friction drag of all airfoils without flow separation has for high Reynolds numbers the following form

$$c_{Df} \sim 2\kappa^2/\ln^2 Re \qquad (27)$$

The turbulence model has an effect only on the function $\hat{C}(X)$. Therefore, it follows from (24) and (26), that the turbulence model has only a small influence on the friction drag of a few percent. In order to evaluate turbulence models, the function $\hat{C}(X)$ should be used rather than $c_f(X)$ for comparison.

## 3 Equilibrium Boundary Layers

When the defect layer equation (16) leads to a self–similar solution, the turbulent boundary layer is called an equilibrium boundary layer. This happens for outer flows of the form

$$U_{10}(X) = B(X - X_0)^m \quad , \qquad (28)$$

see Rotta (1950).

By using the transformation

$$\eta = \hat{y}\ (1 + 3\ m)\ B\ (X - X_0)^{m-1}\ , \quad \hat{u}_2 = -F(\eta), \quad \hat{t}_1 = S(\eta) \qquad (29)$$

it follows from (16)

$$(1 + 2\,\beta)\eta\, F'' + 2\,\beta\, F' = S' \qquad (30)$$

with the boundary conditions

$$\begin{aligned} \eta \to 0: &\quad F'' = 1/\kappa\eta & S = 1\ , &\quad F = 0 \\ \eta = \eta_\delta: &\quad F' = 0\ , & S = 0\ , &\quad F = 1 \end{aligned} \qquad (31)$$

and with the Rotta–Clauser parameter

$$\beta = \frac{\delta_1^*}{\tau_w^*}\frac{dp^*}{dx^*} = -\frac{m}{1 + 3\ m} \quad , \qquad (32)$$

which is a constant for equilibrium boundary layers. Detailed solutions of (30) are described in Gersten, Herwig (1992).

The function $\hat{C}(X)$ according to (18) reads now

$$\hat{C}(X) = \overline{C}(\beta) + \frac{1}{\kappa} \ln \left[(1+3\ m)U_{10}/(X-X_0)\right] \ , \tag{33}$$

where

$$\overline{C}(\beta) = \lim_{\eta \to 0} \left[F'(\eta) + \frac{1}{\kappa} \ln \eta\right] \ . \tag{34}$$

This leads to the final form of the resistance law for equilibrium boundary layers:

$$\sqrt{\frac{2}{c_f}} = \frac{1}{\kappa} \ln \left[Re_x \frac{c_f}{2}\right] + C^+ + \overline{C}(\beta) + \frac{1}{\kappa} \ln (1 + 3\ \beta) \tag{35}$$

which can also be brought into an explicit form by using the function $G(\Lambda;D)$ defined in (22). Again, $c_f$ is a function of $\ln Re_x$ , where $Re_x = U_{10}^* \ x^*/\nu^*$ is the local Reynolds number.

## 4 Extensions to the Basic Problem

### 4.1 Viscous Superlayer

So far only three layers have been distinguished according to Fig. 1. Strictly speaking there are four layers as shown in Fig. 3. But it has been shown by Jeken (1992),

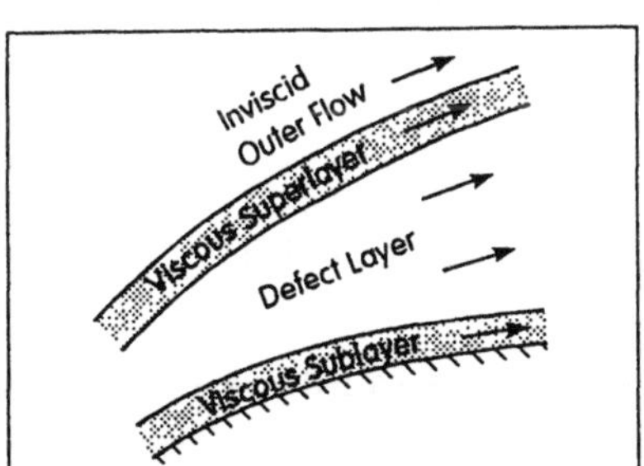


**Fig.3.** Multi–layer structure of turbulent attached boundary layers: inviscid outer flow and three–layer boundary layer

that the viscous superlayer has like the viscous wall layer a universal solution. Turbulence modellers can take advantage of this universal solution, which leads to relations between model constants, see also Gersten (1994).

## 4.2 Curvature Effects

As has been shown by Jeken (1992), turbulence models have to take into account the curvature of the flow even for high Reynolds numbers. Along curved walls one has to distinguish between Reynolds stresses normal to the wall and parallel to the wall, respectively. This can be done only by the Reynolds–stress models, also called second–order closure models.

## 4.3 Flow Separation

In Sections 1 and 2 it was assumed that the wall shear stress is always positive. For flows with separation the analysis has to be extended.

The matching conditions have been generalized such, that they satisfy both laws, i.e. the logarithmic law for $\tau_w^* \neq 0$ and the square–root law for $\tau_w^* = 0$, respectively, see Gersten et al. (1993). The parameter

$$K = \frac{\nu^*}{\rho^*(\tau_w^*/\rho^*)^{3/2}} \frac{dp^*}{dx^*} \tag{36}$$

describes the transition from $\tau_w^* \neq 0$ ($K \to 0$ because of $\nu^* \to 0$) to $\tau_w^* = 0$ ($K \to \infty$). In the literature sometimes the symbol $p^+ = -K$ is used, see Cebeci, Bradshaw (1984, p. 357). The generalized matching law is again universal, but not yet very well known for $K \to \infty$, see Gersten et al. (1993).

## 4.4 Low Reynolds Number Turbulence Models

Contributions in the literature on so–called low Reynolds number turbulence models produce the impression that they deal with higher–order effects with respect to Reynolds number. Unfortunately, this is not true. These contributions practically try to model the wall–layer equations, which then lead to the well–known universal solutions, see also Wilcox (1993, p. 142). An asymptotic higher–order theory for turbulent boundary layers does not exist yet. In many cases its contributions are expected to be negligibly small and not identifiable by experiments, see Gersten, Herwig (1992).

## 4.5 Effects of Suction or Blowing

The viscous wall layer of turbulent boundary layers with continuous suction or blowing has not a universal solution any more, as has been shown by Gersten, Herwig (1992). For small blowing velocities the well–known logarithmic matching condition is extended to

$$\lim_{y^+\to\infty} u^+ = \frac{1}{\kappa}\ln y^+ + C^+ +$$

$$v_w^+\left[\frac{1}{4\kappa^2}(\ln y^+)^2 + \left[\frac{C^+}{2\kappa} + \frac{F_{vo}}{\kappa}\right]\ln y^+ + C_v^+\right] , \qquad (37)$$

where $u^+ = u^*/u_\tau^*$ , $y^+ = y^*\, u_\tau^*/\nu^*$, and $v_w^+ = v_w^*/u_\tau^*$. The value $F_{vo}$ follows from the turbulence model applied to the inviscid turbulent outer layer.

The turbulent Couette flow with suction or blowing have been investigated by Gersten, Herwig (1992, p. 448). They found the following values for $F_{vo}$ :

| | | |
|---|---|---|
| Model by Stevenson | : | $F_{vo} = 0$ |
| k–ε model : | : | $F_{vo} = -2.7$ |
| Model by Rotta | : | $F_{vo} = -4.6$ |

The boundary layer along a flat plate with suction or blowing is again an equilibrium boundary layer, when $v_w^*(X)$ is proportional to the local friction velocity $u_\tau^*(X) = \sqrt{\tau_w^*(X)/\rho^*}$ .

The implicit form of the resistance law for small values of $v_w^+$ can then be written as

$$\frac{2}{v_w^+}\left[\left[1+\frac{v_w^+}{u_\tau}\right]^{1/2}-1\right]=\frac{1}{\kappa}\ln\left(\mathrm{Re}_x\,u_\tau^2\right)+C^++\overline{C}$$

$$+\,v_w^+\left[\frac{F_{v0}}{\kappa}\ln\left(\mathrm{Re}\,u_\tau^2\right)+C_v^++\overline{C}_v-\frac{1}{4}\,C^{+2}\right], \qquad (38)$$

which reduces to (35) ($\beta = 0$) for $v_w \to 0$, see Gersten, Herwig (1992). In addition to $\overline{C}$ also the constants $F_{v0}$ and $\overline{C}$ depend on the turbulence model used.

Unfortunately, experimental data are not available to evaluate the various turbulence models for flows with blowing or suction.

## 5 References


Bush, W.B., Fendell, F.E. (1972): "Asymptotic analysis of turbulent channel and boundary–layer flow", J. Fluid Mech., Vol. 56, pp. 657–681

Cebeci, T., Bradshaw, P. (1984): *Physical and Computational Aspects of Convective Heat Transfer* (Springer–Verlag, New York)

Deriat, E., Guiraud, J.–P. (1986): "On the asymptotic description of turbulent boundary layers", Journal de Mecanique Theorique et Appliquee, Numero special, pp. 109–140

Fendell, F.E. (1972): "Singular perturbation and turbulent shear flow near walls", The Journal of the Astronautical Sciences, Vol. 20, pp. 129–165

Gersten, K. (1987): "Some contributions to asymptotic theory for turbulent flows", in: Proc. 2nd Int. Symp. on Transport Phenomena in "Turbulent Flows", Tokyo, October 25–29, pp. 201–214

Gersten, K. (1994): "What can asymptotic theory do for turbulence modellers?" In Proc. Symp. on Developments in Fluid Dynamics and Aerospace Engineering, Indian Institute of Science, Bangalore, 9–10 December 1993

Gersten, K., Herwig, H. (1992): *Strömungsmechanik. Grundlagen der Impuls–, Wärme– und Stoffübertragung aus asymptotischer Sicht,* (Vieweg, Braunschweig/Wiesbaden)

Gersten, K., Klauer, J., Vieth, D. (1993): "Asymptotic analysis of two–dimensional turbulent separating flows", in K. Gersten (Ed.): *Physics of Separated Flows – Numerical, Experimental, and Theoretical Aspects.* Notes on Numerical Fluid Mechanics, (Vieweg, Braunschweig/Wiesbaden) Vol. 40, pp.125–132

Jeken, B. (1992): "Asymptotische Analyse ebener turbulenter Strömungen an gekrümmten Wänden bei hohen Reynolds–Zahlen mit einem Reynolds–Spannungs–Modell", Dissertation (Ruhr–Universität Bochum), see also: ZAMM, Vol. 72, pp. T308–T312

Kluwick, A. (1989): "Interacting turbulent boundary layers", ZAMM, Vol. 69, pp. T560–T561

Mellor, G.L. (1972): "The large Reynolds number asymptotic theory of turbulent boundary layers", Int. J. Engng. Sci., Vol. 10, pp. 851–873

Rotta, J. (1950): "Über die Theorie der turbulenten Grenzschichten", in Mitteilungen aus dem Max–Planck–Institut für Strömungsforschung, ed. by A. Betz

Schneider, W. (1991): "Boundary–layer theory of free turbulent shear flows", Z. Flugwiss. Weltraumforsch., Vol. 15, pp. 143–158

Yajnik, K.S. (1970): "Asymptotic theory of turbulent shear flows," J. Fluid Mech., Vol. 42, pp. 411–427

# Turbulent Entrance Flow in a Channel: An Asymptotic Approach

Heinz Herwig [1], Marco Voigt [2]

[1] Technische Thermodynamik, TU Chemnitz-Zwickau, D-09107 Chemnitz, Germany
[2] Institut für Thermo- und Fluiddynamik, Ruhr-Universität Bochum, D-44780 Bochum, Germany

**Abstract:** Entrance flow in a channel is investigated for high Reynolds numbers. In the limit $Re \to \infty$ there exists a distinct structure of the flow field which can be taken into account by asymptotic methods. Contrary to the laminar flow situation, there is no need for an extra expansion in the immediate entrance region. The downstream expansion which results from a leading order momentum equation of slender channel type can be also used from the very beginning of the channel.

## 1 Introduction

During the last two decades CFD (computational fluid dynamics) has emerged as an independent branch of fluid mechanics and there is no doubt that its importance will further increase. This rapid development goes hand in hand with a shift in intellectual effort from analysing the physical problem to improving the computer code. But, analytical considerations will always be important to understand the physics underlying the problem under consideration. Asymptotic methods, in this respect, are powerful tools, especially in the turbulent case. Moreover, computational and analytical methods supplement each other as far as the Reynolds number scale is concerned: while CFD solutions are well established in the low Reynolds number range and are gradually extended to higher Reynolds numbers, the asymptotic theory based on a *Re*-linked perturbation parameter is an approximation for $Re \to \infty$, being exact in the limit.

As far as turbulent flows are concerned, CFD solutions incorporate turbulent models from the beginning (when they are based on the Reynolds averaged equations and not DNS-approaches) whereas asymptotic considerations, prior to turbulence modeling, will reveal much of the physics of a problem, as extracted from the basic Navier-Stokes equations in unclosed form.

In this sense, CFD and asymptotic approaches should not be considered to be competing methods but rather complementing approaches. Formally this might be expressed by placing an 'A' in front of CFD which stands for 'analytical' or 'asymptotic'. In this study we will give an example of this ACFD-approach by analysing the turbulent entrance flow into a channel for high Reynolds numbers.

## 2 The laminar case for $Re \to \infty$

Since the special physics of the turbulent entrance flow shall be illustrated as opposed to its laminar counterpart, we first give a brief description of the high Reynolds number structure of the laminar case.

A comprehensive study of laminar entrance flow in the high Reynolds number limit has been performed in the late sixties (Van Dyke, 1970). The asymptotic structure revealed in this study is sketched in figure 1.

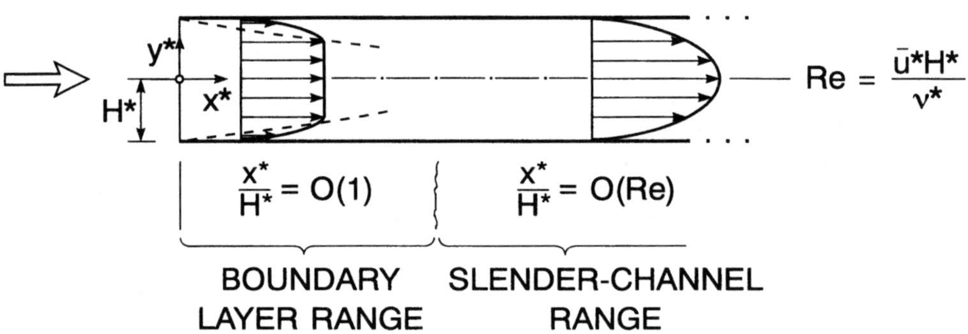


**Fig. 1.** Asymptotic structure and variables in the laminar case

Basically, there are two regions. The first one is the entrance range with streamwise extension of order $O(1)$ in the dimensionless coordinate $x = x^*/H^*$. Here, boundary layer physics describes the gradual penetration of viscous effects into the initially inviscid core region. A second region appears downstream where changes occur due to the slenderness of the geometry on a streamwise coordinate $x$ of order $O(Re)$ only. Both regions have to be matched properly to give a rational description of the flow for high Reynolds numbers.

It turns out that the downstream expansion applied to the whole entrance region, i.e. also in the boundary layer range, is a good (though irrational) approximation, even for rather low Reynolds numbers. As an example, a heat transfer result is shown in figure 2 compared to two Navier-Stokes results for small Reynolds numbers. For a Reynolds number of 25 the asymptotic result ($Re = \infty$) is a very good approximation even for the first data point which is located, less than half a channel height into the channel !

The physics of turbulence suggests that a similar result should be expected in the turbulent case.

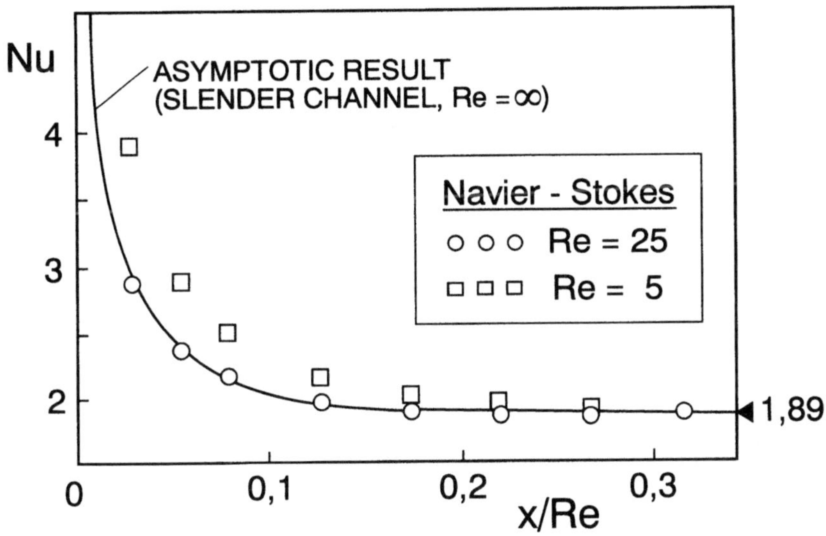


**Fig. 2.** Heat transfer distribution, $T_W^* = \text{const}$, $Pr = 1$ from: Gersten and Herwig (1992)

## 3 The turbulent case for $Re \to \infty$

In figure 3, the structure of the high Reynolds number solution is sketched in the downstream region leaving open, for the moment, what happens in the immediate entrance region. Adjacent to the fully turbulent core region, there are two asymptotically thin sublayers in which viscous effects are of the same order of magnitude as the turbulent transfer processes. This singular structure comes (formally) from an expansion of the time averaged Navier-Stokes equations with respect to the perturbation parameter

$$\epsilon \equiv u_{\tau R} = \frac{u_{\tau R}^*}{\bar{u}^*} = O\left(\frac{1}{\ln Re}\right) . \tag{1}$$

This special choice of the perturbation parameter is not as obvious as in the laminar case. Physically, $1/(\ln Re)$ is related to the skin friction velocity $u_{\tau R}^* = \sqrt{|\tau_{WR}^*|/\rho^*}$ which serves as a scaling quantity in the turbulent case, see Mellor, 1972. This quantity should be looked upon as being given rather than being part of the result. Compared to the laminar case, we thus have a kind of inverse problem: which average velocity $\bar{u}^*$ corresponds to the friction velocity $u_{\tau R}^*$? Once you accept this point of view, you can identify the flow in the viscous sublayers as that part of the flow field which is described by the primary expansion (no information is necessary from the adjacent region to start the expansion, see Schneider, 1978). The core region, as a consequence, is the secondary expansion which, with its leading order term, is already matched to the viscous sublayers.

The viscous sublayer expansions turn out to be the same for all wall bounded flows with non-zero shear stress. Without turbulence modeling, much is already known about

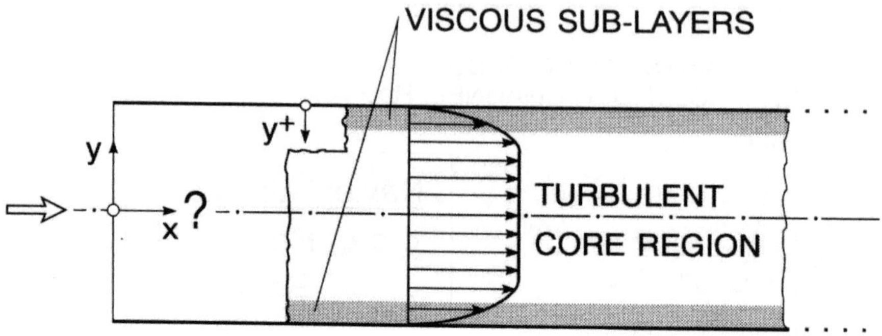


**Fig. 3.** Flow structure in the downstream region

its leading order solution. For example

$$\lim_{y^+\to\infty} u^+ = \frac{1}{\kappa}\ln y^+ + C^+ , \tag{2}$$

the well known logarithmic behaviour, can be deduced without turbulence modeling, see Gersten and Herwig, 1992.

The asymptotic expansion of the core region reveals that (to leading order) equations of the slender channel type appear, for details see Voigt, 1994. In order to get a *Re*-independent solution of these equations, they must be transformed to the variable

$$\tilde{u} = \frac{u_c^* - u^*}{u_\tau^*} + \frac{1}{\kappa}\ln\left(1 - \frac{y^*}{H^*}\right) \tag{3}$$

which is characterized by three features:

1) it is in defect form $u_c^* - u^*$ , $u_c^*$ is the centerline velocity,

2) it is scaled with the (local) skin friction velocity $u_\tau^*$,

3) the singular term $\frac{1}{\kappa}\ln(1-y)$, which comes in by matching with (2), is subtracted.

The momentum equation (leading order equation of the expansion) thus reads

$$\frac{\partial \tilde{u}}{\partial \hat{x}} = \frac{\partial \tau_t}{\partial y} - \left.\frac{\partial \tau_t}{\partial y}\right|_c + O(\epsilon^2) . \tag{4}$$

Here $\tau_t$ is the turbulent shear stress. The independent variables are of slender channel type:

$$\hat{x} = \frac{x^*}{H^*}\epsilon \quad ; \quad y = \frac{y^*}{H^*} . \tag{5}$$

Equation (4) can be solved after a turbulence model is incorporated for $\tau_t$. Before we do this, equation (4) should be analysed according to its behaviour for $\hat{x} \to 0$, i.e. in the immediate entrance region.

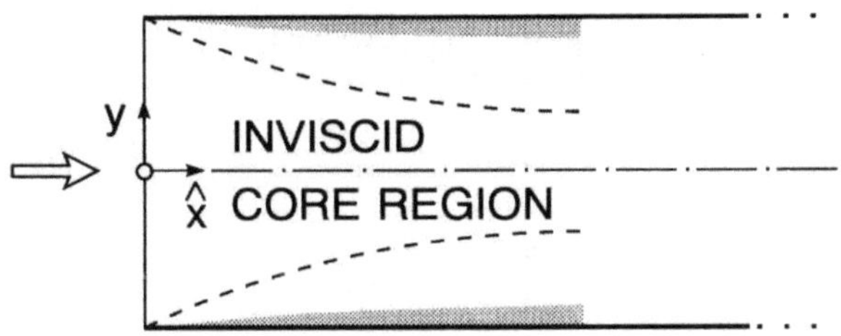


**Fig. 4.** Flow structure in the entrance region

Figure 4 illustrates that, in the entrance region, turbulence is constrained to the wall boundary layers. As a consequence, the term $\partial\tau_t/\partial y|_c$ in equation (4) vanishes if this equation is (formally) applied in this region.

The remaining terms in equation (4) are identical to the terms of a boundary layer asymptotic analysis, though given in slender channel rather than in boundary layer variables. Table 1 compares both approaches.

| | SLENDER CHANNEL | BOUNDARY LAYER |
|---|---|---|
| STREAMWISE COORDINATE | $\hat{x} = \frac{x^*}{H^*}\epsilon$ | $x = \frac{x^*}{H^*}$ |
| NORMAL COORDINATE | $y = \frac{y^*}{H^*}$ | $\hat{y} = \frac{y^*}{H^*}/\epsilon$ |
| LEADING ORDER MOMENTUM EQUATION | $\frac{\partial\tilde{u}}{\partial\hat{x}} = \frac{\partial\tau_t}{\partial y}$ | $\frac{\partial\tilde{u}}{\partial x} = \frac{\partial\tau_t}{\partial\hat{y}}$ |

**Tab. 1.** Turbulent leading order slender channel and boundary layer equations, perturbation parameter: $\epsilon = u_{\tau R}$

From table 1 it is obvious that both equations are equivalent. Multiplying the slender channel equation by $\epsilon$ transfers it formally to the boundary layer equation. This is distinctly different from the laminar situation as will be shown afterwards. As a consequence, equation (4) can be applied in the entrance region (i.e. up to $\hat{x} = 0$) as well as downstream. No extra expansion is necessary in the entrance region.

# 4 Physical interpretation

From the two previous sections, it is obvious that laminar and turbulent flow structures are quite different in the high Reynolds number limit. To be more specific, the laminar case is shown in table 2 in the same way as the turbulent case in table 1.

| | SLENDER CHANNEL | BOUNDARY LAYER |
|---|---|---|
| STREAMWISE COORDINATE | $\hat{\hat{x}} = \frac{x^*}{H^*}\epsilon^2$ | $x = \frac{x^*}{H^*}$ |
| NORMAL COORDINATE | $y = \frac{y^*}{H^*}$ | $\hat{y} = \frac{y^*}{H^*} / \epsilon$ |
| LEADING ORDER MOMENTUM EQUATION | $u\frac{\partial u}{\partial \hat{\hat{x}}} + \hat{\hat{v}}\frac{\partial u}{\partial y} = -\frac{dp}{d\hat{\hat{x}}} + \frac{\partial \tau}{\partial y}$ | $u\frac{\partial u}{\partial x} + \hat{v}\frac{\partial u}{\partial \hat{y}} = -\frac{dp}{dx} + \frac{\partial \tau}{\partial \hat{y}}$ |

**Tab. 2.** Laminar leading order slender channel and boundary layer equations, perturbation parameter: $\epsilon = 1/\sqrt{Re}$

The slender channel equations in the laminar case cannot be formally transferred into the boundary layer equations, basically since $x$ contains $\epsilon^2$, $\hat{y}$ however $\epsilon$ only. Obviously the physics in the entrance region is quite different from that downstream, so that an extra expansion is necessary. But, what is the reason for this? A simple explanation can be given in the following way: in the laminar case, the oncoming inviscid profile is changed downstream by molecular viscosity effects equally distributed over the cross-section, so that a parabolic velocity profile finally forms for $x \to \infty$, irrespective of how high the Reynolds number is.

In the turbulent case, molecular effects are restricted to the wall regions. The turbulent eddy viscosity effects in the core region are much stronger, so that the velocity profile is only slightly changed. As a consequence, the downstream profile is kept similar to the oncoming full profile for high Reynolds numbers.

In figure 5, this difference is illustrated by the form of the profiles upstream and

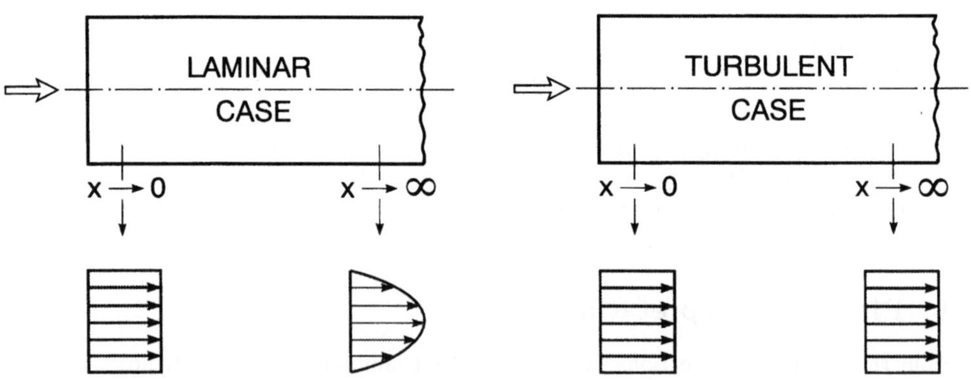


**Fig. 5.** Velocity profiles in the limit of infinitely high Reynolds numbers upstream and downstream

downstream in the high Reynolds number limit. In the turbulent case, the actual profile is a small perturbation of the full profile everywhere, whereas in the laminar case this is true only in the immediate entrance region.

This may elucidate a very important "principle" of asymptotic analysis: never study turbulent flows with a laminar mind !

## 5 Results and comparison with experiment

When a turbulence model is introduced into equation (4), the flow field in the core region can be determined (independent of *Re*). By matching with the viscous wall layers, the skin friction law can be obtained, from which the skin friction distribution (for different Reynolds numbers) emerges. In figure 6, the distribution of $c_f$ is shown for three different Reynolds numbers, based on the $k$-$\epsilon$-model for turbulence (for details, see Voigt, 1994). Even for the lowest Reynolds number $2.5 \cdot 10^4$, the deviations are presumably within the range of experimental uncertainty.

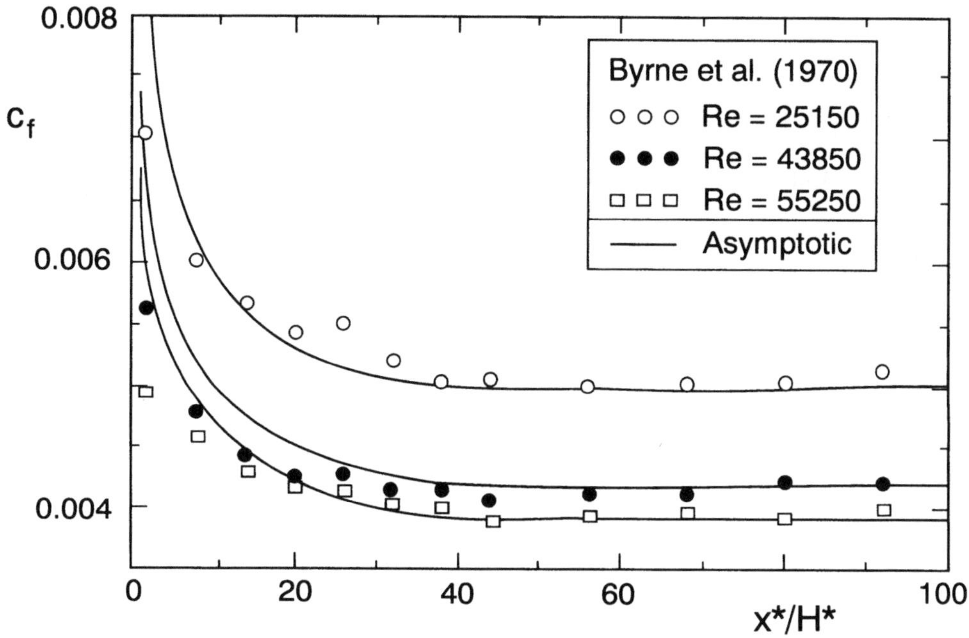


**Fig. 6.** Skin friction distribution from the asymptotic theory compared to experimental results

## 6 Concluding remarks

This study aimed at demonstrating that (and showing how) asymptotic and computational methods may enter into a profitable mutual alliance, especially for turbulent flows.

As an example, turbulent entrance flow was studied in the high Reynolds number limit. This limit solution provides approximate solutions for finite Reynolds numbers which are excellent approximations down to very low Reynolds numbers. Or, in other words: Reynolds numbers for which turbulent flows exist are high Reynolds numbers in an asymptotic sense.

## References

Byrne, J., Hatton, A.P., Marriot, P.G. (1970): "Turbulent flow and heat transfer in the entrance region of a parallel wall passage" in Proc. Instn. Mech. Engrs., Vol. 184, pp. 697–710

Gersten, K., Herwig, H. (1992): *Strömungsmechanik* (Vieweg-Verlag, Braunschweig/Wiesbaden)

Mellor, G.L. (1972): "The large Reynolds number asymptotic theory of turbulent boundary layers", Int. J. Engng. Sci., Vol. 10, pp. 851–873

Schneider, W. (1978): *Mathematische Methoden in der Strömungsmechanik* (Vieweg-Verlag, Braunschweig)

Van Dyke, M. (1970): "Entry flow in a channel", J. Fluid Mech., Vol. 44, pp. 813–823

Voigt, M. (1994): *Die Entwicklung von Geschwindigkeits- und Temperaturfeldern in laminaren und turbulenten Kanal- und Rohrströmungen aus asymptotischer Sicht* (Dissertation, Ruhr-Universität Bochum)

# An efficient wave interaction mechanism within a turbulent boundary layer

Sabine Ortiz [1] , Emmanuel Deriat [2]

[1] Ecole Nationale Supérieure de Techniques Avancées, 32 Bd Victor 75015 Paris
[2] Onera, 29 ave de la Division Leclerc 92320 Châtillon

**Abstract:**The paper deals with the weakly non-linear temporal stability of the mean flow of a turbulent boundary layer. The linear approximation first studied by Stephan had led to weakly damped modes. Here the amplitude equations are derived, the non linearities arising to second order due to the existence of three wave resonance. Some exact solutions of the amplitude equations are examined: they break down after a finite time, yielding an amplification mechanism.

## 1. Introduction

In 1985 Guiraud, in order to gain a better understanding of self-sustained turbulence by using asymptotic methods, proposed to Pace the study of the linear temporal stability of an incompressible turbulent boundary layer mean flow. Stability concepts are well known by those who are interested in transition to turbulence. Guiraud's point of view was to develop similar ideas in fully developped turbulence. The majority of publications on "turbulent stability" has concentrated on pertubations governed by Orr-Sommerfeld-type equation. Reynolds (1972) obtained for an incompressible two-dimensional turbulent wake a stability map for large scale disturbances; he aimed at giving some insight into the generation of coherent structures. Hussain and Reynolds (1972) considered the response of a chanel mean flow to a forced excitation.

The originality of the studies carried out by Pace (1987) and then Deriat (1988) have consisted in taking into account, first the pertubed Reynolds stress tensor through the mixing length model (adopted for simplicity) and then the asymptotic structure of the turbulent boundary layer. Disturbances are taken as two-dimensional waves of small amplitudes. Under the assumption of nearly parallel flows (Bouthier(1972) and Deriat(1988)) the linearized pertubed Reynolds equations lead to an eigenvalue problem governed by an operator which is not the Orr-Sommerfeld one but the **Pace-Deriat operator** as called by Guiraud(1989) .

Pace(1987) showed that the defect layer is the critical layer of our problem in the sense that Reynolds stresses play a main part. The Squire transformation is no longer valid for the new operator; this explains why Stephan (1992) coped with the tedious problem of determining the three-dimensional modes aiming at finding out a linear mechanism of amplification. However the numerical calculation established a general damping. Nevertheless, by considering the energy balance equation of each mode, Stephan (1992) found a positive production term for some of them. This encouraging result and the fact that some modes are weakly damped persuaded us to go on and to look for a weakly non-linear mechanism of amplification by examining their interactions between.

By examining in some details numerical results of Stephan (1992), one can find three resonant waves which are weakly damped. We call a resonant triad, three waves such that two of them can interact to second order to excite the third one; the sum of their wave vectors equals the wave vector of the third one, and the sum of their wave speeds equals the third one.

When resonance occurs in weakly non-linear theory, amplitude equations exhibit quadratic non linearities which dominate cubic non linearities. In fact, for a wide variety of flows, without resonant interactions, the symmetry of the problem leads to cubic non linearities, the amplitude equation being a Landau-type equation.

The first to investigate a three wave resonance was Phillips(1960) for gravity waves. Within the framework of laminar boundary layer, Craik(1971) tackled Tollmien-Schlichting wave interactions. He underlined the importance of the critical layer in energy transfer. The purpose of this paper is to devise a weakly non-linear theory, in the light of Craik's approach, which can be adapted to *any shear flow at large Reynolds numbers in which suitable wave triad may exist* (Craik(1971)), with the hope of bringing to the fore an amplification mechanism.

## 2. Scope of the Study

### 2.1 assumptions

When using a weakly non-linear theory one assumes the linear approximation to be the leading term. This section is devoted to linear stability, studied by Stephan(1992). The turbulent mean flow given by $U_B = (u_B, v_B, 0, p_B)$ , is an incompressible, two-dimensional and steady flow, governed by Reynolds equations. The longitudinal velocity component is denoted by $u_B$, the normal one by $v_B$ and $p_B$ stands for the mean pressure. We also assume that $\|u_B\| >> \|v_B\|$, in order to suppose that the flow is nearly parallel (Bouthier(1972) and Deriat (1988)). The pertubation, whose amplitude gauge is $\eta$, is denoted by $\eta U(\chi, y)$ where $y$ is defined by $Y = \sqrt{C_f}y$, $Y$ being the normal coordinate and $\sqrt{C_f}$ standing for the gauge of the boundary layer thickness. In order to break the thin layer structure of the turbulent boundary layer, we suppose that the perturbation depends upon a rapid variable, $\chi$ (Deriat (1988)), that means:

$$\begin{cases} O(\frac{\partial \chi}{\partial X}) = \left(\sqrt{C_f}\right)^{-1}, \\ O(\frac{\partial \chi}{\partial Z}) = \left(\sqrt{C_f}\right)^{-1}, \\ O(\frac{\partial \chi}{\partial T}) = \left(\sqrt{C_f}\right)^{-1}. \end{cases} \tag{1}$$

where $X$, $Z$ and $T$ stand for the longitudinal coordinate, the transversal coordinate and time. The components of the pertubation are denoted by $u$, $v$, $w$ and $p$. One originality of the linear stability study is to take into account the pertubed Reynolds stresses. For simplicity, we choose the mixing length model as a closure model, the dimensionless mixing length is named $\Lambda$. The unpertubed traceless Reynolds tensor, $\tau^d(U_B)$, is given hereafter:

$$\tau^d(U_B) = 2\sqrt{2}C_f\Lambda^2 Tr(D_B^2)^{1/2}\, D_B, \tag{2}$$

where $Tr$ and $D_B$ stand for the trace and the strain rate tensor. The pertubed Reynolds tensor can be expanded in a formal series (Ortiz(1994), appendix A); we show such a series below, truncated to third order (the hereafter indices 1, 2 or 3 represent respectively the linear, the quadratic and the cubic approximation):

$$\tau^d(U_B + \eta U) = \tau^d(U_B) + \eta\sqrt{C_f}\tau_1^d + \eta^2\tau_2^d + \frac{\eta^3}{\sqrt{C_f}}\tau_3^d \quad . \tag{3}$$

## 2.2 pertubed Reynolds equation

Under the above assumptions, the pertubed Reynolds equations can be expressed as is shown now (Ortiz (1994)):

$$\underbrace{\frac{\partial}{\partial T}MU + LU}_{\mathbf{I}} + \underbrace{\eta_{\neq}L_{\neq}U}_{\mathbf{II}} = \frac{\eta}{\sqrt{C_f}}\underbrace{\left(\nabla\tau_2^d - \nabla(MU)MU\right)}_{\mathbf{III}} + \underbrace{\frac{\eta^2}{C_f}\left(\nabla\tau_3^d\right)}_{\mathbf{IV}} \quad . \tag{4}$$

In this equation, **I** refers to the linearized problem where $M$ and $L$ are linear operators, **III** is the quadratic non-linear term, **IV** corresponds to the cubic non linearity, finally **II** is induced by non-parallel terms of the mean flow, typically one of those terms is $\eta u\frac{\partial u_B}{\partial X}$. Term **II** is liable to arise in equations to second order.

## 2.3 linear approximation

For a small pertubation ($\eta << 1$), equation (4) reduces to $\frac{\partial}{\partial T}MU + LU = 0$. The corresponding solution, $U_0$, is sought as a normal mode:

$$\begin{cases} U_0 = \tilde{U}exp(i\chi) = \tilde{U}(y)exp(i\alpha x + i\gamma z - i\sigma t), \\ \text{where } (x, y, z, t) = \sqrt{C_f}\,(X, Y, Z, T), \end{cases} \tag{5}$$

the longitudinal and transversal wave numbers $\alpha$ and $\gamma$ are reals and the frequency $\sigma$ is a complex number (temporal stability). It may be seen, taking into account the asymptotic structure of the turbulent boundary layer, that the problem in the wall layer yields explicit solutions which give boundary conditions to the problem set in the defect layer, by matching in an overlap domain.

In the defect layer, in order to prevent a non significant degeneracy of the problem, Stephan(1992) showed that $\sigma/\alpha$ must have the same structure as the longitudinal velocity,

$u_B = 1 + \sqrt{C}_f u_2$ ($u_2$ stands for the *defect velocity*). Hence the complex frequency can be put in the form:

$$\sigma = \alpha + \sqrt{C_f}\,(\sigma_r + i\sigma_i) \quad \text{where} \quad i^2 = -1 \quad . \tag{6}$$

pace*.72cmhspace*.72cmA numerical computation was necessary to obtain the mode shapes $\tilde{U}(y)$ and the complex frequency $\sigma$. Stephan(1992) showed that $\sigma$ is a multiform function of $\alpha$ and $\gamma$. All the modes are damped ($\sigma_i < 0$). In the figure hereafter, we have represented four branches of solutions.

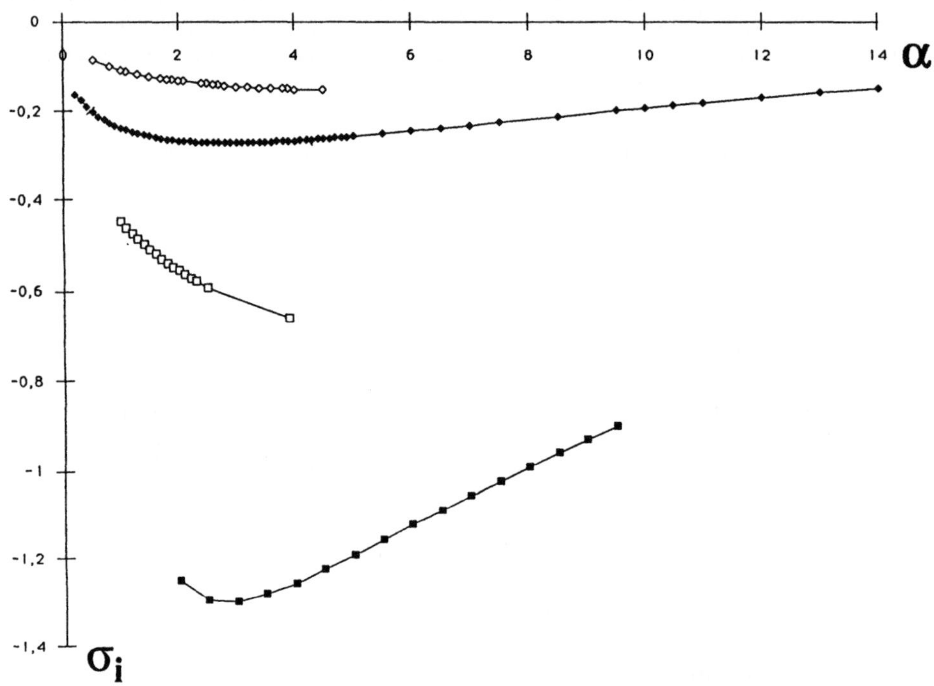


fig1: $\sigma_i$ (defined in (6)) as a function of $\alpha$ for $\gamma = 0.5$ according to Stephan (1992)

Modes, whose frequency belongs to one of the first two branches, are considered weakly damped; therefore it is set:

$$\sigma_i = \epsilon\tilde{\sigma}_i \quad \text{where} \quad \epsilon << 1 \quad , \tag{7}$$

where the weak damping gauge $\epsilon$ refers to our numerical data (Stephan (1992)).

## 3. Weakly non-linear theory

Ortiz(1994) showed that non-linear terms in (4) can be neglected in the wall layer, therefore the weakly non-linear problem, based on the weak damping and on the exis-

tence of resonant triads, will be set in the sole defect layer. In the introduction, we already stressed the importance of the critical layer for energy transfer as underlined by Craik(1971).
Now we precise the conditions satisfied by three resonant waves (constituying a resonant triad):

$$\begin{aligned} \textbf{(a)} \quad \alpha_1 &= \alpha_2 + \alpha_3 \ , \\ \textbf{(b)} \quad \gamma_1 &= \gamma_2 + \gamma_3 \ , \\ \textbf{(c)} \quad \sigma_{r1}^{s_1} &= \sigma_{r2}^{s_2} + \sigma_{r3}^{s_3}. \end{aligned} \tag{8}$$

The frequency evaluated for $\alpha_i, \gamma_i$ and belonging to the $s_i$ branch (see figure 1) is denoted by $\sigma_{ri}^{s_i}$, where $s_i$ is a numbering of weakly damped branches on which one can find resonant waves.

The amplitude of each mode is undetermined by the linear theory. It can be considered a slowly varying function of time. So we supposed that the amplitude depends on a **non-linear time** $T_1$, defined by:

$$T_1 = \epsilon T. \tag{9}$$

Our purpose is to bring to the fore an amplification mechanism by deriving the amplitude equations from (4). The first step is to defin correctly $U$ which is governed by (5) to leading order. The gauge of the non-parallel term in (4) $\eta_{\neq}$ equals $\eta\sqrt{C_f}$; we choose the amplitude gauge such that in (4) **III** dominates the term **II**. The cubic term **IV** in (4) is disregarded, which leads us to set:

$$C_f << \eta << \sqrt{C_f} \quad . \tag{10}$$

### 3.1 choice of the pertubation

The problem is to select $U$ which is governed to leading order by (5) . The first idea coming to mind is that $U$ is a continuum of modes, each mode having an amplitude $A(\alpha, \gamma, T_1)$, undetermined by linear theory ($T_1$ is defined in (9)). Unfortunately, deriving the amplitude equations from (4) is a very difficult problem which could be dealt with by using statistical property, Benney and Saffman(1966). We have decided to adopt a simpler point of view relying on the concept of extended resonance (Ortiz (1994))

Each mode (5) can be represented as a point $M$ in the $(\alpha, \gamma)$ plane and the correponding frequency $\sigma_r$ . We consider three domains $\Delta_1$, $\Delta_2$ and $\Delta_3$ containing $M_1(\alpha_1, \gamma_1)$, $M_2(\alpha_2, \gamma_2)$ and $M_3(\alpha_3, \gamma_3)$, those points being associated respectively with the following frequencies: $\sigma_{r1}^{s_1}$, $\sigma_{r2}^{s_2}$ and $\sigma_{r3}^{s_3}$. They correspond to the resonant triad (8).
By definition, three modes associated, in the $(\alpha, \gamma)$ plane, with $M$, $M'$ and $M''$ belonging to $\Delta_1$, $\Delta_2$ and $\Delta_3$, resonate in our extended sense when their wave numbers fullfil the (8a) and (8b) requirements , and when:

$$O\left(\sigma_r - \sigma_r' - \sigma_r''\right) = \epsilon \ , \tag{11}$$

where $\epsilon$ is the weak damping gauge (6), and where $\sigma_r$, $\sigma'_r$ and $\sigma''_r$ are the frequencies associated with $M$, $M'$ and $M''$. The resonant triad, associated with $M_1$, $M_2$ and $M_3$ satisfies (8c) which does not invalidate (11). We can define now the perturbation by its inverse Fourier transform:

$$\begin{cases} U = \int_{R^2} \left(\dot{U} + \dot{W}\right) e^{i\alpha x + i\gamma z} d\alpha d\gamma \ , \\ \text{where} \quad \dot{U} = \sum_{j=1}^{nb} A^j(\alpha,\gamma)\tilde{U}^j(y)e^j \ + \ \sum_{\mu=1}^{\mu=3} B^{s_\mu}(\alpha,\gamma,T_1) 1_{\Delta_\mu}(\alpha,\gamma)\tilde{U}^{s_\mu} \ e^{s_\mu} \ , \\ \text{and } O\left(\dot{U}\right) \ >> \ O\left(\dot{W}\right) \ . \end{cases} \tag{12}$$

The first summation is over $nb$ branches of solutions (see fig.1), the second is over three waves which resonate in the extended sense; $1_{\Delta_\mu}(\alpha,\gamma)$ is a function which equals one if $(\alpha,\gamma)$ belongs to $\Delta_\mu$, zero if not and $e^j$ stands for $exp(-i\sigma^j t)$. We seek some evolution equations for $B$, assumed to vary rapidly on each domain, this can be expressed by the following property:

$$\forall\, (\mu,\lambda) \in \{1,2,3\}^2 \ \left|\int_{\Delta_\lambda} A^j(\alpha,\gamma) d\alpha d\gamma\right| \ << \ \left|\int_{\Delta_\mu} B^{s_\mu}(\alpha,\gamma,T_1) d\alpha d\gamma\right| . \tag{13}$$

The term $\dot{W}$ is induced by $\dot{U}$ self-interaction.

### 3.2 amplitude equations

By applying the Fourier transform to (4) taking into account (8a), (8b), (9), (10) and (12), the non-linear equation can be written as shown now (see Ortiz(1994) chapter IV):

$$\eta\epsilon e^{s_1} \frac{\partial B^{s_1}}{\partial T_1} M\tilde{U}^{s_1} + \left[\frac{\partial}{\partial T} M + \mathcal{L}_{(\alpha,\gamma)}\right] \dot{W} = \mathcal{R} + \mathcal{NR} \ , \tag{14}$$

where $\mathcal{L}_{(\alpha,\gamma)}$ is the Fourier transform of $L$ (4). The non-linear (quadratic) term can be divided into two components: the first one $\mathcal{R}$ comes from resonant interactions (in the extended sense) of linear modes whose amplitudes are $A^j$ or $B^{s_\mu}$. The second term comes from non resonant terms. Therefore, $\dot{W}$ can be expressed as the sum of two terms, $\dot{W}_R$ and $\dot{W}_{R^c}$ (for more details, see Ortiz (1994) chapter IV). The latter leads to a problem whose righthand member is $\mathcal{NR}$, which can be solved (Ortiz (1994) p.94). The former can be, as suggested by the above equation, written in the form:

$$\dot{W}_R \ = \epsilon\, \eta \ e^{s_1} \ B^{s_\mu} U_2^{s_1} \ , \tag{15}$$

where $U_2^{s_1}$ is an unknown function. The assumption made in (13) shows that, among the interactions between modes, only those correponding to modes of the second summation in (12) are significant, the leading term, $\mathcal{R}_0$ of $\mathcal{R}$ is as follows:

$$\mathcal{R}_0 = \int_{R^2} B^{s_2}(\alpha',\gamma',T_1) B^{s_3}(\alpha'',\gamma'',T_1) e^{s_2} e^{s_3} 1_{\Delta_2}(\alpha',\gamma') 1_{\Delta_3}(\alpha'',\gamma'') \mathcal{Q}(\tilde{U}_2^s,\tilde{U}_3^s) d\alpha' d\gamma' \ , \tag{16}$$

where $\mathcal{Q}(\tilde{U}_2^s,\tilde{U}_3^s)$ arises when taking the inverse Fourier transform of the quadratic non linearities in (4) relying on the particular expression of the inverse Fourier transform of

$U$ (12). Therefore, the non-linear equation (14) becomes:

$$e^{s_1} \frac{\partial}{\partial T_1} B^{s_1}(\alpha,\gamma,T_1) M\tilde{U}^{s_1} + B^{s_1} \left[\frac{\partial}{\partial T} M + \mathcal{L}_{(\alpha,\gamma)}\right] (e^{s_1} U_2^{s_1}) = \mathcal{R}_0 \ . \tag{17}$$

Furthemore, one can convince himself that each of the terms in (17) has the same structure: a term rapidly varying upon $\Delta_\mu$ ($B^{s_\mu}(\alpha,\gamma,T_1)$) times another one slowly varying. By integrating (17) over $\Delta_1$ we obtain by setting:

$$\forall \ \mu \ \in \ \{1,2,3\} \ \ a_\mu \ = \ e_\mu^{s_\mu} \int_{\Delta_\mu} B^{s_\mu}(\alpha,\gamma,T_1) d\alpha d\gamma \ , \tag{18}$$

a non-linear inhomogeneous equation for $U_2^{s_1}$ (in the above equation $e_\mu^{s_\mu}$ is the value of $e^{s_\mu}$ for $\alpha$ equals $\alpha_\mu$ and $\gamma$ equals $\gamma_\mu$) . It is well known by those who deal with weakly non-linear theory, that the solvability condition gives the amplitude equations. It is necessary to derive the adjoint operator associated with the homogeneous linearized problem studied by Stephan(1992) (Ortiz (1994) chapter III). Finally, we obtain the amplitude equations for each wave of the triad:

$$\begin{cases} \frac{d}{dT_1} a_1 &= \tilde{\sigma}_i^{s_1} a_1 + \Lambda_1 a_2 a_3 \ , \\ \frac{d}{dT_1} a_2 &= \tilde{\sigma}_i^{s_2} a_2 + \Lambda_2 a_1 a_3^* \ , \\ \frac{d}{dT_1} a_3 &= \tilde{\sigma}_i^{s_3} a_3 + \Lambda_3 a_1 a_2^* \ . \end{cases} \tag{19}$$

where $*$ stands for the complex conjugate, and $\tilde{\sigma}_i^j$ is defined in (7). The coefficients $\Lambda_i$ arising in (19) come from the inner product of (17) with an adjoint normal mode.

### 3.3 amplification mechanism

The normal shapes $\tilde{U}$ (5) and the adjoint normal shapes are known (Stephan(1992) and Ortiz (1994)). We can, for a given triad, compute the interaction coefficients appearing in the above equations. Since they are complex numbers, (19) is in a non-conservative form Craik(1986). The solution of (19) for a particular set of initial conditions is obtained by using a particular case of Craik's method (Craik (1986)). By doing this, we find (Ortiz (1994) chapter V):

$$|a_\mu| \ \sim \ \frac{C}{|1 - K exp(-\nu T_1|} \ , \tag{20}$$

where $\nu$ is the sum of the $\tilde{\sigma}_i^j$. The amplitude is unbouded after a finite time. The numerical values of $\tilde{\sigma}_i^j$ (Ortiz(1994) chater V) imply that the blow-up time is, if $K$ is chosen smaller than one, roughly $T$. This is the amplification mechanism within the turbulent boundary layer that we claimed, which could explain how the mean flow feeds the turbulence through pertubations. It seems that this blow-up could be obtained even if we had used an another closure model, on condition that weakly damped resonant triad still exists (Ortiz(1994)).

Other solutions of (19) could be sought, corresponding to an another set of initial conditions. After a finite time the amplitude are not bounded, cubic non linearities cannot be neglected in (4). The analysis of cubic non linearities could bring a better understanding of this amplification mechanism.

## References

Benney, D.J., Saffman, P.G (1966): "Non linear interactions of random waves in a dispersive medium", Proc. Roy. Soc. Lond, A, 389, pp. 301-320

Bouthier, M. (1972): "Stabilité linéaire des écoulements presque parallèles Partie II. La couche limite de Blasius", J. Mécanique, 12, pp. 75-95

Craik, A.D.D. (1971): " Non linear resonant instability in boundary layers", J. of Fluid Mech., 50 pp. 393-413

Craik, A.D.D.(1986):"Exact solution of non conservative equations for three waves and second harmonic resonance", Proc. Roy. Soc. Lond, A, 406, pp. 1-12

Deriat, E. (1988): *Contributions à l'étude de la structure asymptotique de la turbulence de couche limite* (Phd thesis Univ. P & M Curie, Paris)

Deriat, E., Pace, J.M., Guiraud, J.P. (1989): " Asymptotic theory of unstable normal modes of pertur bations on a turbulent boundary layer velocity profile", GAMM conference, Vienne, ZAMM, Vol. 69 n.6, pp. 558-560

Ortiz, S. (1994): *Interaction de modes tridimensionnels résonants au sein d'une couche limite turbulente* (Phd thesis Univ. P & M Curie, Paris)

Pace, J.M. (1987): *Mécanisme d'instabilité du profil de vitesse moyenne dans une couche limite turbulente* (Phd thesis Univ. P & M Curie, Paris)

Phillips, O.M. (1960): " On the dynamics of unsteady gravity waves of finite amplitude", J. of Fluid Mech., 9 part 1, pp. 193-217

Reynolds, W.C. (1972): "Large scale instabilities of turbulent wakes", J. of Fluid Mech., 54, pp. 481-488

Reynolds, W.C., Hussain, A.K.M.F (1972): "The mechanics of an organized wave in turbulent shear flow, Part 3 Theoretical models and comparison with experiments", J. of Fluid Mech., 54, pp. 263- 288

Stephan, F. (1992): *Etude théorique linéaire et recherche de modes propres tridimensionnels de perturbation dans une couche limite turbulente de plaque plane* (Phd thesis Univ. P & M Curie, Paris)

# Part II

# Hydrodynamic Instability

# Bifurcations of Capillary–Gravity Interfacial Waves

Frédéric Dias

Institut Non-Linéaire de Nice, UMR 129 – CNRS & UNSA
1361 route des Lucioles, F-06560 Valbonne

**Abstract:** Bifurcations of solitary capillary–gravity water waves are reviewed. A generalization to interfacial waves is provided. The case of solitary waves bifurcating at the minimum of the dispersion curve is chosen to illustrate the analysis, which makes use of tools of dynamical system theory. It is shown that there is a critical density ratio which strongly affects the type of solitary waves which may bifurcate. Depending on the density ratio, one can obtain "bright" solitary waves or "dark" solitary waves.

# 1 Introduction

Although the problem of water waves has been studied by many researchers, it is still a fascinating topic. Recently, there has been active research in analyzing the steady water-wave problem as a spatial Hamiltonian or reversible system. Iooss & Kirchgässner ([8],[9]) for example have revealed new solutions to the water-wave problem, especially in the context of solitary waves. If one is only interested in studying waves which travel uniformly at a constant speed $c$, one can remove the time from the equations by reformulating the problem in a frame of reference moving with the wave and by looking for steady solutions. The resulting spatial problem can be formulated as a dynamical system in space, the "evolution" variable being the horizontal coordinate $x$. Once in dynamical system form, the problem can be reduced by using for example the center manifold theorem, combined with the use of normal forms. The spectrum of the linearized operator $L$ depends strongly on the parameters of the problem, which we choose to be the Froude number $F$ and the Weber number $W$ defined as

$$F = \sqrt{\frac{c^2}{gh}} \quad \text{and} \quad W = \frac{\sigma}{\rho h c^2}, \tag{1.1}$$

where $g$ denotes the acceleration due to gravity, $\rho$ the fluid density, $\sigma$ the surface tension coefficient, $h$ the mean depth of the fluid. Of interest from the bifurcation point of view are the cases when $L$ possesses a double zero or double imaginary eigenvalues.

A physical interpretation of the meaning of such double eigenvalues can be given by looking at the dispersion relation for linearized capillary–gravity progressive waves. If

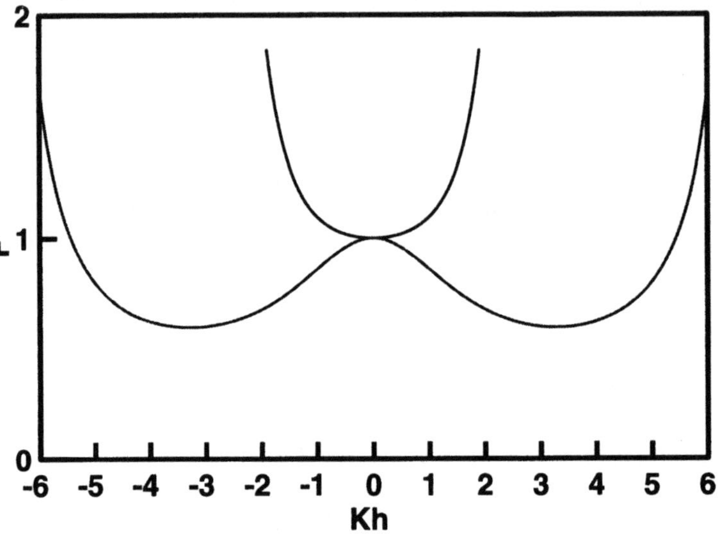


Figure 1: Dispersion relation for $W < 1/3$ (lower curve) and for $W > 1/3$ (upper curve)

$\kappa = Kh$ denotes the dimensionless wavenumber, the dispersion relation can be written as

$$D(\kappa, F, W) \equiv F^2 - \frac{\tanh \kappa}{\kappa - W\kappa^2 \tanh \kappa} = 0\,. \tag{1.2}$$

First note that, if $\kappa$ satisfies (1.2), so does $-\kappa$. Depending on the values of $W$ and $F$, equation (1.2) can have zero, one or two positive roots, as it can clearly be seen in Figure 1.

There are essentially three cases to consider:
(i) $W - 1/3 > 0$: The only double root is at $\kappa = 0$. The corresponding value for $F$ is one. A family of solitary waves of depression bifurcates from a uniform flow. Their existence has been proved by Amick & Kirchgässner [2] and by Sachs [13]. These waves have been computed numerically by Hunter & Vanden–Broeck [7]. A good approximation to these waves is given by the solitary-wave solution of the Korteweg–de Vries equation for $W > \frac{1}{3}$.
(ii) $W - 1/3 < 0$: There is a double zero root at $\kappa = 0$ and nonzero double roots at the minima of the dispersion curve. Note that the bifurcations occurring at the double zero root are more complex than in the previous case because there are also nonzero roots corresponding to $F = 1$. Iooss & Kirchgässner [9], Beale [3] and Sun [14] have proved existence results for generalized solitary waves consisting of a single crest plus a train of periodic waves of much smaller amplitude which persists at infinity. A good approximation to the shape of the single crest is given by the solitary-wave solution of the Korteweg–de Vries equation for $W < \frac{1}{3}$. Numerical computations by Vanden–Broeck [15] have produced particular waves for which the amplitude of the small oscillations is zero within numerical accuracy. Lombardi [4] mathematically proved that the amplitude can be extremely small. But whether or not the amplitude at infinity can be exactly zero remains an open

problem. At the minima of the dispersion curve, Iooss & Kirchgässner [8] proved the existence of solitary waves with damped oscillations ("bright" solitary waves), bifurcating from a progressive wave with wavenumber $K$ and with zero amplitude. Vanden–Broeck & Dias [16] computed such waves numerically. Dias & Iooss [5] computed such waves analytically.
(iii) $W - 1/3 = 0$: There is a quadruple zero root. The bifurcations occurring at $F = 1$ are complex and will not be described here.

The problem of interfacial waves has not been studied as much as the problem of water waves. Although many results on water waves can be extended to interfacial waves without leading to significant qualitative differences, there are some wave phenomena for which the density ratio can play an important role. In particular, it is shown that certain types of solitary waves ("dark" solitons), which are not possible in the context of water waves, are possible in the context of interfacial waves (see [6]).

# 2 The physical problem

The propagation of capillarity–gravity waves on the surface of a perfect fluid layer with constant density is considered. The layer is two-dimensional and bounded below by a horizontal wall. The flow is supposed to be irrotational. The coordinates are $x$ in the horizontal direction and $y$ along the vertical direction, with $y = 0$ representing the free surface at rest. The free surface is described by $y = \eta(x)$. The velocity components are $u$ and $v$. Dimensionless variables are introduced by taking the unit length to be $\sigma/\rho c^2$ and the unit velocity to be $c$.

The continuity equation and the irrotationality condition can be written in the dimensionless form

$$\left.\begin{array}{rcl} u_x + v_y & = & 0 \\ u_y - v_x & = & 0 \end{array}\right\} . \tag{2.1}$$

The boundary conditions at $|x| \to \infty$ are

$$u \to 1, \quad v \to 0 . \tag{2.2}$$

The kinematic condition on the free surface can be written as

$$u\eta_x - v = 0 , \qquad \text{at } y = \eta(x) . \tag{2.3}$$

Using Bernoulli's equation and eliminating the pressure at the free surface, one can write the dynamic condition in the form

$$\tfrac{1}{2}(u^2 + v^2) + \alpha\eta - \frac{\eta_{xx}}{(1 + \eta_x^2)^{3/2}} = \tfrac{1}{2} \quad \text{at } y = \eta(x) , \tag{2.4}$$

where the dimensionless number $\alpha$ is defined by

$$\alpha \equiv \frac{W}{F^2} = \frac{g\sigma}{\rho c^4} . \tag{2.5}$$

The water-wave problem is a nonlinear problem, the nonlinearities being present in the boundary conditions (2.3) and (2.4) at the free surface. The problem can be reformulated as a dynamical system in space (see [5] for details). Using $x$ and the streamfunction

$\psi$ (with $\psi = 0$ on the free surface) as independent variables makes the analysis somewhat easier. It is also preferable to use the variables

$$w_1(x,\psi) = \tfrac{1}{2}(u^2 + v^2 - 1)\,, \qquad w_2(x,\psi) = \frac{v}{u}\,, \tag{2.6}$$

rather than the velocity components $u$ and $v$. The dynamical system formulation is

$$\begin{aligned} w_{0x} &= (1+w_0^2)^{3/2}\,(w_1(\cdot,0) + F^{-2}\,[u^{-1}] - F^{-2}) \\ w_{1x} &= v w_{1\psi} - u^3 w_{2\psi} \\ w_{2x} &= w_{1\psi} u^{-1} + v w_{2\psi}\,, \end{aligned} \tag{2.7}$$

where $[\cdot]$ denotes the mean value of $\cdot$ over the interval $[-\frac{1}{W}, 0]$, together with the boundary conditions

$$\begin{aligned} w_2(\cdot,0) &= w_0 \\ w_2(\cdot,-W^{-1}) &= 0\,. \end{aligned} \tag{2.8}$$

Let $\mathbf{w}$ be the vector of components $(w_0, w_1, w_2)^T$. The system (2.7)–(2.8) to be solved can be written as

$$\frac{\partial \mathbf{w}}{\partial x} = L\mathbf{w} + N(\mathbf{w})\,, \tag{2.9}$$

where $L$ is the linearization about $\mathbf{w} = 0$ and $N$ a nonlinear operator. It is easy to show that the eigenvalues of $L$, denoted by $s$, must satisfy the relation

$$s\cos\left(\frac{s}{W}\right) = (\alpha - s^2)\sin\left(\frac{s}{W}\right)\,. \tag{2.10}$$

A complete analysis of the spectrum was done by Kirchgässner [11]. When $s = \pm\mathrm{i}W\kappa$, equation (2.10) becomes the well-known dispersion relation (1.2).

# 3 Analysis near the 1:1 resonance

We consider the case when $L$ has double imaginary eigenvalues. We denote by $\alpha_c$ the corresponding value for the parameter $\alpha$. These eigenvalues are non semi-simple. One can use the center manifold theorem (which is essentially a separation of variables) to show that all bounded solutions in a neighborhood of the 1:1 resonance ($\alpha$ near $\alpha_c$) can be written as

$$\mathbf{w}(x) = A(x)\varphi_0^+ + B(x)\varphi_1^+ + \overline{A(x)}\varphi_0^- + \overline{B(x)}\varphi_1^- + \text{higher-order terms}\,, \tag{3.1}$$

where $\varphi_0^+$ (resp. $\varphi_1^+$) is an eigenvector (resp. a generalized eigenvector) corresponding to the double imaginary root $s = \mathrm{i}k_c$ of (2.10). The bifurcation parameter is chosen to be $\mu = \alpha_c - \alpha$. The amplitudes $A$ and $B$, which only depend on the spatial variable $x$, are complex. One easily finds that

$$\varphi_0^+ = \begin{pmatrix} \mathrm{i}\tanh\kappa_c \\ -(\cosh k_c\psi + \sinh k_c\psi\tanh\kappa_c) \\ \mathrm{i}(\sinh k_c\psi + \cosh k_c\psi\tanh\kappa_c) \end{pmatrix}, \tag{3.2}$$

$$\varphi_1^+ = \begin{pmatrix} 0 \\ \mathrm{i}\psi(\sinh k_c\psi + \cosh k_c\psi \tanh \kappa_c) + \mathrm{i}\cosh k_c\psi(\tanh \kappa_c - \coth \kappa_c)/W \\ \psi(\cosh k_c\psi + \sinh k_c\psi \tanh \kappa_c) + \sinh k_c\psi(\tanh \kappa_c - \coth \kappa_c)/W \end{pmatrix}, \tag{3.3}$$

and

$$\varphi_0^- = \overline{\varphi_0^+}, \quad \varphi_1^- = \overline{\varphi_1^+}.$$

The normal form of the dynamical system, which allows to express the higher-order terms in (3.1) in a rather simple form, is (see [10])

$$\begin{aligned} A_x &= \mathrm{i}k_c A + B + \mathrm{i}AP\left(\mu; |A|^2, \tfrac{1}{2}\mathrm{i}(A\overline{B} - \overline{A}B)\right), \\ B_x &= \mathrm{i}k_c B + \mathrm{i}BP\left(\mu; |A|^2, \tfrac{1}{2}\mathrm{i}(A\overline{B} - \overline{A}B)\right) + AQ\left(\mu; |A|^2, \tfrac{1}{2}\mathrm{i}(A\overline{B} - \overline{A}B)\right), \end{aligned} \tag{3.4}$$

where $P$ and $Q$ are real polynomials

$$\begin{aligned} P(\mu; U, V) &= -p_1\mu + p_2 U + p_3 V + O(|\mu| + |U| + |V|)^2, \\ Q(\mu; U, V) &= -q_1\mu + q_2 U + q_3 V + O(|\mu| + |U| + |V|)^2. \end{aligned} \tag{3.5}$$

The coefficients $p_1$ and $q_1$ can be easily computed from (2.10). The other coefficients (see details in [5] on how to compute them) require more work.

The next step is to find the solutions to the normal form (3.4). The analysis is greatly simplified by using the fact that the system (3.4) is integrable, with integrals

$$K = \tfrac{1}{2}\mathrm{i}(A\overline{B} - \overline{A}B) \quad \text{and} \quad H = |B|^2 - \int_0^{|A|^2} Q(\mu; s, K)\,\mathrm{d}s.$$

A complete analysis of (3.4) was provided in [10]. The sign of the coefficient $q_2$ plays a key role in the structure of the solutions. For water waves, $q_2$ is always negative. For interfacial waves, the sign of $q_2$ depends on the density ratio.

# 4 Interfacial waves

We now consider the propagation of waves at the interface between two superimposed layers of perfect fluids. The formulation is the same as before except that the governing equations must be written in each fluid layer. There are two additional parameters: the ratio of depths and the ratio of densities. The analysis in arbitrary depth has not been performed yet. The analysis in infinite depth has been done formally [6]. It is appropriate here to mention the fact that, although all the expressions given in the previous section have a limit that makes sense in infinite depth, the rigorous arguments (in particular the application of the center manifold theorem) fail. Therefore the results given below have not yet been fully justified from a mathematical point of view. In infinite depth, the expression for $q_2$ is

$$q_2 = -\frac{11 - 42R + 11R^2}{8(1+R)^2}, \tag{4.1}$$

where the parameter $R$ is the density ratio. There is a critical value $R = R_c = (21 - 8\sqrt{5})/11$ of the density ratio for which the coefficient $q_2$ is zero. A precise study of the solutions in a neighborhood of that critical value would require the computation of higher-order terms and will be the subject of future work. Here we consider the cases $R > R_c$ and $R < R_c$, with $R$ not too close to $R_c$.

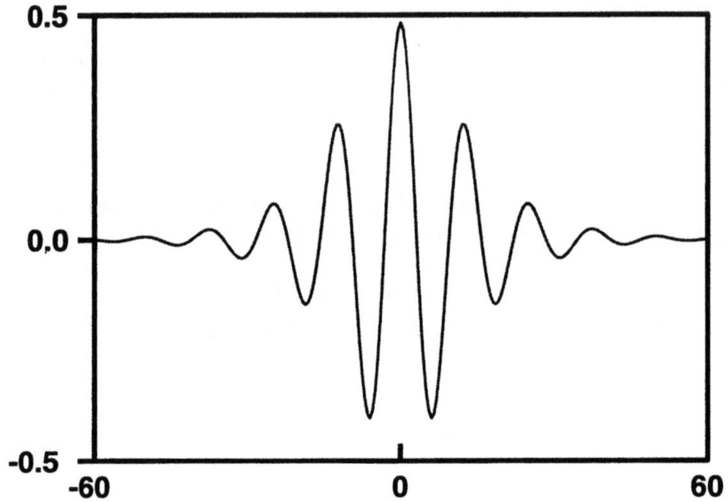


Figure 2: Bright solitary wave bifurcating at the minimum of the dispersion curve when the density ratio is small

## 4.1 Small density ratio

This case corresponds to $R < R_c$ (that is to say $q_2 < 0$). When $\mu > 0$, the solutions are periodic or quasi-periodic. When $\mu < 0$, the solutions are periodic, quasi-periodic or homoclinic. The homoclinic solutions correspond to solitary waves of the type shown in figure 2. They look like wave packets. Their profile at leading order is given by

$$\eta(x) = \pm\frac{4}{1+R}\sqrt{\frac{2|\mu|q_1}{q_2}}\frac{1}{\cosh\left(\sqrt{|\mu|q_1}x\right)}\cos\left(\tfrac{1}{2}(1+R)x\right). \tag{4.2}$$

## 4.2 Density ratio close to one

This case corresponds to $R > R_c$ (that is to say $q_2 > 0$). When $\mu > 0$, the solutions are periodic, quasi-periodic or homoclinic. The homoclinic solutions correspond to solitary waves of the type shown in figure 3. They look like dark solitons. They are connected to the same periodic wave at plus and minus infinity, with a phase shift. In the central part, the amplitude is small. Their profile at leading order is given by

$$\eta(x) = \pm\frac{4}{1+R}\sqrt{\frac{\mu q_1}{q_2}}\tanh\left(\sqrt{\frac{\mu q_1}{2}}x\right)\sin\left(\tfrac{1}{2}(1+R)x\right). \tag{4.3}$$

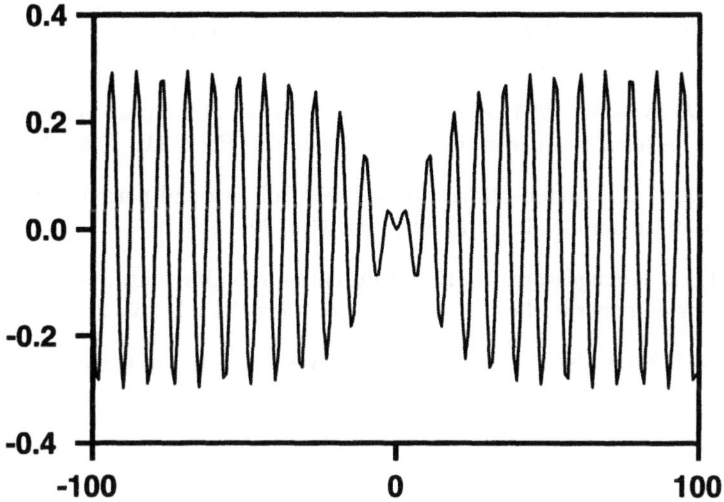


Figure 3: Dark solitary wave bifurcating at the minimum of the dispersion curve when the density ratio is close to one

## 4.3 Physical interpretation

There is an interesting link between these solitary waves which bifurcate at the minimum of the dispersion curve and the stationary solutions of the nonlinear Schrödinger equation (NLS), which describes the modulations of a wave train [1]. For surface waves, the negative value of $q_2$ corresponds to the focusing region of the NLS equation. When $q_2$ is positive, one is in the defocusing region. The corresponding solutions in the form of hyperbolic tangent are well-known in nonlinear optics and are called dark solitary waves because in their central part the wave amplitude is small.

# References

[1] Akylas, T. R. 1993 Envelope solitons with stationary crests. *Phys. Fluids A* **5**, 789–791.

[2] Amick, C. J. & Kirchgässner, K. 1989 A theory of solitary water-waves in the presence of surface tension. *Arch. Rational Mech. Anal.* **105**, 1–49.

[3] Beale, J. T. 1991 Exact solitary water waves with capillary ripples at infinity. *Comm. Pure and Appl. Math.* **44**, 211–257.

[4] Lombardi, E. 1992 Bifurcation d'ondes solitaires à oscillations de faible amplitude à l'infini, pour un nombre de Froude proche de 1. *C. R. Acad. Sci. Paris* **314 I**, 493–496.

[5] Dias, F. & Iooss, G. 1993 Capillary–gravity solitary waves with damped oscillations. *Physica D* **65**, 399–423.

[6] Dias, F. & Iooss, G. 1994 Ondes solitaires "noires" à l'interface entre deux fluides en présence de tension superficielle. *C. R. Acad. Sci. Paris* **319 I**, 89–93.

[7] Hunter, J. K. & Vanden–Broeck, J.-M. 1983 Solitary and periodic gravity–capillary waves of finite amplitude. *J. Fluid Mech.* **134**, 205–219.

[8] Iooss, G. & Kirchgässner, K. 1990 Bifurcation d'ondes solitaires en présence d'une faible tension superficielle. *C. R. Acad. Sci. Paris* **311 I**, 265–268.

[9] Iooss, G. & Kirchgässner, K. 1992 Water waves for small surface tension: an approach via normal form. *Proc. R. Soc. Edinburgh A* **122 A**, 267–299.

[10] Iooss, G. & Pérouème, M.-C. 1993 Perturbed homoclinic solutions in reversible 1:1 resonance vector fields. *J. Diff. Eq.* **102**, 62–88.

[11] Kirchgässner, K. 1988 Nonlinearly resonant surface waves and homoclinic bifurcation. *Advances Appl. Mech.* **26**, 135–181.

[12] Mielke, A. 1991 Homoclinic and heteroclinic solutions in two-phase flow. Universität Stuttgart, preprint.

[13] Sachs, R. L. 1991 On the existence of small amplitude solitary waves with strong surface tension. *J. Diff. Equ.* **90**, 31–51.

[14] Sun, S. M. 1991 Existence of a generalized solitary wave solution for water with positive Bond number less than 1/3. *J. Math. Anal. Appl.* **156**, 471–504.

[15] Vanden–Broeck, J.-M. 1991 Elevation solitary waves with surface tension. *Phys. Fluids A* **3**, 2659–2663.

[16] Vanden–Broeck, J.-M. & Dias, F. 1992 Gravity–capillary solitary waves in water of infinite depth and related free surface flows. *J. Fluid Mech.* **240**, 549–557.

# The effect of a low-frequency modulation on some codimension 2 bifurcations

M.N. Ouarzazi and P.A. Bois

Laboratoire de Mécanique de Lille
U.S.T.L., Bât. M3, F-59655 Villeneuve d'Ascq

**Abstract.** - The interactions between several types of intabilities lead to bifurcations of multiple codimension. The convective flow of a binary mixture in a porous medium is a good example of such an interaction of two instabilities. These instabilities depend on two parameters: the Rayleigh number *Ra* and the separation ratio $\psi$. The corresponding neutral curves intersect, in the $(Ra, \psi)$ plane, at a polycritical point. We describe, in this paper, the influence, on this bifurcation, of a small low-frequency time variation of the thermal boundary conditions. The reduced nonlinear Mathieu equation depends on a parameter $\varepsilon$, which characterizes the low frequency modulation. For $\varepsilon = 0$ the Mathieu equation possesses, in the phase plane, heteroclinic and periodic solutions. For $\varepsilon \neq 0$, the system may exhibit a chaotic régime of Smale horseshoe type. The onset of the chaotic regime as well as the curve at which a saddle-node bifurcation occurs can be estimated by means of Melnikov's technique. Numerical simulations agree with the results of Melnikov's theory. This problem can be considered to be a model for other instability problems arising in other domains of the Physics, such as thermosolutal convection and convection in magnetic fields or under rotation.

## 1. Introduction

It is well known, in problems dealing with the thermal convection in buoyant viscous fluids, that the analytical modelling of the stationary convective instability is entirely characterized by the Rayleigh number: an unstable state must set in when this number just reaches a *critical value*. If the Rayleigh number increases beyond the critical value, it is no longer possible to describe analytically the successive bifurcations toward a convective state depending on time, except for particular cases such as the one of low Prandtl number (Busse, 1972). Many numerical studies were performed a long time ago (see, e.g., Manneville, 1991, and references cited herein), in order to predict several scenarii leading to chaos.

Motivated by the hope of finding, through analytical studies, the complex dynamical behaviour near the chaos, nonlinear approaches in the neighbourhood of higher codimension singularities have been popular for a few years.

The study of the convective instability of a binary mixture in a porous medium belongs to this class of problems: a stationary instability and an oscillatory instability can branch off, starting from the state of rest of the fluid. The interaction between these instabilities leads to a complex dynamics of the medium. This problem has been considered, with constant prescribed thermal boundary conditions, by Brand & Steinberg (1983a), then Brand & *al.* (1984) and others. In the present paper we assume that the thermal boundary data are periodically varying with respect to time. The physical meaning of this assumption is the presence of some irregularities which can appear in experiments. Other theoretical works aimed at modelling these irregularities have considered other possible mechanisms such as spatially inhomogeneous thermal boundary conditions for the convection of a simple fluid (Walton, 1982) or a laser system with inhomogeneous pumping (Ouarzazi & *al.*, 1994c).

The aim of the present paper is to present, by applying Melnikov's method, an analytical determination of such bifurcations. We first recall the results of a linear study (Ouarzazi & Bois, 1994a, later referred as O.& B.): for the present work, the most important result is the existence, in some parameter plane, of so-called *Arnold's tongues* which define zones of mode-locked motions (sect. 2). Inside such a tongue, the frequency of the excitator is weak and is characterized by some small parameter $\varepsilon$. After some calculations the equation for the amplitude of the motion can be reduced to a nonlinear Mathieu equation (sect. 3) depending on $\varepsilon$. For $\varepsilon = 0$, this equation possesses, in the phase plane, heteroclinic and periodic solutions. For $\varepsilon \neq 0$ it is possible to apply Melnikov's technique in order to estimate the onset of the threshold of the chaotic régime: the results are matched with direct numerical simulations (sect. 4).

## 2. Hydrodynamic equations and linear stability

We begin by briefly recalling some results of the linear stability of a fluid mixture in a porous medium. We refer the reader to (O.& B.) for more details. Consider a mixture of three constituents: a solid phase (porous medium) and a fluid phase, which is itself a mixture of two fluids. The temperature is denoted $T^*$, the concentration of one fluid in the other is $C^*$. The velocity is denoted $\mathbf{u}$, the vertical velocity $w$. The medium occupies the two-dimensional domain $(\Omega)$: $0 \leq x \leq a$, $0 \leq z \leq a$. The system is heated periodically below, and $T^* = T_0$ above. Denote $T = T^* - T_0$ the temperature difference, $C = C^* - C_0$ the concentration difference. The equations of motion read:

$$\rho = \rho_0(1 - \alpha(T^* - T_0) - \alpha'(C^* - C_0)), \qquad \nabla.\mathbf{u} = 0,$$

$$\frac{1}{\rho_0}\nabla p + \{\alpha(T^* - T_0) + \alpha'(C^* - C_0)\}\mathbf{g} + \frac{\nu}{K}\mathbf{u} = 0,$$

$$\frac{\partial T^*}{\partial t} + \mathbf{u}.\nabla T^* = \kappa\, \Delta T^*, \tag{1}$$

$$\frac{\partial C^*}{\partial t} + \mathbf{u}.\nabla C^* = \mathrm{D}\{\Delta C^* + \frac{k_\mathrm{T}}{T^*}\Delta T^*\},$$

and we assume the following boundary conditions:

$$T = \delta(T_0 - T_1)\cos \omega t \ \text{ at } z = 0, \qquad T = T_1 - T_0 \text{ at } z = a, \tag{2}$$

$$\frac{\partial C}{\partial x} = -\frac{k_T}{T_0}\frac{\partial T}{\partial x}, \quad u = 0, \text{ at } x = 0, \ x = a,$$

$$\frac{\partial C}{\partial z} = -\frac{k_T}{T_0}\frac{\partial T}{\partial z}, \quad w = 0, \text{ at } z = 0, \ z = a. \tag{3}$$

Eqs. (1) to (3) are very classical (see e.g. O. & B. or Brand & Steinberg). The parameters $\alpha$, $\alpha'$, $\kappa$, $\nu$, $k_T$, $K$ characterize the different constituents.

Taking in account the boundary conditions (2) and (3), we obtain the following solution of (1) describing an hydrostatic state $\{\mathbf{u} = 0, T = T_h(z, t), C = C_h(z, t)\}$:

$$T_h(z,t) = (T_0 - T_1)\left\{-\frac{z}{a} + \delta Re\{[a(\lambda)\, e^{\lambda z/a} + a(-\lambda)\, e^{-\lambda z/a}]\, e^{i\omega t}\}\right\},$$

$$C_h(z,t) = \frac{k_T}{T_0}(T_0 - T_1)\left\{\frac{z}{a} + \frac{\delta}{\kappa - D} Re\{[D\, a(\lambda)\, e^{\lambda z/a} + D\, a(\lambda)\, e^{-\lambda z/a} \right. \tag{4}$$

$$\left. + K\, b(\lambda,\mu)\, e^{\mu z/a} + K\, b(\lambda,-\mu)\, e^{-\mu z/a}]\, e^{i\omega t}\}\right\}.$$

The quantities $\lambda$, $\mu$, $a(\lambda)$ and $b(\lambda, \mu)$ are some complex combinations of $\omega$, $a$, $\kappa$ and D (see O. & B.).

We now look for a perturbation of this state. We introduce nondimensional time and lengths $t' = \kappa t/a^2$ and $(x', z') = (x/a, z/a)$ (in the sequel, omitting the primes, again denoted $t$, $x$ and $z$). By introducing the nondimensional perturbation variables $\mathbf{v}$, $\theta$, $c$:

$$\mathbf{u} = \frac{\kappa}{a}\mathbf{v}, \quad T = T_h(z,t) + \frac{\nu\,\kappa}{\alpha\, g\, K\, a}\,\theta, \quad C = C_h(z,t) - \frac{\nu\,\kappa\, k_T}{\alpha\, T_0\, g\, K\, a}\, c, \tag{5}$$

we then obtain for these variables after some calculations the system:

$$\Delta w - \frac{\partial^2\theta}{\partial x^2} - \psi\frac{\partial^2 c}{\partial x^2} = 0, \qquad \nabla.\mathbf{v} = 0,$$

$$\frac{\partial\theta}{\partial t} - \Delta\theta - Ra\, w = -\delta\, Ra\, f(z,t,\omega) w - \mathbf{v}.\nabla\theta, \tag{6}$$

$$\frac{\partial c}{\partial t} - \Lambda\,\Delta c + \Lambda\,\Delta\theta - Ra\, w = \delta\, Ra\, g(z,t,\omega)\, w - \mathbf{v}.\nabla c,$$

Note that, in the first of (6), because of D is a very small coefficient, the term including the time derivative $\partial w/\partial t$ has been neglected, as usually in the Darcy equation.

The two functions $f(z, t, \omega$ and $g(z, t, \omega)$ in (6) are related to the hydrostatic state, and their form is given in (O. & B.). There appear, in addition to $\delta$, three numbers characterizing the problem: the filtration Rayleigh number $Ra$, the separation ratio $\psi$, and the Lewis number $\Lambda$. These numbers are defined with respect to the physical characteristics of the problem by:

$$Ra = \frac{\alpha(T_0 - T_1) g\, K\, a}{\nu\,\kappa}, \qquad \psi = -\frac{\alpha'\, k_T}{\alpha\, T_0}, \qquad \Lambda = \frac{D}{\kappa}. \tag{7}$$

For small $\delta$ and for a given value of $\Lambda$, we then look for a solution of (6) in the form:

$$w = w_0 + \delta\, w_1 + \delta^2\, w_2 + \dots\,, \qquad Ra = R_0 + \delta\, R_1 + \delta^2\, R_2 + \dots \tag{8}$$

The instability thresholds $(w_0, R_0)$, corresponding to the unmodulated boundary conditions, are well known (Brand and Steinberg, 1983a). For a solution of the form:

$$w_0 = \tilde{w}_0\, e^{\sigma t} \cos \pi x \sin \pi z, \qquad \sigma = \sigma_r + i\, \sigma_i,$$

the marginal states are defined by $\sigma_r = 0$, and there appear two types of unstable modes when this state is crossed. A first one, corresponding to $\sigma_i = 0$ when $\sigma_r = 0$, leads to a bifurcation towards a steady state. The corresponding neutral curve, in the $(R_0, \psi)$ plane, is defined by the equation $R_0 = R_{cs}(\Lambda, \psi)$. A second one, corresponding to $\sigma_i = \Omega(\Lambda, \psi)$ when $\sigma_r = 0$, leads to a Hopf bifurcation towards an oscillating mode. The corresponding neutral curve is

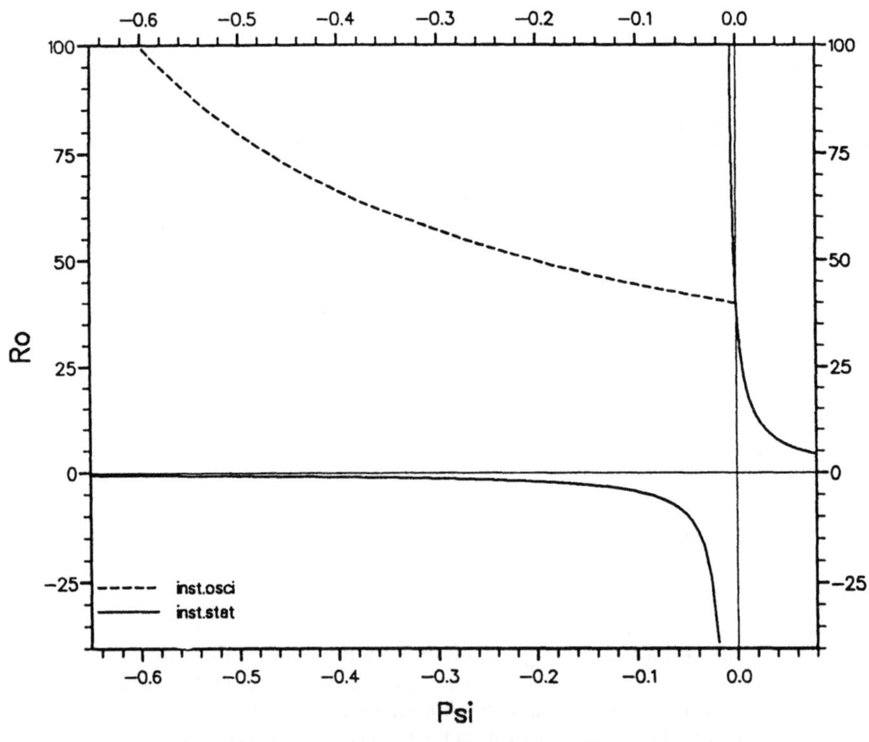


**Fig.1.** Stability curves in the $(Ra, \psi)$ plane for unmodulated prescribed temperature at the boundaries (results of Brand & Steinberg, $\Lambda = 10^{-2}$).

defined by the equation $R_0 = R_{co}(\Lambda, \psi)$. These curves (figure 1) intersect at a polycritical point $(\psi_{pc}, Ra_{pc})$:

$$\psi_{pc} = -\frac{\Lambda^2}{1 + \Lambda + \Lambda^2}, \qquad Ra_{pc} = 4\pi^2(1 + \Lambda + \Lambda^2). \tag{9}$$

## 3. Nonlinear study: amplitude equation

We now consider, in the $(Ra, \psi)$ plane, in the unstable region, a neighbourhood of the polycritical point. In order to study this neighbourhood, we follow an analysis analogous to the one used by Knobloch & Proctor (1981) for a similar problem dealing with the thermohaline convection. For convenience we denote by $\varepsilon^2$ the order of magnitude of this neighbourhood. Hence we expand $Ra$ and $\psi$ as:

$$Ra = Ra_{pc} + \varepsilon^2\mu, \qquad \psi = \psi_{pc} - \varepsilon^2\,\overline{\psi}. \tag{10}$$

The singularities at the polycritical point occur when $\omega$ is of order $\varepsilon$ (see O.& B.). Hence we set $\omega = \varepsilon\,\tilde{\omega}$, so that $\omega t = \tilde{\omega}\tau$, where $\tau$ is the slow time $\varepsilon t$. In the same manner, the case of interest occurs when the time modulation is $O(\varepsilon^2)$, so that we assume $\delta = \varepsilon^2\,\bar{\delta}$. Moreover we expand $w$, $\theta$ and $c$ in the form:

$$w = \varepsilon\, W_1(\tau)\sin\pi z\cos\pi x + O(\varepsilon^3), \quad \theta = \varepsilon\,\theta_1(\tau)\sin\pi z\cos\pi x + \varepsilon^2\theta_2(\tau)\sin 2\pi z + O(\varepsilon^3),$$
$$c = \varepsilon\, C_1(\tau)\sin\pi z\cos\pi x + \varepsilon^2 C_2(\tau)\sin 2\pi z + O(\varepsilon^3). \tag{11}$$

By inserting the expressions (11) in (6) and expanding $f(z)\sin\pi z$ and $g(z)\sin\pi z$ as Fourier series, we obtain a system of fifth order which can itself be reduced to a unique equation satisfied by $W_1$:

$$\frac{d^2W_1}{d\tau^2} = \mu_1 W_1 + f_1 W_1^3 + \varepsilon(\mu_2 - f_2 W_1^2)\frac{dW_1}{d\tau}$$
$$+ \bar{\delta} f_3 W_1 \cos\tilde{\omega}\tau - \varepsilon\,\bar{\delta}\left(f_4 W_1 \sin\tilde{\omega}\tau - f_5\frac{dW_1}{d\tau}\cos\tilde{\omega}\tau\right) + O(\varepsilon^2), \tag{12}$$

with:

$$\mu_1 = \frac{4\pi^4\Lambda}{Ra_{pc}}\,\frac{Ra - R_{cs}}{\varepsilon^2}\,\overline{\psi}, \qquad \mu_2 = \frac{2\pi^2}{Ra_{pc}}\,\frac{Ra - R_{co}}{\varepsilon^2}\,(1+\Lambda)\,\overline{\psi},$$
$$f_1 = \frac{\pi^2}{4}, \quad f_2 = \frac{1+\Lambda}{4\Lambda}, \quad f_3 = 4\pi^4\Lambda, \quad f_4 = \frac{\pi^4\Lambda}{3} + 2\pi^2(1+\Lambda), \quad f_5 = 2\pi^2(1+\Lambda). \tag{13}$$

Eq. (12) is a inhomogeneous form of a nonlinear Mathieu equation. In order to study the interaction between the dissipative terms, the periodic fluctuations and the nonlinearities, we consider in the sequel the simplified case where $\bar{\delta}$ is itself of order $\varepsilon$. Hence we assume that $\bar{\delta} = \varepsilon \tilde{\delta}$, and the simplified corresponding Mathieu equation reads:

$$\frac{d^2 W_1}{d\tau^2} = \mu_1 W_1 + f_1 W_1{}^3 + \varepsilon(\mu_2 - f_2 W_1{}^2) \frac{dW_1}{d\tau} + \varepsilon \tilde{\delta} f_3 W_1 \cos \tilde{\omega}\tau + o(\varepsilon). \qquad (14)$$

## 4. Solution of the amplitude equation by Melnikov's method

The Melnikov method allows to determine the conditions under which chaotic behaviours and subharmonic resonances appear. These determination is performed by using the so-called Melnikov's function $M(\tau_0)$ and subharmonic Melnikov's function $M^{m/n}(\tau_0)$ (see e.g. Guckenheimer & Holmes, 1983, ch. 4, §§ 4.5 and 4.6: these functions represent good measures of the distance between stable and unstable manifolds).The simple zeros of $M(\tau_0)$ correspond to transverse intersections of stable and unstable manifolds while those of $M^{m/n}(\tau_0)$ give rise to subharmonic orbits. For the present case we obtain the following conclusions:

(i) *Smale horseshoe chaos.*

The Melnikov's function, for the present problem, reads:

$$M(\tau_0) = \frac{2\sqrt{2}}{3}\left[\mu^*{}_2 - \frac{f^*{}_2}{5}\right] - \frac{\delta^* \pi\, \omega^2}{\mathrm{sh}(\pi\omega/\sqrt{2})} \sin \tilde{\omega}\tau_0 = \mathrm{B}(1 - \mathrm{A} \sin \tilde{\omega}\tau_0),$$

$$\delta^* = -\frac{1}{\mu_1}, \qquad \mu^*{}_2 = \sqrt{\delta^*}, \qquad f^*{}_2 = \mu^*{}_2 \frac{f_2}{f_1}. \qquad (15)$$

If $|\mathrm{A}| > 1$, then $M(\tau_0)$ possesses simple zeros, and, in the phase plane, there exist chaotic behaviours of Smale horseshoe type. If $|\mathrm{A}| < 1$, no intersection can occur and the whole system is not chaotic. Hence the limit $|\mathrm{A}| = 1$, which corresponds to a value $\delta^* = \delta^*{}_c$, characterizes the onset of the chaotic régimes. Such régimes and onsets can be exhibited for realistic numerical values of the several parameters (see Ouarzazi *& al.*, 1994b). The theoretical value of the corresponding onset has been matched with the result of a direct numerical simulation, and gives a good agreement.

(ii) *Nonlinear periodic resonances.*

The subharmonic Melnikov's function for resonant periodic orbits inside heteroclinic cycles reads:

$$M^{m/n}(\tau_0) = \mu^*{}_2 I_2(m, n) - f^*{}_2 I_1(m, n) - \delta^* I_3(m, n) \sin \tilde{\omega}\tau_0, \qquad (16)$$

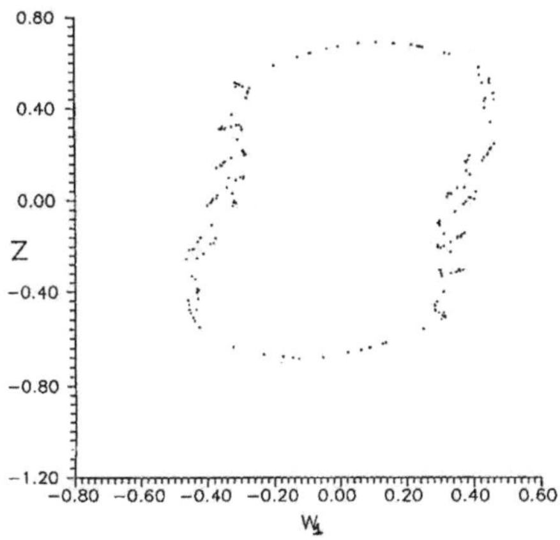


**Fig. 2.** The onset of the chaotical behaviour as predicted by Melnikov's method for $\varepsilon\,\mu^*_2 = 0.1$, $\varepsilon f^*_2 = 1$ (figs. 2 and 3 are extracted from O.& *al.*, 1994b).

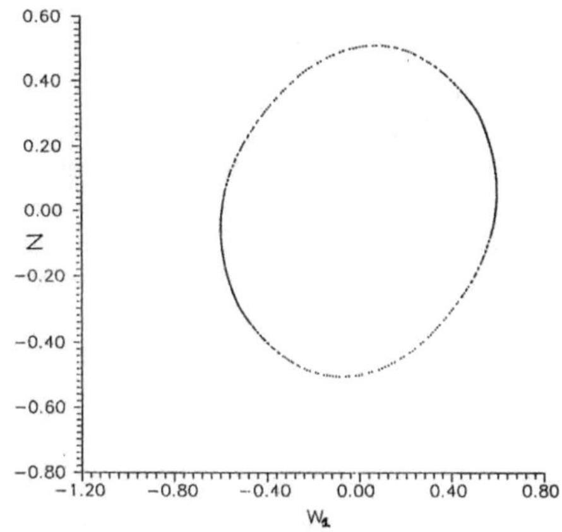


**Fig. 3.** Saddle-node bifurcation of the periodic orbit of modulus k = 1/2: transition from a quasiperiodic state to a phase-locked state of order 2 ($\varepsilon\,\mu^*_2 = 0.08$, $\varepsilon f^*_2 = 1$).

where $m$, $n$ are integers satisfying a resonance relation, and $I_1$, $I_2$, $I_3$ are some combinations of Jacobi elliptic integrals. We obtain the following conclusion: for $n \neq 1$ and/or m odd, there arise only second order effects. For n = 1 and m even, if:

$$\delta^* > \left| \frac{\mu^*_2\, I_2(m,\,1) - f^*_2\, I_1(m,\,1)}{I_3(m,\,1)} \right| = \delta^*_c(m,\,1), \tag{17}$$

then $M^{m/1}(\tau_0)$ has simple zeros, and the equation (14) possesses a subharmonic orbit of period $T = 2\pi m/\omega$. Moreover the curve $\delta^* = \delta^*_c(m, 1)$ possesses a saddle-node bifurcation. These results can be confirmed by direct numerical simulations (see figs. 2 and 3).

## 5. Conclusion

The interesting dynamics found in the binary fluid mixture problem owes its existence to the competition between a destabilizing temperature gradient and a stabilizing concentration distribution. A result of this competition, is that a chaotic régime of Smale horseshoe type can arise from small boundary thermal fluctuations. We have used Melnikov's analysis in order to evaluate the threshold of the appëarance of the chaotic régime, as well as the one of saddle-node bifurcation. Other mechanisms have been proposed by Rucklidge (1992) and by Knobloch *& al.* (1992), who showed that double-diffusive systems exhibit chaotic solutions associated with heteroclinic orbit linking a pair of saddle-foci with eigenvalues q, $-p \pm i\omega$ ($p > 0$, $q > 0$, $\omega \neq 0$) which satisfy Shilnikov's criterion $p/q < 1$.

It will be interesting to consider effects due to spatial variations of the thermal boundary conditions. We expect that, in some parameter range, a similar behaviour will be identified in various problems. Such an example is the one of the inhomogeous pumping of a laser system (Ouarzazi *& al.*, 1994c). The dynamics of this system is entirely described by an universal equation of Ginzburg-Landau type with complex coefficients varying in space.

## References

Brand H.R., Hohenberg P.C. & Steinberg V., 1984. "Codimension 2-bifurcations for convection in binary fluid mixtures". - *Phys. Rev.*, **30 A**, 2548-2561.

Brand H.R. & Steinberg V., 1983a: "Convective instabilities in binary mixtures in a porous medium". - *Physica*, **119 A**, 327-338.

Brand H.R. & Steinberg V., 1983b. "Nonlinear effects in the convective instability of a binary mixture in a porous medium near threshold". - *Physics letters*, **93 A**, 333-336.

Busse F.H., 1972. "The oscillatory instability of convection rolls in a low Prandtl number fluid". - *J. Fl. Mech.*, **52**, 97-112.

Guckenheimer J. & Holmes P.J., 1983. "*Nonlinear oscillations, dynamical systems and bifurcations in vector fields*". Springer, Berlin.

Knobloch E. & Proctor M.R.E., 1981. "Nonlinear periodic convection in double-diffusive systems". - *J. Fl. Mech.*, **108**, 291-316.

Knobloch E., Proctor M.R.E. & Weiss N.O., 1992. "Heteroclinic bifurcations in a simple model of doubel-diffusive convection". *J. Fl. Mech.*, **239**, 273-292.

Manneville P., 1991. "*Structures dissipatives, chaos et turbulence*". Aléa, Saclay.

Ouarzazi M.N. & Bois P.A., 1994a. "Convective instability of a fluid mixture in a porous medium with time-dependent temperature gradient". - *Eur. J. Mech. (fluids)*, **13**, 275-298.

Ouarzazi M.N., Bois P.A. & Taki M., 1994b. "Nonlinear interaction of convective instabilities and temporal chaos of fluid mixture in porous medium". - *Eur. J. Mech. (fluids)*, **13**, 423-438.

Ouarzazi M.N., Bois P.A. & Taki M., 1994c. "Transverse structures and inhomogeneous pumping in laser systems". - *in preparation*.

Rucklidge A.M., 1992. "Chaos in models of double convection". - *J. Fl. Mech.*, **237**, 209-229.

Walton J.C., 1982. "The effects of slow spatial variations on Bénard convection". - *Quart. J. Mech. Appl. Math.*, **35**, 33-48.

# Finite amplitude waves in a boundary layer flow over compliant walls

**Maurice Rossi**[1], **Uwe Ehrenstein**[2]

[1]Laboratoire de Modélisation en Mécanique
URA CNRS 229 Université Paris 6, case 162
4 Place Jussieu, 75252 PARIS CEDEX 5. FRANCE
e.mail: maur@ccr.jussieu.fr
[2] Laboratoire de Mécanique de Lille
URA CNRS 1441 Université de Lille
59665 Villeneuve d'Ascq CEDEX FRANCE
e.mail: uwe.ehrenstein@univ-lille1.fr

**Abstract:** In the present work, we study the influence of a compliant wall on the two-dimensional evolution of instability waves occuring in Plane Poiseuille flow. Contrary to most studies, which determine how linear stability characteristics are altered by compliance, we have chosen to focus on finite amplitude travelling waves. Moreover we dwell upon the role of a single parameter coefficient modeling the elasticity of the wall. These finite amplitude patterns are interesting objects since they exist below the linear instability threshold down to a nonlinear critical Reynolds number closed to the observed transitional Reynolds numbers. Gradually changing the compliance, we observe how the set of finite amplitude wave solutions is transformed. It appears that there is a substantial increase in the nonlinear critical Reynolds number when the compliance is high enough. The examination of streamlines attests as well the qualitative change occuring when the compliance is increased. Finally the skin friction is computed and shows a decrease with increasing compliance.

## 1. Introduction

Since Kramer(Kramer,1962) described his famous experiment showing considerable drag reductions for an object covered with a flexible material, a lot of theoretical studies have been based on the idea that the use of a compliant wall provides a practical way of delaying the appearence of turbulence in a boundary layer flow. Explanations which account for this delayed transition are various. One obvious possibility is related to a change in linear stability characteristics of the boundary layer flow(Carpenter P. W.,1990). More specifically a shift of the neutral curve towards higher Reynolds number or else an overall decrease of perturbations growth rates may cause a change in the transitional process. In this approach, several works have described, since the seminal contributions by Benjamin(Benjamin,1960) and Landhal(Landhal,1962), the compliant wall problem as a case where a potential is present for different type of instabilities. One mainly differentiates between three types of instabilities. A first set corresponds to the so-called modified Tollmien-Schlichting waves: these are the usual instability waves which originate because of an irreversible transfer of energy from the basic flow towards wave perturbations via the Reynolds tensor action. Indeed due to the presence of a viscous critical layer and a friction wall layer, there exists the possibility of a nonnul Reynolds tensor contribution. When the wall is compliant, boundary conditions at the wall are significantly modified: this affects in particular the friction wall layer hence the stability characteristics of these waves which are essentially flow based waves. Since a compliant wall is a wave bearing medium, it possesses as well its own free surface waves. When modified by the fluid pressure, these neutral waves can be destabilized through a mechanism very similar to Miles mechanism which explains how waves are produced by the action of wind. This second set which does not exist when the wall is rigid, hence contains surface based solutions called Flutter waves. Finally a purely inviscid process arises for sufficiently high flow speed: this is equivalent to the celebrated Kelvin-Helmholtz instability of shear layers.

The purpose of most articles is then an overall description of the stability of each type of instability waves in a multiparameter space where the parameters can be respectively inertia, damping,flexural rigidity, tension and spring stiffness of the compliant surface. If one compares experimental and theoretical datas, this approach has not yet provided a clear cut picture. This may be caused by various factors in particular by the neglect of nonlinear terms in the transition process. Our standpoint here is precisely to put the emphasis on such nonlinearities. In this research area, just a few recent works (Rotenberry,1993)(Thomas,1992)(Gajjar,1990) took into account this effect and all of them — as far as as the author knows— dealt with the weakly nonlinear approximations. Since many nonlinear problems are still to be investigated, it is not useless to start analyzing the simplest one of them. Following the theoretical studies on the rigid case we hence picked up the Poiseuille flow problem. As in the rigid case, this is a somewhat academic problem because it is difficult to achieve on an experimental basis. This example is however worth studying since it is an analytic solution of the Navier-Stokes equations which contains— it is known from our rigid case experience— many effects observed in more complex boundary layer profiles.

Nonlinear effects might change drastically the behavior of a flow: this is particularly obvious for subcritical instabilities —this is the case of Poiseuille flow over a rigid wall— where the flow is linearly stable in some regions in parameter space but destabilized by finite amplitude perturbations in the same regions. This remark is true for supercritical flows as well since secondary instabilities described by Bayly(Bayly,1986) which arise on top of the finite amplitude waves, account for the appearence of small scale structures during transition. For the compliant wall case, nonlinearities are present not only under the form of the usual convective term in Navier-Stokes equation but also and mostly because of interactions between the fluid and the moving boundary of the compliant wall. There exists diverse ways to introduce nonlinearities. A first one is through direct numerical simulations. A second approach —the one we choose here— relies on bifurcation theory: one numerically follows nonlinear equilibria bifurcating from the infinitesimal linear wave solutions. It is our purpose to describe how such finite amplitude waves evolve when the wall becomes compliant.

Contrary to the multiparameter analysis done in linear studies, we have reduced our description of the compliant surface to a single parameter characterizing the compliance: a spring constant. That is we impose the wall displacement to be simply proportionnal to the fluid pressure generated on the surface. This implies that we neglect other effects such as damping or tension of the compiant membrane or that we disregard the viscous component of the normal stress which is known to be small. These various simplifications are not compulsory and a more elaborate structure with inertia, damping,flexural rigidity, tension and spring stiffness can be theoretically introduced at no cost. However it should be kept in mind that the nonlinear domain is much more demanding as regards numerical computations than the linear approximation. Mainly when one computes a point in the mutiparameter space one should simply determine the spectrum of a linear operator — the Orr-Sommerfeld operator with the appropriate boundary conditions— while one should define a continuous family of solutions of a nonlinear eigenproblem for the nonlinear problem. This is the reason why we started looking at the first set of instability waves with a single parameter.

In the first section we provide the compliant wall model and the basic equations. Therafter we introduce a mapping which makes nonlinearities due to the fluid structure interaction obvious. We then reduce the problem via numerical discretization to a large nonlinear algebraic equation depending on several parameters. The second section discusses the numerical results.

## 2. Description of the Model and Basic Equations

### 2.1 The wall model

We consider here the celebrated case of a channel flow between two parallel walls but where the lower one is covered with a compliant material. This compliant wall is modelized using a surface based model. In such cases, the dynamics is quite simple since it simply depends on the time evolution of the graph which defines the interface fluid-compliant material. More specifically, let us nondimensionalize using the fluid density $\rho$, the centerline velocity U, viscosity $\mu$ and half channel width h as dimensional units. the upper wall is then located at $y = 1$ while the compliant wall is located at $y = -1 + \eta(x,t)$. Moreover the compliant wall dynamics is then given by

$$\frac{\partial \eta}{\partial t} = v(x, -1 + \eta(x,t), t), \tag{1a}$$

$$u(x, -1 + \eta(x,t), t) = 0, \tag{1b}$$

$$-\delta p(x, -1 + \eta(x,t), t) = \frac{\chi}{Re^2}\eta \tag{1c}$$

The kinematical condition(1a) relates the surface motion with the fluid normal velocity at the interface. Equation(1b) expresses the no-slip condition on the compliant medium for the horizontal velocity u(x,y,t). It implies that only vertical displacements are allowed for the compliant wall. This characteristics often ameliorates the efficiency of experimental compliant surfaces. Finally the last equation is the dynamical description of the surface by a Hooke's Law where the spring constant $\chi$ is the single parameter of the flexible wall and $-\delta p$ the pressure perturbation.These equations are already in dimensionless form, the quantities in equation (1c) being related to the physical quantities(especially the dimensional compliance $\chi^*$) by the relations

$$Re = \frac{U_\infty L}{\nu}, \quad \chi = \frac{\chi^* h^3 \rho}{\mu^2} \tag{2}$$

Let us mention that the expressions in (2) are unambiguous and quite different from that obtained for a boundary layer evolving along the streamwise coordinate e.g. Blasius flow. In such a case, the Reynolds number appears at the nominator in (1c). Finally let us mention that for Poiseuille flow we have to prescribe a physical parametrization: we work here at constant pressure gradient and the Reynolds number we are using is such that $U = -\frac{h^2 P}{2\nu}$ with $P$ the constant mean pressure gradient.

Among nonlinear states, the simplest we can think of are two-dimensional travelling wave solutions of the form

$$u(x,y,t) = u(x',y), \quad v(x,y,t) = v(x',y), \quad p(x,y,t) = p(x',y), \quad \eta(x,t) = \eta(x') \tag{3}$$

where $x' = x - ct$ and c is the phase speed of the wave. The problem we have to solve is no more an initial value problem since time does not show up. It is rather a nonlinear eigenvalue problem where phase speed c is the unknown eigenvalue to be determined. Such eigensolutions even if they are unstable may be worth studying. Indeed Saffman(Saffman P.G.,1983),( Milinazzo F.A, Saffman P. G., (1985) proposed the idea that such states are related to the coherent structures seen in experiments. Moreover they are possible new basic states from which secondary instabilities may arise. In this work we precisely analyze the existence or nonexistence of such states and their physical characteristics with respect to the spring constant. If one defines the stream function as

$$u = \varepsilon \frac{\partial \psi}{\partial y} \quad v = -\varepsilon \frac{\partial \psi}{\partial x'} \tag{4}$$

where $\varepsilon$ is the wave amplitude, these progressive waves satisfy Navier stokes equation written in vorticity streamfunction formulation as

$$-c\frac{\partial \omega}{\partial x'} - \frac{1}{Re}\Delta\omega + (1-y^2)\frac{\partial \omega}{\partial x'} + 2y\frac{\partial \psi}{\partial x'} + \varepsilon\left(\frac{\partial \psi}{\partial y}\frac{\partial \omega}{\partial x'} - \frac{\partial \psi}{\partial x'}\frac{\partial \omega}{\partial y}\right) = 0 \tag{5}$$

$$\omega = -\Delta\psi \tag{6}$$

At the upper wall we impose the no-slip condition on the rigid wall.

$$\frac{\partial \psi}{\partial x'}(x',1) = 0, \quad \psi(x',1) = 0 \tag{7}$$

At the lower wall the exact kinematical conditions become

$$-c\frac{\partial \eta}{\partial x'} = v(x', -1 + \eta(x',t), t), \tag{8a}$$

$$u(x', -1 + \eta(x')) = 0, \tag{8b}$$

The dynamical equation can be written as

$$\frac{\chi}{Re^2}\frac{d\eta}{dx'} = \frac{1}{Re}\frac{\partial\omega}{\partial y}(x', -1+\eta(x')) - \frac{\varepsilon}{Re}\frac{d\eta}{dx'}\frac{\partial\omega}{\partial x}(x', -1+\eta(x')) + \varepsilon c^2\frac{d\eta}{dx'}\frac{d^2\eta}{dx'^2} \quad (8c)$$

### 2.2 Mapping Transformation and Numerical Methods

The above dynamical system is highly nonlinear because we prescribe the kinematical and dynamical conditions at an interface the location of which is an unknown of the problem. This clearly appears when we perform a transformation which maps the physical domain $-1+\eta(x') \leq y \leq 1$ into the interval $-1 \leq y \leq 1$

$$\tilde{x} = x'; \quad \tilde{y} = (1-\gamma)y + \gamma \quad with \quad \gamma = \frac{\eta(x')}{\eta(x') - 2} \quad (9)$$

This coordinate change which is performed to impose the boundary condition on fixed boundaries, engenders new nonlinearities for instance equation(6) becomes

$$\tilde{\Delta}\psi + A(\eta)\frac{\partial^2\psi}{\partial\tilde{y}^2} + B(\eta)\frac{\partial^2\psi}{\partial\tilde{y}\partial\tilde{x}} + C(\eta)\frac{\partial\psi}{\partial\tilde{y}} = -\omega \quad (10)$$

where $A$,$B$,$C$ are functions of the graph $\eta(x')$. Using Chebyshev polynomials in the normal coordinate and fourier modes in the streamwise coordinate, we then discretize the overall system using a collocation method(For details see (Ehrenstein and Rossi,1993)). In order to get a good numerical approximation for these states it is necessary to use at least $K = 44$ collocation points in the y direction and $N = 3$ fourier modes. The system takes then the form of a large system of nonlinear algebraic equations(more than 900 equations) that depends on three parameters the wave number $\alpha$ of the progressive periodic wave, $\chi$ and $Re$. The phase velocity $c$ and amplitude $\varepsilon$ of this wave are then obtained starting from a bifurcation point solution —that corresponds to an infinitesimal neutral instability wave— and solving this system using a Newton-Raphson iteration in conjunction with Keller'pseudo-arclength continuation procedure where the continuation parameter is one of the three parameters $\alpha$, $\chi$ and $Re$. For instance, to each wave length corresponds a continuous curve of travelling waves in the plane Reynolds number—amplitude.

## 3. Numerical results

For the rigid wall case, the set of all possible solutions in the three dimensional parameter space $\alpha$, $Re$, $\varepsilon$ form an ellipsoidal surface with a turning point at a Reynolds number equal to $Re = 2911.36$. We determine the distorsion of this equilibium solution set due to flexibility. To do so we gradually reduce, at a fixed value of the wave number $\alpha$ the spring stiffness $\chi$. Figure 1 shows the $\alpha = 1.21$ cut in the Reynolds number-amplitude plane for different values of the compliance. When $\chi$ is large enough ( continuous curve in Figure 1),one recovers the rigid case curve.On the contrary when flexibility is increased(dashed curves), the equilibrium surface evolves. One striking difference is that some finite amplitude solutions get even disconnected from the Reynolds number axis. One hence obtained isolated solution which are not related to the solutions from the linear analysis.

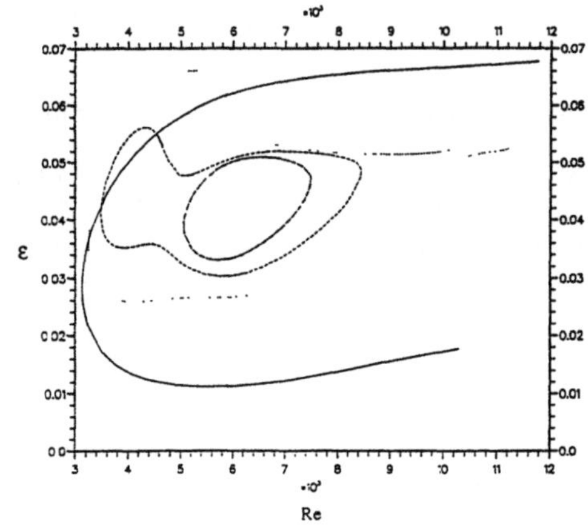


**Figure 1** $\alpha = 1.21$ cuts through various equilibrium surfaces in the Reynolds number Re and $\varepsilon$ amplitude plane with K=44 and N=3. $\chi = 4.10^7$ plain line, $\chi = 7.10^6$....., $\chi = 5.10^6$ - - - -, $\chi = 4.10^6$ _._.

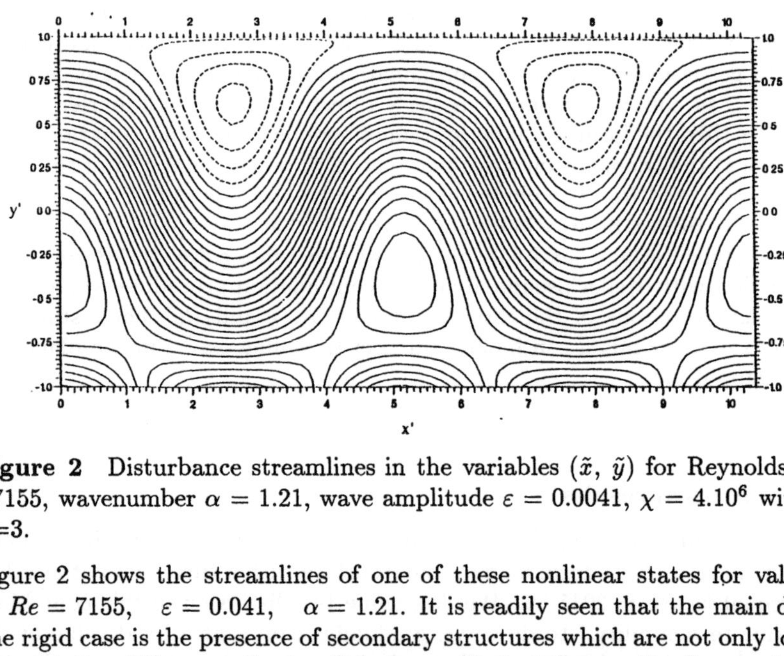


**Figure 2** Disturbance streamlines in the variables $(\tilde{x}, \tilde{y})$ for Reynolds number $Re = 7155$, wavenumber $\alpha = 1.21$, wave amplitude $\varepsilon = 0.0041$, $\chi = 4.10^6$ with K=44 and N=3.

Figure 2 shows the streamlines of one of these nonlinear states for values $\chi = 410^6$, $Re = 7155$, $\varepsilon = 0.041$, $\alpha = 1.21$. It is readily seen that the main difference with the rigid case is the presence of secondary structures which are not only located in the very vicinity of the compliant wall but are able to influence the flow up to a finite distance. In Figure 3, we illustrate the change in the nonlinear critical Reynolds number which corresponds to the turning point below which no travelling wave exists. We cut the surface at three different Reynolds numbers $Re = 5080$, $Re = 4065$, $Re = 3905$ for $\chi = 4.10^6$. According to the narrowing of the bandwith of possible wavenumbers, there is evidence that the turning point lies around $Re = 3905$ which should be compared to the usual $Re = 2911.36$ for the rigid wall. This indeed is an indication of a possible transition delay if transition is linked in any way to the existence of these travelling wave solutions as was assumed by Saffman.

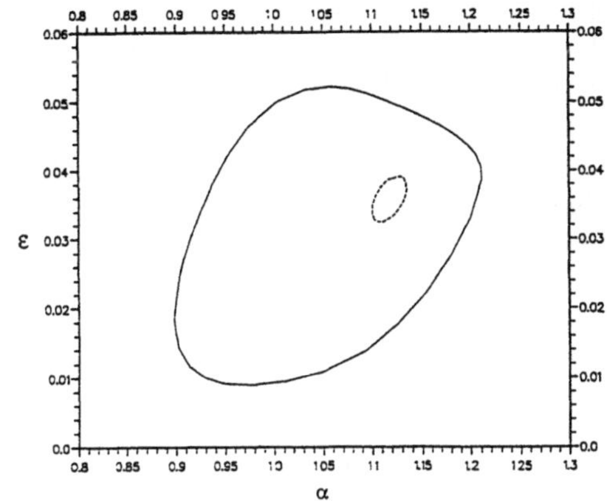


**Figure 3** Various Re=const.-cuts through equilibrium surface for $\chi = 4.10^6$ in the wavenumber $\alpha$ and amplitude $\varepsilon$ plane with K=44 and N=3. $Re = 5080$ plain line, $Re = 4065$ ...., $Re = 3905$ – – – – –

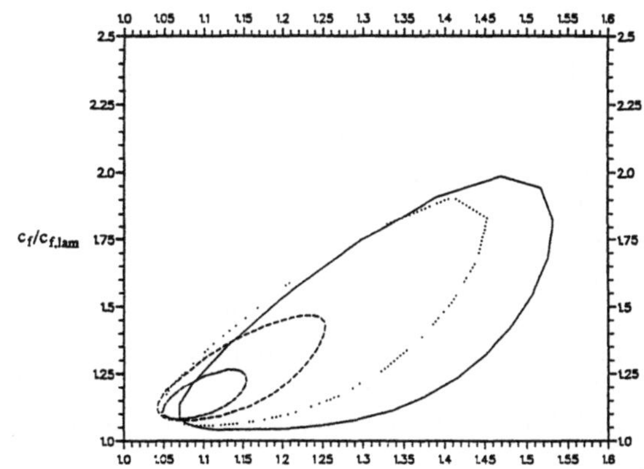


**Figure 4** Normalized friction factor $\frac{c_f}{c_{f,lam}}$ as function of the wavenumber $\alpha$, for Reynolds number $Re = 4065$ with K=44 and N=3. $\chi = 10^8$ plain line, $\chi = 10^7$ ...., $\chi = 510^6$ – – – – , $\chi = 410^6$ -.-.

Finally a skin friction coefficient associated with the new state formed by the Poiseuille flow plus the progressive wave, is computed. In Figure 4 appears $\frac{c_f}{c_{f,lam}}$ the friction factor normalized with respect to the laminar skin friction coefficient. This shows that decreasing flexibility not only does lead to an increase in bandwith of possible wavenumbers but also to an increase of the maximal normalized skin friction. This correponds to an effective drag reduction at least for these states.

## 4. Conclusion

In this work, we have shown how nonlinear travelling wave solutions are modified by the presence of compliance. For each travelling wave, streamline patterns show that they are influenced by compliant wall up to a finite distance from it. Moreover the nonlinear critical Reynolds number below which no such solutions are possible seem to be shifted towards higher values. Finally skin friction associated with these states decrease with increasing compliance. The overall picture for Poiseuille flow is thus pretty encouraging. We are presently analyzing along the same lines the more practical but more complex problem of a Blasius boundary layer. Preliminary results indicate that nonlinearities may change the results obtained with the linear approximation despite Blasius is a less subcritical instability than Poiseuille flow.

### Acknowledgement

This work was was supported by DRET( 91 1198A )

### References

Bayly B.J.: "Three-dimensional Instablility of elliptical flow", *Physical Review Letters* Vol 57, Number 17, 2160-62.

Benjamin T.B.(1960): "Effects of a flexible boundary on hydrodynamic stability", *J. Fluid Mech.*, **9**(4), 513-32.

Carpenter P. W.(1990): "Status of transition delay using compliant walls in Viscous Drag Reduction in Boundary Layers",*Progress in Astronautics and Aeronautics*, vol.123 (ed. Bushnell D. M. and Hefer J. N.), 79-113, AIAA.

Ehrenstein U., Rossi M.(1993): "Nonlinear Tollmien-Schlichting waves with compliant walls", *Eur.J.Mech., B/ Fluids*, **12**, 789-810.

Gajjar J. S. B.(1990): "Nonlinear stability of flow over compliant surfaces", in Proc. 5th. European Drag Reduction Meeting, London, November 1990, Kluwer.

Kramer M. O.(1962): "Boundary layer stabilization by distributed damping", *J. AM. Soc. Naval Engrs*, **74**, 341-348.

Landhal M.T.(1962): "On the stability of a mlaminar incompressible boundary layer over a flexible surface",*J. Fluid Mech.*, **13**(4), 609-32.

Milinazzo F. A., Saffman P. G.(1985): "Finite-amplitude steady waves in plane viscous shear flows", *J. Fluid Mech.*, **160**, 281-295.

Rotenberry J. M.(1993): "Finite amplitude steady waves in the Blasius boundary layer", *Phys. Fluids* A 5, 1840-1842.

Saffman P. G.(1983): "Vortices-stability and turbulence", *Ann. N. Y. Acad. Sci.*, **404**, 12-24.

Thomas M. D., (1992): "The nonlinear stability of flows over compliant walls", *J. Fluid Mech.*, **239**, 657-670.

# RECENT DEVELOPMENTS IN THE THEORY OF THE NON-LINEAR STABILITY OF HIGH REYNOLDS NUMBER FLOWS

Andrew G. Walton

Department of Mathematics, Imperial College of Science, Technology and Medicine
180 Queen's Gate, London SW7 2BZ, U.K.

**Abstract** The aim of this communication is to summarize some of the progress that has been made concerning the problem of modelling the phenomenon of transition to turbulence using a high Reynolds number asymptotic approach. In recent years a number of weakly and strongly nonlinear theories have been developed and their effects investigated in a variety of flow situations. The particular path to transition taken in a practical situation will depend on the nature of the disturbance environment. Particular mention will be made of the theory of vortex-Tollmien-Schlichting-wave interaction (VWI) in which a tiny wave disturbance is able, through longscale-short-scale interplay, to alter the basic flow at leading order.

## 1. The linear stability of a decelerating boundary-layer

Let us begin by considering the linear stability of a steady two-dimensional boundary-layer on a flat plate subject to an adverse pressure gradient. We assume the presence of small amplitude Tollmien-Schlichting (TS) disturbances proportional to $exp(i\alpha x + i\beta z - i\Omega t)$. We define the following non-dimensional parameters :

$$F = \Omega \nu / U_e^2, \mathrm{Re} = \mathrm{U}_e L / \nu$$

where $U_e$ is a typical external velocity, $L$ is distance along the plate and $\nu$ represents kinematic viscosity. A typical neutral stabilitty curve for this type of flow is shown in figure 1 in which we plot the frequency parameter $F$ versus the Reynolds number $Re$ . The flow is unstable inside the curve and stable outside. This result is found by solving the Orr-Sommerfeld equation which can be derived from the Navier-Stokes equations under the assumption that the basic flow is parallel. To give an idea of the size of the parameters involved here, the instability is encountered in a region of parameter space in which $F$ is typically of order $10^{-4}$, while the critical value of $Re$ at which the flow first becomes unstable is of order $10^4$. The values of these parameters sugget that an asymptotic approach to the problem is relevant : indeed it is only at asymptotically large values of *Re* that the parallel-flow assumption used in deriving the Orr-Sommerfeld equation is a valid leading-order approximation. An important point to note is that the Reynolds number increases as the disturbance propagates downstream. If we consider the disturbance to have a fixed frequency $\Omega$, then its evolutionary path on the stability diagram will be a horizontal line such as the dashed line on figure 1. Typically, such a disturbance would be expected to experience a period of decay until it reaches the distance along the plate corresponding to the lower branch neutral point (*A*) ; this is followed by growth until the upper branch neutral point is reached further downstream, beyond which the wave will once again decay. This behaviour was first observed experimentally by Schubauer & Skramstad (1947). However the scenario envisaged above can only be realised provided the amplitude of the wave remains sufficiently small throughout, ensuring that linear theory is valid. Clearly if the wave is amplified to such an extent that it can no longer be regarded as a small perturbation to the basic flow, then its subsequent development must be described by a nonlinear theory. As discussed above, the assumptions upon which linear theory is based are violated unless *Re* is asymptotically large. Thus in order to extend the linear theory into the nonlinear regime in a mathematically consistent manner we must take $\mathrm{Re} \gg 1$ throughout the analysis. Using this approach a number of nonlinear theories have been advanced : the differences between them arise essentially from assumptions made concerning the form of the disturbance environment and the location of the disturbance relative to the neutral points mentioned above. A part from the region near the lower branch, where $\mathrm{Re} \approx O(F^{-4/3})$ which we will concentrate on in the next section, there are two other asymptotic regions upon which attention can be focussed : for Reynolds numbers of order $F^{-3/2}$, close to the point *B* on figure 1, the disturbance modes can be described by a five-layer asymptotic structure, while closer to the upper branch neutral point, at distances for which $\mathrm{Re} \approx \mathrm{F}^{-2}$ , (point *C*) the wavelength of the modes becomes comparable with the boundary-layer thickness and the instability becomes primarily inviscid in nature. As remarked by Cowley & Wu (1993) a case can be made for the study of each of these asymptotic regions of the stability curve. For a more detailed discussion of these regions the reader is referred to the aforementioned paper and references therein.

In the remainder of this communication I will be concentrating on following the evolution of disturbances that are introduced slightly upstream of the lower branch neutral point and are amplified to a nonlinear state while still within the vicinity of this location. Non-parallel effects can then be taken into account by use of the lower branch triple-deck scalings for the TS wave (Smith 1979). In the next section we will see that this wave need only be of a small amplitude to produce a leading order correction to the basic flow via nonlinear effects - the so-called vortex - TS wave interaction.

## 2. Vortex - TS wave interaction (VWI)

In this section we present a simple order of magnitude argument that indicates how the scalings are chosen to maximize the strength of the interaction between a three-dimensional TS wave and the basic flow in the boundary-layer. We take $u, v, w, p$ to represent the streamwise ($x$), normal ($y$) and spanwise ($z$) velocity components and the pressure. The proximity of the disturbance to the lower branch neutral point implies that the TS wave can be represented in the form

$$\begin{bmatrix} u_w \\ v_w \\ w_w \\ p_w \end{bmatrix}(x,\bar{y},\bar{z})\exp\left[i\mathrm{Re}^{3/8}\int \alpha(x)dx - i\mathrm{Re}^{1/4}\Omega t\right] + c.c. \ ; \ z = \mathrm{Re}^{-3/8}\bar{z}, y = \mathrm{Re}^{-1/2}\bar{y}$$

(with $p_w$ independent of $\bar{y}$) where we have taken the scaling in the spanwise direction to be of the same order as the wavelength of the TS wave, and *c.c.* denotes complex conjugate. The boundary-layer flow has the form :

$$[u_m(x,\bar{y},\bar{z}), v_m(x,\bar{y},\bar{z}), w_m(x,\bar{y},\bar{z}), p_m(x,\bar{z})] \ ,$$

the subscript $m$ denoting "mean". We expect the incoming boundary-layer flow to be two-dimensional with the spanwise variations induced by the introduction of the wave. Under the assumption that the streamwise boundary-layer flow $u_m$ is of $O(1)$ we have, from continuity :

$$\partial u_m / \partial x \approx \partial w_m / \partial z \ ,$$

implying that the spanwise mean flow component must be of order $\mathrm{Re}^{-3/8}$ in view of the short spanwise scale. This spanwise flow is induced by the self-interaction of the wave disturbance within the lower deck of the triple-deck structure, where $y = O(\mathrm{Re}^{-5/8})$ . In this region there is an inertial-viscous balance of terms in the Navier-Stokes equations. In particular, from the spanwise momentum equation we have :

$$w_w \frac{\partial w_w^*}{\partial z} + w_w^* \frac{\partial w_w}{\partial z} \approx \frac{1}{\mathrm{Re}} \frac{\partial^2 w_m}{\partial y^2}$$

with $*$ denoting complex conjugate. Using the scalings deduced above, this balance is achieved provided the spanwise component of the wave velocity $w_w$ is of order $\mathrm{Re}^{-1/4}$ , remarkably tiny in practical terms. It can be shown that the other velocity components and pressure are similarly small. In more detail, allowance must be made for the logarithmic growth of the spanwise mean velocity within the lower deck and this leads to the inclusion of a logarithmic factor in the above analysis. For more information regarding these scales the reader is referred to Smith & Walton (1989), Hall & Smith (1991). On the basis of these scalings, we can derive from the Navier-Stokes equations the "wide-vortex/TS" interaction equations (Hall & Smith 1991). The mean flow is found to be governed by the 3D boundary-layer equations

$$u_x + v_{\bar{y}} + w_{\bar{z}} = 0 \ , \qquad (1)$$

$$uu_x + vu_{\bar{y}} + wu_{\bar{z}} = u_e u_{ex} + u_{\bar{y}\bar{y}}, \qquad (2)$$

$$uw_x + vw_{\bar{y}} + ww_{\bar{z}} = w_{\bar{y}\bar{y}}, \qquad (3)$$

subject to the boundary conditions

$$u = v = 0,\ w = -\frac{1}{\alpha^2\lambda^2}\frac{\partial}{\partial z}\left(\alpha^2|p|^2 + |p_{\bar{z}}|^2\right) \text{ on } \bar{y} = 0\ , \tag{4a-c}$$

$$u \to u_e, w \to 0 \quad \text{as } \bar{y} \to \infty\ . \tag{5}$$

The velocities and pressure have been rescaled with respect to the Reynolds number in the manner suggested above, *e.g.* we have written $u_m = u, v_m = \mathrm{Re}^{-1/2} v, w_m = \mathrm{Re}^{-3/8} w$ with $(u,v,w)$ of $O(1)$. In (4c), $p$ represents the wave pressure which is governed by the equation

$$\begin{gathered} p_{\bar{z}\bar{z}} - \frac{\lambda_{\bar{z}}}{\lambda}\left[\frac{3}{2} + \frac{\xi}{2Ai(\xi)}(\xi\kappa(\xi) + Ai'(\xi))\right]p_{\bar{z}} - \alpha^2 p = (i\alpha\lambda)^{5/3}\frac{Ai'(\xi)}{\kappa(\xi)}A(x,\bar{z})\ , \\ \text{with } \lambda = u_{\bar{y}}\big|_{\bar{y}=0},\ \xi = -\frac{i^{1/3}\Omega}{(\alpha\lambda)^{2/3}},\ \kappa(\xi) = \int_{\xi}^{\infty} Ai(q)dq\ , \end{gathered} \tag{6}$$

where $Ai$ denotes the Airy function. Equation (6) arises from the analysis of the flow in the lower deck of the triple-deck structure where the 3D TS waves are driven by the unknown shear $\lambda(x,\bar{z})$. To close the system we require a relationship between the wave pressure $p$ and the wave displacement $-A(x,\bar{z})$. This law arises from considering the linearized potential flow in the upper deck where the TS waves are small perturbations to the external flow $u_e$.

The equations for the mean and wave flows are strongly coupled : the spanwise slip condition (4c) represents the nonlinear effect of the wave on the mean flow, while the back effect of the mean flow on the wave is transmitted via the appearance of the mean-flow skin-friction $\lambda$ in (6), both explicitly and via the function $\xi$.

For given starting conditions the system can be solved under the assumption that the wave remains neutral as it evolves in a nonlinear fashion. The wavenumber and amplitude of the wave adjust in such a way that the forced mean flow maintains a skin-friction consistent with a neutral wave. Owing to the complexity of the numerical problem, few solutions of the full equations have been obtained (Hall & Smith 1991, Walton & Smith 1992). In the next section we investigate a simplified set of equations which results from assuming the incoming boundary-layer to have a primarily two-dimensional, two-tier structure.

## 3. Vortex - TS wave interaction in a strong adverse pressure gradient

Consider once again the 2D decelerating boundary-layer flow past a flat surface. At the point of zero wall skin-friction, the classical boundary-layer equations exhibit a square-root singularity (Goldstein 1948). However if a low amplitude 3D TS wave is input slightly upstream of the singularity location, the theory of § 2 applies under the assumption that the input position is close to the lower branch neutral point. In view of the simplified form of the boundary-layer flow, the mean flow equations (1-3) of § 2 simplify to the single equation

$$\lambda\lambda_x = -1 + w_{\bar{z}}\big|_{\bar{y}=0} \tag{7}$$

subject to the condition (4c), while equation (6) for the wave remains unchanged. For details of the derivation of this system the reader is referred to Walton, Bowles & Smith (1994). We note that in the absence of an interaction, the cross-flow term on the right hand side of (7) is zero and the equation reduces to that of Goldstein (1948). The starting condition we impose on the system is as follows. At a given location $x=-1$, say, upstream of the classical Goldstein singularity, we assume that a 3D wave is introduced into the 2D boundary layer with a pressure distribution of the form $p(-1,\bar{z})=Q\cos\beta\bar{z}$. The requirement that such a starting condition be compatible with the interaction equations (4c), (6), (7) determines the amplitude $Q$ for a given spanwise wavenumber $\beta$. The equations are then integrated from this starting condition for different values of $\beta$. The results of three such calculations are shown in figure 2. We follow the evolution of the wave pressure amplitude at one particular spanwise location ($\bar{z}=0$) as we march downstream. On the scale in the figure, the Goldstein singularity would be expected to occur at $x=0$. However, upstream of this location nonlinear effects lead to a rapid growth in the wave amplitude and an apparent singularity. The next step is to determine how the character of the interaction changes beyond this point : terms previously neglected in the formulation must now center the reckoning in such a way that the singularity is removed. Another area of interest concerns the possibility of initiating the interaction from exponentially small disturbances upstream. On the streamwise length scale considered here, a feature of the interaction is the requirement of a finite value for the wave pressure amplitude at the location of wave input $x=-1$ (see figure 2 for example). On a shorter length scale around this location the amplitude of the imposed wave can no longer be assumed to be independent of downstream distance. In this localized region, the wave pressure can be shown to satisfy an integro-differential equation similar to that studied by Smith, Brown & Brown (1993) in the context of vortex-Rayleigh wave interactions. Currently work is in progress at Imperial College investigating the above two areas in association with Ms R.A. Patel.

## References

Cowley, S.J., Wu, X. (1993) : "Asymptotic approaches to transition modelling", in A.G.A.R.D.

Goldstein, S. (1948) : "On laminar boundary-layer flow near a point of separation", *Q.J. Mech. Appl. Math. Vol. I*, pp. 43-69.

Hall, P., Smith, F.T. (1991) : "On strongly nonlinear vortex/wave interactions in boundary layer transition", *J. Fluid Mech. Vol 227*, pp. 641-666.

Schubauer, G.B., Skramstad, H.K. (1947) : "Linear boundary-layer oscillations and transition on a flat plate", *J. Res. NAT. Bur. Stand. Vol 38*, pp. 251-292.

Smith, F.T. (1979) : "On the non-parallel flow stability of the Blasius boundary-layer", *Proc. R. Soc. Lond. A Vol 366*, pp. 91-109.

Smith, F.T., Walton, A.G. (1989) : "Nonlinear interaction of near-planar TS waves and longitudinal vortices in boundary-layer transition", *Mathematika Vol 36*, pp. 262-289.

Smith, F.T., Brown, S.N, Brown, P.G. (1993) : "Initiation of three-dimensional nonlinear transition paths from an inflectional profile", *Euro. J. Mech. B-Fluids Vol 12*, pp. 447-473.

Walton, A.G., Smith, F.T. (1992) : "Properties of strongly nonlinear vortex/Tollmien-Schlichting-wave interactions", *J. Fluid Mech. Vol 244*, pp. 649-676.

Walton, A.G., Bowles, R.I., Smith, F.T. (1994) : "Vortex/wave interaction in separating flows", to appear in *Euro. J. Mech. B-Fluids.*

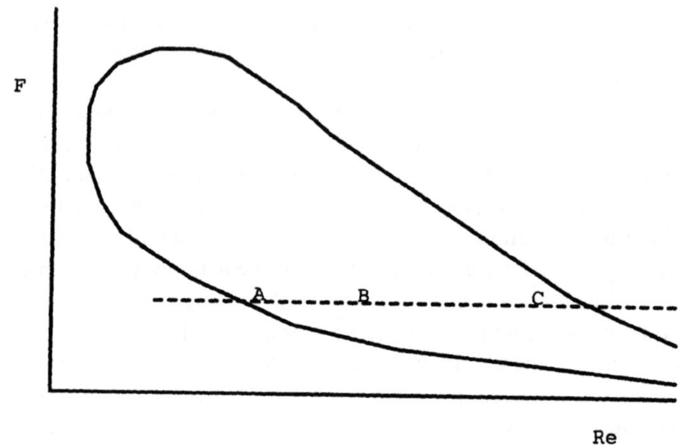


Figure 1

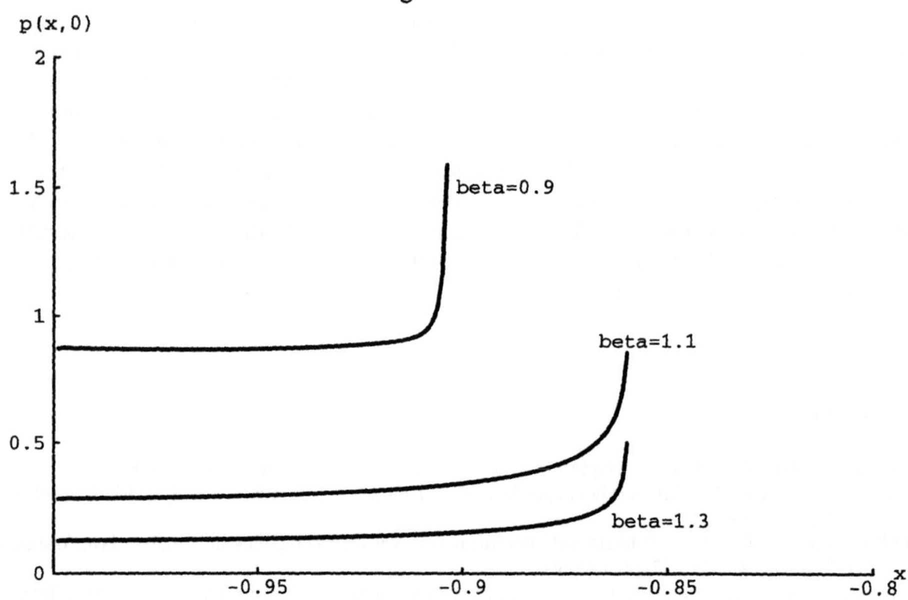


Figure 2

# Weakly non linear instability of the laminar boundary layer calculated by a parabolic system of equations

Ch. AIRIAU, G. CASALIS

CERT/ONERA, 2 av. Ed. Belin, B.P. 4025, 31055 Toulouse Cedex (France)

**Abstract:** The classical form of normal mode for the Tollmien-Schlichting wave can be generalized by taking into account the weak streamwise evolution of a steady boundary layer flow. Furthermore, the corresponding stability equations can be parabolized in the streamwise direction which leads to a simple marching procedure for the numerical resolution. This approach, called PSE, has the great advantage of allowing non linear models. For an incompressible flow and a two-dimensional mean flow, some results are given hereafter and compared with experiments, secondary instability theory and direct numerical simulations.

# 1 Introduction

The linear development of Tollmien-Schlichting (TS) waves in a two-dimensional incompressible boundary layer is now well known. Experiments are in good agreement with the classical linear stability theory (LST) which uses the parallel approximation, with non parallel approaches such as multiple scales analysis, and with Direct Numerical Simulation (DNS). However, after the linear stage, the TS-amplitudes reach a finite level for which the linear approximations are no longer valid. Some experiments in controlled conditions on a flat plate boundary layer (Klebanoff *et al* 1962; Kachanov and Levchenko 1984) have shown that transition to turbulence occurs after resonance between a two-dimensional TS wave (primary mode) and a pair of symmetrical three-dimensional waves (secondary mode).

This paper is devoted to the study of these weakly non linear phenomena using the interactive resonant wave concept with a new stability approach : the Parabolized Stability Equations (PSE). This approach was first proposed by Herbert and Bertolotti in 1987 to investigate the non parallel effects in a Blasius boundary layer. Detailed discussions on the linear PSE can be found in some articles (Bertolotti *et al* 1992; Airiau and Casalis 1993; Airiau 1994), the first part of the present paper only gives the main points. However, the PSE has the great advantage of allowing weak non linear models as described in the second part. Finally some applications for similar mean flow with and without pressure gradient are given and the results are compared with experiments, secondary instability theory (SIT) and DNS.

# 2 Linear Theory

Let be $x$ the streamwise direction, $y$ the direction normal to the wall and $z$ the spanwise direction. The lengths and velocities are written in non dimensional form using a constant length $\delta$ (boundary layer thickness) and the constant upstream velocity. The two-dimensional steady mean flow is given by $\vec{Q}_B$ :

$$\vec{Q}_B(x,y) = (U(x,y), V(x,y), 0, P(x,y)) \tag{1}$$

This basic flow is assumed to be solution of the Navier-Stokes equations. Then a three-dimensional unsteady disturbance is superimposed to the previous flow ; this is given by $Re(\tilde{q}(x,y,z,t))$ and has a single normal mode form :

$$\tilde{q}(x,y,z,t) = q(x,y) \exp i \left( \int_{x_0}^{x} \alpha(\xi)\, \mathrm{d}\xi + \beta z - \omega t \right) \tag{2}$$

This form is similar to the corresponding one in the parallel theory : $q(x,y)$ is an amplitude function and the exponential term is a wave like function. We used the spatial theory : $\alpha$ is a complex wavenumber, the real part $\alpha_r$ is a streamwise wavenumber, the imaginary part $\alpha_i$ is a spatial growth rate. Both quantities and the mean flow quantities (1) are assumed to have a weak dependence of $x$ such that terms like $\frac{\partial^2 U}{\partial x^2}$, $\frac{\partial^2 q}{\partial x^2}$, $\frac{\partial V}{\partial x}$, $\frac{\mathrm{d}\alpha}{\mathrm{d}x}\frac{\partial q}{\partial x}$, ... can be neglected in the equations. On the other hand, $\beta$ and $\omega$, which represent a spanwise wave number and a circular frequency, are real and constant. With these assumptions, a system of partial differential equations for the amplitude function $q(x,y)$ can be obtained from the perturbated linearized Navier-Stokes equations :

$$Lq + M\frac{\partial}{\partial x}q = 0 \tag{3}$$

This system is parabolic in the streamwise direction only under some conditions which are assumed to be satisfied hereafter (Airiau and Casalis 1993; Airiau 1994). In equation (3), $L$ contains the Orr-Sommerfeld operator and some "non parallel" terms. The unknowns are $q$ and $\alpha$, but to determine this latter, there is no dispersion relation as in the parallel theory. In fact a closure relationship is needed because the $x$-dependence between the amplitude function $q$ and the exponential term in (2) is ambiguous. As proposed by Herbert (Herbert and Bertolotti 1987), a normalization for the amplitude function is introduced in order to limit the $x$-dependence in this amplitude function. Several normalizations can be adopted, they have no influence upon the numerical value of $\tilde{q}$.

In non parallel studies using the multiple scales analysis, it is classical to introduce a small parameter expansion based on $\varepsilon_0 = 1/\sqrt{R_L}$, where $R_L$ is a Reynolds number calculated with a constant streamwise length $L$. Then a hierarchy of equations is written, but in order to have the Orr-Sommerfeld equation, first order terms (the viscous ones) are kept in the zero order equation. However, in the first order equation, the same viscous terms, which are now second order terms, are kept. In 1986, Itoh (Itoh 1986) proposed to write simultaneously the zeroth and first orders terms in a single equation, this equation is parabolic in $x$. But he didn't think to add a normalization and to solve the corresponding system. In 1987, Herbert (Herbert and Bertolotti 1987) considered simultaneously the zeroth and first order equations of the multiple scale analysis, the corresponding equation is parabolic and he had the idea to introduce a normalization in order to solve the system, finally called PSE. Comparing both parabolic equations (Itoh and Herbert), there are two extra terms in the PSE which are second order terms, these terms are included in operator $M$ of equation (3). But the PSE can be easily generalized for more complex flows (compressible and/or three-dimensional effects).

The parabolic system of equation (3), a normalization, the classical boundary conditions and an initial condition (the parallel theory solution is often chosen at the first value of $x$) are strictly speaking the Parabolized Stability Equations. These latter are solved by a marching procedure coupled with a predictor-corrector method between the normalization and equation (3). Different comparisons between DNS and LST, and calculations on real configurations (Airiau and Casalis 1993; Airiau 1994) have been made. The non parallel effect is found to be weak in a two dimensional boundary layer. It increases when the flow (the mean flow and/or the disturbance) becomes three-dimensional.

# 3 Non linear PSE

In controlled experiments, two routes to turbulence generated by three-dimensional resonances have been observed : the subharmonic one (Kachanov and Levchenko 1984; Saric *et al* 1987, are two examples) and the fundamental one (Klebanoff it et al 1962 for example). Both routes can be modelled by two ways. The secondary instability theory (SIT) is a first one : the total flow, composed of the mean flow and the TS wave, is very unstable to three-dimensional waves. Stability analysis is achieved using Floquet's theory. Secondly, there is the interactive resonant wave concept, firstly developed by Craik, which is used in the present paper. In the subharmonic route, there is a resonance between a TS wave (frequency $\omega$) and two symmetrical oblique waves (frequency $\omega/2$), this generates three-dimensional staggered $\Lambda$-shape vortices just before transition. For the fundamental resonance, the oblique waves have the frequency $\omega$ of the primary mode and the $\Lambda$-shape vortices are aligned.

## 3.1 Assumptions-Fluctuations

In the usual interactive resonant wave concept, the disturbances are written as infinite sum of discrete normal modes. In the PSE approach, the disturbance is searched in the following form :

$$\tilde{q}(x,y,z,t) = \sum_{n=-\infty}^{+\infty} \sum_{m=-\infty}^{+\infty} q_{n,m}(x,y) A_{n,m}(x)\, \chi_{n,m}(x,z,t) \tag{4}$$

where $A_{n,m}$ and $\chi_{n,m}$ are given by :

$$A_{n,m} = \exp\left[-\int_{x_0}^{x} Im(\alpha_{n,m}(X))\, \mathrm{d}X\right] \tag{5}$$

and :

$$\chi_{n,m} = \exp\left[i\left(\int_{x_0}^{x} Re(\alpha_{n,m}(X))\, \mathrm{d}X + m\beta z - n\frac{\omega}{2}t\right)\right] \tag{6}$$

Each $(n,m)$-mode looks like a linear PSE mode with a phase speed equal to $c_{n,m} = n\omega/(2.Re(\alpha_{n,m}))$. Let us consider a triad : a $(k,l)$-mode and a $(r,s)$-mode amplify a $(n,m)$-mode by non linear interaction. A perfect resonance appears when :

$$\chi_{n,m} = \chi_{k,l}.\chi_{r,s} \quad , \quad n = k + r \, , \quad m = l + s \tag{7}$$

A non perfect resonance means :

$$(Re(\alpha_{n,m}), m\beta, n\frac{\omega}{2}) = (Re(\alpha_{k,l}), l\beta, k\frac{\omega}{2}) + (Re(\alpha_{r,s}), s\beta, r\frac{\omega}{2}) + (\Delta\alpha, \Delta\beta, \Delta\omega) \tag{8}$$

$(\Delta\alpha, \Delta\beta)$ gives a slow spatial shift (discorded phase) and $\Delta\omega$ a temporal phase shift (detuned modes) in the resonant triad. In some experiments, phase speeds have been measured for different modes ; until strong resonance, the phase speed has been found to be the same for each mode, moreover tuned modes appeared to be the most amplified ones.

So we assume : $\forall n,m \;\; c_{n,m} = \omega/\alpha$, where $\alpha = Re(\alpha_{2,0})$ is the streamwise wavenumber of the primary TS mode. In equation (4), the $(0,0)$ and $(0,m)$ modes are respectively two-dimensional and three-dimensional mean flow distorsions. Applying the same assumptions as used in the linear case for each $(n,m)$-mode leads to a set of non linear coupled partial differential equations :

$$L_{n,m} q_{n,m} + M_{n,m} \frac{\partial}{\partial x} q_{n,m} = \sum_{r=-\infty}^{+\infty} \sum_{s=-\infty}^{+\infty} \frac{A_{r,s} A_{n-r,m-s}}{A_{n,m}} NL_{r,s}(q_{r,s}, q_{n-r,m-s}) \tag{9}$$

These equations are parabolic in the streamwise direction, the left hand side terms are the linear PSE for the $(n,m)$-mode (identical as equation (3)) and the right hand side terms represent the Non Linear interactions ($NL_{r,s}$) of the other modes upon the $(n,m)$-mode.

The boundary conditions are the same as for the linear PSE, except for the $(0,0)$-mode : this latter represents a steady distorsion of the mean flow, classical boundary layer conditions are applied for it. Moreover, normalizations are written for each mode in order to determine the remaining unknowns : $\alpha$ and the different growth rates $a_{n,m} = Im(\alpha_{n,m})$. The initial conditions are a primary TS-mode with a given initial amplitude $A$ and a secondary mode with a given initial amplitude $B$. This mode is the $(1,1)$-mode in the subharmonic resonance case and the $(2,1)$-mode in the fundamental resonance case. A special procedure (Airiau 1994) provides the initial conditions for the other generated modes. The figure below shows these generated modes for the two resonances ; due to symmetry reasons, only the $(n,m)$-modes with $n \geq 0$ and $m \geq 0$ are represented.

## 3.2 Non linear results

For the subharmonic resonance of the flat plate Blasius boundary layer, we found, as other authors (Bertolotti *et al* 1992 for example), very good agreement with PSE, DNS and Kachanov's experiments (Airiau 1994). Figure 2 shows an original comparison with experimental values obtained by Saric (Saric *et al* 1987). Only the evolution of the primary mode is represented for two different initial amplitudes ; for the lowest value, the evolution corresponds to the linear stability, for the highest value, there is an important non linear growth which is well reproduced by the PSE.

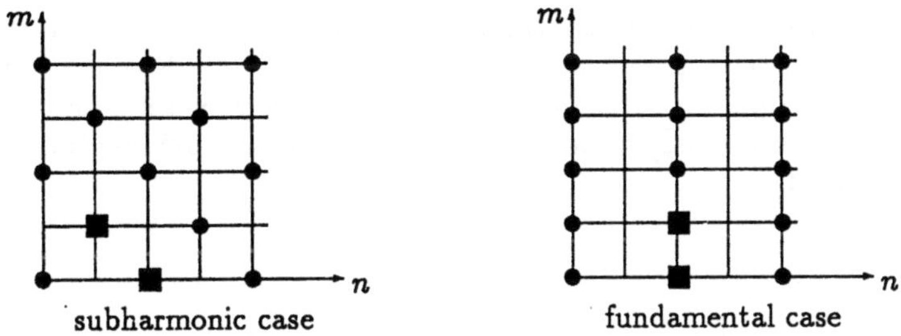


Figure 1: Squares : imposed modes, circles : generated modes

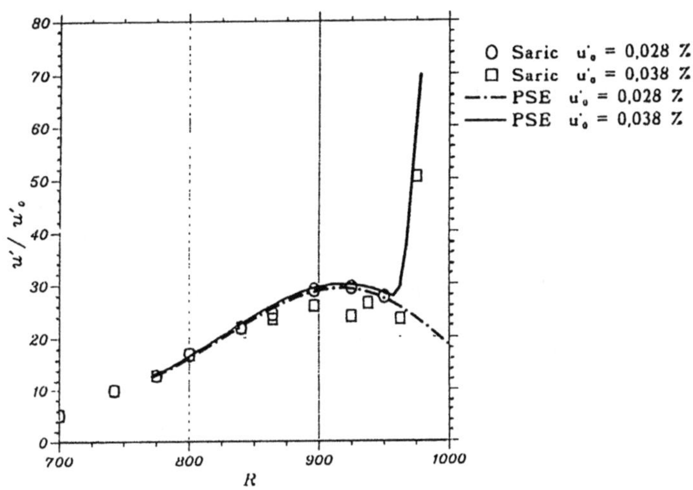


Figure 2: C-type resonance, comparisons with Saric et al. experiments

The PSE is in agreement with the secondary instability theory, as far as the evolution of the secondary mode is linear as it is assumed for the SIT. Figure 3 represents the streamwise velocity amplitude of both primary and secondary modes as functions of a local Reynolds number ($R = \sqrt{R_x}$) for a case which is similar to Kachanov's one (Kachanov and Levchenko 1984).

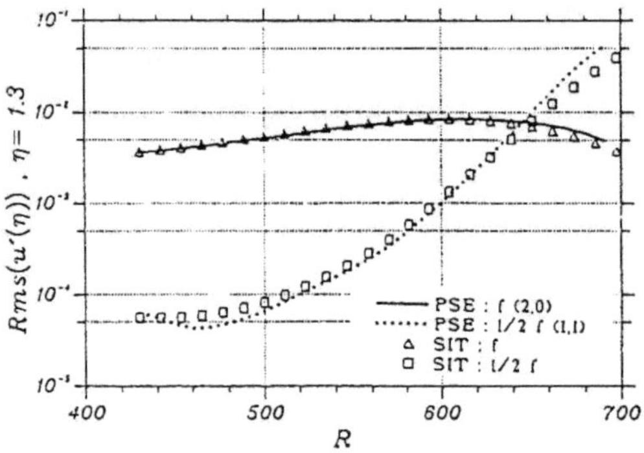


Figure 3: Comparisons with Secondary Instability Theory

For flows with strong positive pressure gradient, resonances are more difficult to observe because the mean flow is much more unstable with respect to the TS waves. However, it has been possible to reach the onset of subharmonic and fundamental resonances for a similar Falkner-Skan-profile close to separation. Figure 4 gives different amplitudes evolution in comparison with DNS done by Kloker (Kloker 1993).

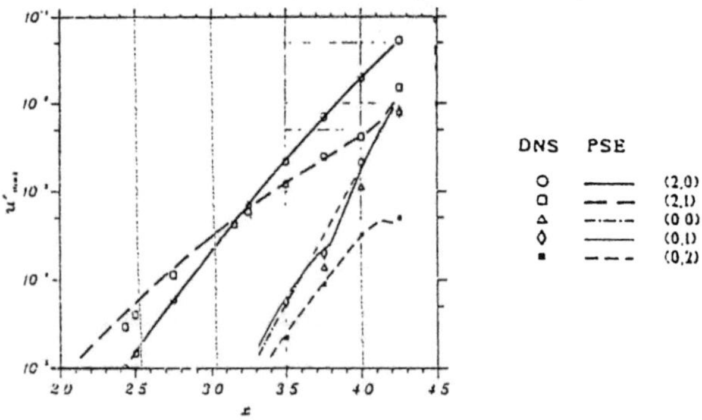


Figure 4: Strong positive pressure gradient, comparisons with DNS

Next figure (5) shows the influence of the initial amplitude $B$ of the secondary mode for a given initial amplitude $A$ of the primary mode. When the Reynolds number is close to 1.300, three-dimensional resonance occurs, but if $B$ is weak there is no non linear interactions upon the TS mode. On the other hand, for higher values of $B$, the TS wave reaches an amplitude saturation state, the steady $(0,0)$-mode becomes important as it is shown in figure 6. The skin-friction coefficient begins to increase, indicating thereby the

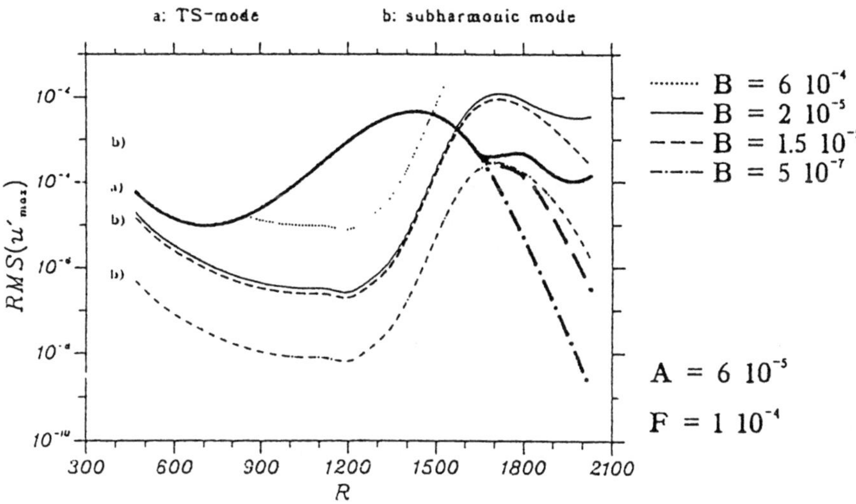


Figure 5: Influence of the amplitude $B$

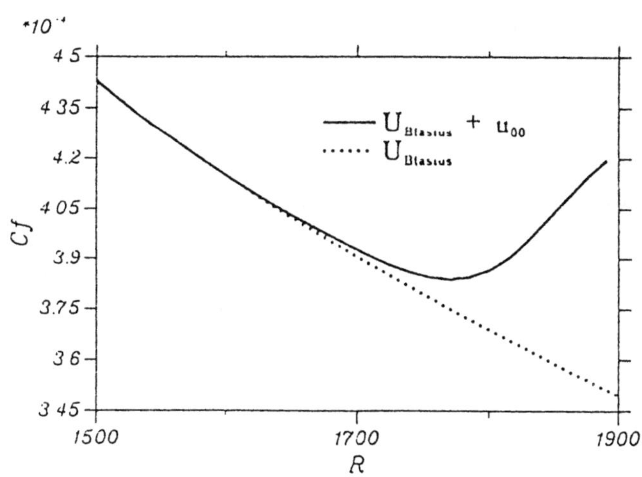


Figure 6: Skin-friction evolution

onset of transition. With the PSE, it is therefore possible for this given case to predict theoretically the transition location.

# 4 Conclusion

The results are in excellent agreement with DNS, but with an enormous time ratio 1 to 1000, the PSE approach appears thereby useful for the study of the non linear interactions which occur just upstream the transition. Furthermore, concerning the transition location prediction, this approach may become an interesting tool.

# References

Airiau Ch. (1994): "Stabilité linéaire et faiblement non-linéaire d'une couche limite laminaire incompressible par un système d'équations parabolisé", thèse ENSAE.

Airiau Ch. & Casalis G. (1993): "Stabilité linéaire de la couche limite par un système d'équations parabolique", La Recherche Aérospatiale, n° 5.

Bertolotti F.P., Herbert Th. & Spalart P.R. (1992): "Linear and non linear stability of the Blasius boundary layer", J.F.M., vol. 242.

Herbert Th. & Bertolotti F.P. (1987): "Stability analysis of nonparallel boundary layers", Bull. Am. Phys. Soc., vol 32.

Itoh N. (1986): "The origin and subsequent development in space of Tollmien-Schlichting waves in a boundary layer", Fluid Dyn. Res. 1, pp 119-130.

Kachanov Yu S. & Levchenko V. Ya. (1984): "The resonant interaction of disturbances at laminar-turbulent transition of a boundary layer", J.F.M., vol. 138, p. 43-74.

Klebanoff P.S., Tidstrom K.D. & Sargent L.M. (1962): , "The three dimensional nature of boundary layer instability", J.F.M., vol. 12, part 1.

Kloker M. (1993): "Direkte numerische Simulation des laminar-turbulenten Strömungsumschlages in einer stark verzögerten Grenzschicht", Dissertation, Universität Stuttgart.

Saric W.S., Levchenko V.Yu & Kozlov V.V. (1987): "Forced and unforced subharmonic resonance in boundary-layer transition", AIAA-84-0007.

# Applying a condition of observability to constitutive equations of laminar motion

Louis Brun

Commissariat à l'Energie Atomique, Centre d'Etudes de Vaujours-Moronvilliers ,
BP N°7 - 77181 Courtry - France

**Abstract:** The constitutive law of the material element, be it a fluid in laminar motion or a solid, is subject to two sets of fundamental requirements. The first set pertains to the Thermodynamics of Irreversible Processes, the second to stability. Stability considerations concerning the fluid generally confine, *via* Gibbs' postulate, to the case of the equilibrium, to the effect of specifying the signs of the specific heat and compressibility coefficient. Do such *observability* conditions somehow extend to the out -of - equilibrium motion of the fluid element? The question is given hereafter a positive answer by reducing it to the point of deciding whether the step-by-step determination of the motion in the vicinity of the element constitutes a well-posed problem. This approach ends up in two distinct necessary observability conditions, one for motion, one for equilibrium, the expressions of which involve the same dispersion function systematically built upon the constitutive law. Applications are made to the three most common models of fluid behaviour.

## Introduction

Despite the acknowledged prevalence of turbulent motion in nature, laminar motion and specifically stability within the thermomechanical context of laminar motion is an everlasting subject of sustained studies. The stability concept we refer to concerns systems in a state of equilibrium or , more generally, steadily moving with respect to an inertial or uniformly rotating frame. The conventional linear stability study involves a family of infinitesimal perturbations in the form of modes obeying the linearized equations of motion with space- dependent coefficients and homogeneous boundary conditions. The family is complete in a mathematical sense and such that each mode uniformly grows in an exponential way. The modes, which convey overall information on the basic motion, are essentially non-zero almost everywhere.

Consider instead at time $t_0$ a regular perturbation localized at point $\underline{x}_0$. By localized we mean that the perturbation decreases rapidly, along with its derivatives present in the equation, outside a neighbourhood of $\underline{x}_0$ with arbitrary small dimensions. One expects that the perturbation will remain localized and be approximately governed, all over its support and close to $t_0$, by the linearized equations, the coefficients of which have been fixed at their values at $\underline{x}_0$.

Observe that the property of the original motion of being stationary plays no role in the analysis and that the latter can be extended to any nonstationary motion provided it be locally regular. The system of linear equations with coefficients accordingly frozen admits a family of solutions, each scalar component of which is of the form :

$$f = \mathcal{R}_e\{f^* \exp(st - i\underline{k}.\underline{x})\}. \tag{1}$$

The growth rate $s$ and vectorial wave number $\underline{k}$ jointly satisfy a certain dispersion relation :

$$\mathcal{D}(s,\underline{k})_{\underline{x},t} = 0, \tag{2}$$

a power function of $s$ and $\underline{k}$, whose coefficients depend on the point and time considered (with zero subscripts dropped). Any solution of the frozen system may be represented by a Fourier integral of (1) over the real $\underline{k}$ - space. The integral corresponding to a localized perturbation necessarily comprises wave numbers with unbounded amplitude $k$. To each $\underline{k}$ there generally correspond several roots $s_1(\underline{k})$ , $s_2(\underline{k})$,... of the dispersion relation. If for $k$ tending to infinity the real part of one $s_i$ at least increased without limit, the localized perturbation would " explode" at once. But, as Hadamard 1964 puts it in his attempt to apply to PDEs of evolution Poincaré's ideas which are at the root of the theory of dynamic systems, "if the slightest error sufficed to change drastically the course of the phenomenon studied (...), this last would appear as governed not by precise laws but by chance" . The problem of computing step by step the solution through time would fail to satisfy a " particularly incisive requirement" resorting to stability which, next to the two requirements of existence and uniqueness, Courant and Hilbert 1966 consider to be "necessary if the mathematical formulation is to describe observable natural phenomena".

The foregoing considerations prompt us to lay down a conjecture in the form of what we shall term an

**Observability condition** *For the mathematical description of the physical system to make sense, it is necessary that, for each i,*

$$\lim \mathcal{R}_e\{s_i(\underline{k})\} < \infty. \tag{3}$$

The implementation of (3) will show that the conjecture eventually brings about limitations on the constitutive equations of the fluid themselves. Three models of fluid behaviour will be considered in turn : viscous incompressible, perfect compressible, viscous compressible. Each one is supposed beforehand to be thermodynamically admissible in the sense that it complies with the prescriptions of the TIP.

## 1 The viscous incompressible fluid

In the sequel $\underline{u}, \underline{\underline{d}}, \rho, p, T$, denote the velocity, the symmetric part of the velocity gradient or strain-rate tensor, the mass density, pressure and temperature. Since $\rho$ is supposed to be invariable, the properties of the fluid are specified by functions of $T$ only : the shear modulus $\mu$, the specific heat $C$ and the conductivity $K$.

The equations of motion, including the heat equation, read :

$$\begin{aligned} &\frac{\partial u_j}{\partial x_j} = 0, \\ &\rho \dot{u}_i + \frac{\partial p}{\partial x_i} - \mu(T)\Delta u_i = 0, \\ &\rho C(T)\dot{T} - K(T)\Delta T - 2\mu(T) d_{ij} d_{ij} = 0, \end{aligned} \tag{4}$$

where the overdot refers to the material derivation.

Let $\delta f$ be the perturbation of any quantity $f$. With the local dispersion relation (2) in mind, we take advantage of galilean invariance to cancel the local $\underline{u}$ in the expression of $\delta\dot{f}$, which accordingly simplifies to :

$$\delta\dot{f} = \frac{\partial}{\partial t}\delta f + \frac{\partial f}{\partial x_j}\delta u_j. \tag{5}$$

Further, in the operation of letting $k \to \infty$, any spatial derivative of $\delta f$ will become negligible next to higher order spatial derivatives of $\delta f$ occurring in the same equation.

Denoting by a prime the derivative with respect to $T$, the perturbed equations of motion thus become :

$$\begin{aligned} &\frac{\partial}{\partial x_j}\delta u_j = 0, \\ &\rho \frac{\partial}{\partial t}\delta u_i + \rho \frac{\delta u_i}{\partial x_j}\delta u_j - \mu \Delta \delta u_i + \frac{\partial}{\partial x_i}\delta p - \mu' \Delta u_i \delta T = 0, \\ &\rho C \frac{\partial}{\partial t}\delta T - K\Delta \delta T + \rho C \frac{\partial T}{\partial x_j}\delta u_j - 4\mu d_{hj}\frac{\partial}{\partial x_h}\delta u_j = 0. \end{aligned} \tag{6}$$

Substitution of expressions (1) for $\delta\underline{u}$, $\delta p$ and $\delta T$ results in :

$$\begin{aligned} &k_j u_j^* = 0, \\ &\left(\rho s + \mu k^2\right) u_i^* + \rho \frac{\partial u_i}{\partial x_j} u_j^* - i k_i p^* - \mu' \Delta u_i T^* = 0, \\ &\left(\rho C s + K k^2\right) T^* + \left(\rho C \frac{\partial T}{\partial x_j} + 4 i \mu d_{jh} k_h\right) u_j^* = 0. \end{aligned} \tag{6*}$$

Eliminating $p^*$ from $(6)_2^*$ thanks to $(6)_1^*$, and further $T^*$ between the equation obtained and $(6)_3^*$, leaves us with:

$$\left(\rho C s+K k^2\right)\left(\rho s+\mu k^2\right)u_i^* +\left(\rho C s+K k^2\right)A_{ij}u_j^* - A_i\left(\rho C\frac{\partial T}{\partial x_j}+4i\mu d_{jh}k_h\right)u_j^* = 0,$$

where the coefficients $A_{ij} = \rho\left(\partial u_i/\partial x_j - \alpha_i\alpha_h\,\partial u_h/\partial x_j\right)$ and $A_i = \mu'\left(\alpha_i\alpha_j\Delta u_j - \Delta u_i\right)$ only depend on the direction $\underline{\alpha} = \underline{k}/k$ of the wave number. The above system is best rewritten in terms of the quantity $\overline{\overline{s}} = s/k^2$ as :

$$\left(\rho C\overline{\overline{s}}+K\right)\left(\rho\overline{\overline{s}}+\mu\right)u_i^* +\left(\rho C\overline{\overline{s}}+K\right)\frac{A_{ij}}{k^2}u_j^* - \frac{A_i}{k^2}\left(\frac{\rho C}{k^2}\frac{\partial T}{\partial x_j}+4i\mu d_{jh}\frac{\alpha_h}{k}\right)u_j^* = 0. \qquad (7)$$

The polynomial of the local dispersion relation, that is the determinant of the system, is the sum :

$$\mathscr{D}\left(\overline{\overline{s}},\underline{k}\right) \equiv \left(\rho C\overline{\overline{s}}+K\right)^3\left(\rho\overline{\overline{s}}+\mu\right)^3 + d\left(\overline{\overline{s}},\underline{k}\right)$$

of two polynomials, the first one of degree 6 in $\overline{\overline{s}}$, the second of degree 5 with all its coefficients tending to zero when $k \to \infty$. There follows at once that none of the six roots $\overline{\overline{s}}_i\left(\underline{k}\right)$ tends to infinity and that their limiting values necessarily cancel either $\rho C\overline{\overline{s}}+K$ or $\rho\overline{\overline{s}}+\mu$.

Since, besides, thermodynamic admissibility amounts to the inequalities $\mu > 0$ and $K > 0$, the observability condition then implies the sole thermal restriction :

$$C > 0. \qquad (8)$$

## 2 The perfect compressible fluid

Here the pressure becomes a thermodynamic quantity, which will be conveniently considered as a function of $v=1/\rho$ and $\eta$, the specific volume and entropy. There being neither viscous stress nor heat conduction, the flow is locally isentropic and the equations of motion reduce to :

$$\begin{aligned} &\rho\dot{v} - \frac{\partial u_j}{\partial x_j} = 0\ , \\ &\rho\dot{u}_i + \frac{\partial p}{\partial v}\frac{\partial v}{\partial x_i} + \frac{\partial p}{\partial \eta}\frac{\partial \eta}{\partial x_i} = 0, \qquad (9) \\ &\dot{\eta} = 0. \end{aligned}$$

Considering the two remarks which flank Eq. (5), the perturbed equations together with their space-harmonic counterpart successively become :

$$\rho\frac{\partial}{\partial t}\delta v + \dot{\rho}\delta v - \frac{\partial}{\partial x_j}\delta u_j = 0,$$

$$\rho\frac{\partial}{\partial t}\delta u_i+\rho\frac{\partial u_i}{\partial x_j}\delta u_j+\frac{\partial p}{\partial v}\frac{\partial}{\partial x_i}\delta v+\frac{\partial p}{\partial \eta}\frac{\partial}{\partial x_i}\delta\eta=0, \tag{10}$$

$$\frac{\partial}{\partial t}\delta\eta+\frac{\partial \eta}{\partial x_j}u_j^*=0.$$

$$(\rho s+\dot\rho)v^*+ik_ju_j^*=0,$$

$$\rho s u_i^*+\rho\frac{\partial u_i}{\partial x_j}u_j^*-i\frac{\partial p}{\partial v}k_i v^*-i\frac{\partial p}{\partial \eta}k_i\eta^*=0, \tag{10)*$$

$$s\eta^*+\frac{\partial \eta}{\partial x_j}u_j^*=0.$$

Elimination of v* and $\eta^*$, provided $s(\rho s+\dot\rho)\neq 0$, ends up in a system which, with the help of the new variable $\bar s = s/k$, is rewritten as :

$$\rho\bar s^2\left(\rho\bar s+\frac{\dot\rho}{k}\right)u_i^*+\left[\left(\rho\bar s+\frac{\dot\rho}{k}\right)\left(\frac{\rho}{k}\bar s\frac{\partial u_i}{\partial x_j}+\frac{i}{k}\frac{\partial p}{\partial \eta}\frac{\partial \eta}{\partial x_j}\alpha_i\right)-\bar s\frac{\partial p}{\partial v}\alpha_i\alpha_j\right]u_j^*=0. \tag{11}$$

The determinant of the system is here the sum

$$\mathscr{D}(\bar s,\underline{k})\equiv\rho^6\bar s^9-\rho^4\frac{\partial p}{\partial v}\bar s^7+d(\bar s,\underline{k}),$$

with $d$ is a polynomial of degree 7 whose coefficients tend to zero with $1/k$. There follows that none of the roots $\bar s_i(\underline{k})$ tends to infinity and that some of them at least tend to a non-zero value given by : $\bar s^2 - v^2\,\partial p/\partial v = 0$.

In the absence of any restriction from thermodynamics, for the observability condition to be satisfied, one should have the single mechanical restriction :

$$\frac{\partial p}{\partial v}(v,\eta)<0. \tag{12}$$

# 3 The viscous compressible fluid

The stress tensor $\underline{\underline{\sigma}}$ here includes a viscous component $\underline{\underline{\sigma}}^v$, while the heat flux $\underline{q}$ no longer vanishes. The pressure and entropy both derive from the specific free energy $\varphi$, a natural function of volume and temperature. The constitutive equations accordingly read :

$$\sigma_{ij}(v,T,\underline{\underline{d}})=-p\delta_{ij}+\sigma_{ij}^v$$

$$\sigma_{ij}^v=\lambda(v,T)\mathrm{tr}.\underline{\underline{d}}\,\delta_{ij}+2\mu(v,T)d_{ij},\quad q_i=-K(v,T)\frac{\partial T}{\partial x_i}, \tag{13}$$

$$p=-\frac{\partial\varphi}{\partial v},\quad \eta=-\frac{\partial\varphi}{\partial T}.$$

They are to be supplemented by the thermodynamic restrictions :

$$\lambda \geq 0, \quad 3\lambda + 2\mu \geq 0, \quad K \geq 0. \tag{14}$$

The mass conservation equation $(9)_1$ for the perfect compressible fluid applies unchanged. The two other equations of motion assume the form :

$$\begin{aligned} &\rho \dot{u}_i - \frac{\partial \sigma_{ij}}{\partial x_j} = 0, \\ &\rho T \dot{\eta} + \frac{\partial q_j}{\partial x_j} - \sigma_{ij}^v \frac{\partial u_i}{\partial x_j} = 0, \end{aligned} \tag{15}$$

where use has been made of (13) to force the entropy rate into the energy equation.

Let $C = T\,\partial\eta/\partial T$ be the specific heat and $\ell = T\,\partial\eta/\partial v$ the latent heat. We rewrite the equations in order to exhibit the dependence upon the unknown fields $v, T, \underline{u}$ :

$$\begin{aligned} &\rho \dot{u}_i - \frac{\partial \sigma_{ij}}{\partial v}\frac{\partial v}{\partial x_j} - \frac{\partial \sigma_{ij}}{\partial T}\frac{\partial T}{\partial x_j} - (\lambda+\mu)\frac{\partial^2 u_j}{\partial x_i \partial x_j} - \mu \Delta u_i = 0, \\ &\rho C \dot{T} - K\Delta T + \ell \rho \dot{v} - \sigma_{ij}^v \frac{\partial u_i}{\partial x_j} - \left(\frac{\partial K}{\partial v}\frac{\partial v}{\partial x_j} + \frac{\partial K}{\partial T}\frac{\partial T}{\partial x_j}\right)\frac{\partial T}{\partial x_j} = 0. \end{aligned} \tag{16}$$

Linearizing equations (9) and (16) along the simplifying steps described and brought into play before results in the system :

$$\begin{aligned} &\rho \frac{\partial}{\partial t}\delta v + \dot{\rho}\delta v - \frac{\partial}{\partial x_j}\delta u_j = 0, \\ &\rho \frac{\partial}{\partial T}\delta u_i - \frac{\partial \sigma_{ij}}{\partial v}\frac{\partial}{\partial xj}\delta v - \frac{\partial \sigma_{ij}}{\partial T}\frac{\partial}{\partial x_j}\delta T - (\lambda+\mu)\frac{\partial^2}{\partial x_i \partial x_j}\delta u_j - \mu \Delta \delta u_i = 0, \\ &\rho C \frac{\partial}{\partial t}\delta T - K\Delta \delta T + (\ell - 2\lambda \rho \dot{v})\dot{\rho}\frac{\partial}{\partial t}\delta v - 4\mu d_{ij}\frac{\partial}{\partial x_j}\delta u_i - \frac{\partial K}{\partial v}\frac{\partial T}{\partial x_j}\frac{\partial}{\partial x_j}\delta v = 0. \end{aligned} \tag{17}$$

Let $s' = \rho s/k^2$ and $k_i' = k_i/k^2$. Eliminating $\underline{u}^*$ between the equations of the space-harmonic counterpart of (17) leaves us with :

$$\left[\left(s' + \frac{\dot{\rho}}{k^2}\right)(s' + \lambda + 2\mu) + \frac{\partial \sigma_{ij}}{\partial v}k_i' k_j'\right]v^* + \frac{\partial \sigma_{ij}}{\partial T}k_i' k_j' T^* = 0, \tag{18}$$

$$\begin{aligned} &\left\{(s'+\mu)\left[(\ell - 2\lambda\rho\dot{v})s' + i\frac{\partial K}{\partial v}\frac{\partial T}{\partial x_j}k_j'\right] + \underline{4\mu d_{ij}k_j'\left[(\lambda+\mu)\left(s' + \frac{\dot{\rho}}{k^2}\right)k_i' + \frac{\partial \sigma_{ih}}{\partial v}k_h'\right]}\right\}v^* \\ &\qquad + \left[(s'+\mu)(Cs' + K) + \underline{4\mu d_{ij}\frac{\partial \sigma_{ih}}{\partial T}k_j' k_h'}\right]T^* = 0. \end{aligned} \tag{19}$$

Further simplifications can still be achieved in the last equation by observing that the underlined terms become comparatively negligible in the limit. The asymptotic dispersion relation follows accordingly :

$$(s'+\mu)\left\{(Cs'+K)\left[\left(s'+\frac{\rho}{k^2}\right)(s'+\lambda+2\mu)+\frac{\partial\sigma_{ij}}{\partial v}k_i'k_j'\right]\right.$$
$$\left.-\left[(\ell-2\lambda\rho\dot{v})s'+i\frac{\partial K}{\partial v}\frac{\partial T}{\partial x_j}k_j'\right]\frac{\partial\sigma_{ij}}{\partial T}k_i'k_j'\right\}=0.$$

After the braces have been developped with respect to $s'$ and considered in the limit $k'\to 0$, it is seen that the dispersion relation splits up into either relation :

$$\begin{cases} s'+\mu=0, \\ \left(s'+\dfrac{\dot{\rho}}{k}\right)(s'+\lambda+2\mu)(Cs'+K)+K\dfrac{\partial\sigma_{ij}}{\partial v}k_i'k_j'=0. \end{cases} \tag{20}$$

Coming back to the original variables, the four roots are respectively :

$$s_1=-\frac{\mu}{\rho}k^2+\mathrm{O}(1),\qquad s_2=-\frac{\lambda+2\mu}{\rho}k^2+\mathrm{O}(1),\qquad s_3=-\frac{K}{\rho C}k^2+\mathrm{O}(1),$$
$$s_4=\underbrace{-\frac{\dot{\rho}}{\rho}-\frac{1}{\rho(\lambda+2\mu)}\frac{\partial\sigma_{ij}}{\partial v}\alpha_i\alpha_j}_{n(\underline{\alpha})}+\mathrm{O}\left(\frac{1}{k^2}\right). \tag{21}$$

The leading term $n(\underline{\alpha})$ of $s_4$, where the partial derivative of $\sigma_{ij}$ refers to expression (13) of the constitutive law, is finite. On account of inequalities (14), the observability condition is accordingly found to be satisfied by each mode, including the third one, provided only :

$$C>0. \tag{22}$$

The above analysis was performed on the generic assumption $\partial K/\partial v\neq 0$. In case of equality, the corresponding contribution to the first brackets in (19) would be replaced by $\rho C\dot{T}/k^2$, without impairing the expressions for the $s_i$ and the ensuing conclusion.

## 4 The equilibrium

The observability condition in the form of inequality (3) applied up to here is supposed to concern any motion. If however a given local state is to keep unchanged all the time, requiring of a localized infinitesimal perturbation (see the introduction) that it doesn't explode will but partially reflect the stability property of the underlying state :the perturbation should stay infinitesimal all the time as well and none of its Fourier components at vanishingly small wavelengths be allowed to grow. There naturally follows the

**Observability condition of equilibrium** *For a local state of equilibrium to be observable, it is necessary that, for each i,*

$$\lim_{k\to\infty} \mathcal{R}_e\{s_i(\underline{k})\} \leq 0. \tag{23}$$

The application to the three models is straightforward. The following set of conditions is obtained according as the fluid is :

$$\begin{aligned} &\text{Viscous : } \ldots\ldots\ldots\ldots\ldots\ldots\ \ C>0 \\ &\text{Perfect compressible : } \ldots\ldots\ \partial p(v,\eta)/\partial v<0 \\ &\text{Viscous compressible : } \ldots\ldots\ \partial p(v,T)/\partial v<0\,, \qquad C>0. \end{aligned} \tag{24}$$

As known (see e.g. Callen 1960), inequalities $(24)_3$ amount to demanding that the specific internal energy $e$ be convex with respect to its natural argument $v,\eta$ at the state considered or equivalently that $\eta$ be concave with respect to $v,e$. These are the famous Gibbs' conditions of stability. Of course, $(24)_2$ and $(24)_3$ are compatible.

## Conclusion

Changing the vocabulary of stability for the one of observability has made it possible to extend , along a line of thought initiated by Poincaré, Gibbs' equilibrium inequalities based on considerations of pure thermostatics into limitations on the constitutive laws for any kind of laminar motion.

Three different laws have been considered. Whenever conduction is effective, the specific heat must be positive. When compressibility comes into play, the positiveness of the former is required in the absence of viscosity (this agrees with the system of equations being necessarily hyperbolic in this case). On the other hand, in the presence of viscosity a negative compressibility doesn't lead to the breakdown of the model. It should be kept in mind however that our necessary conditions are only necessary. If they happened to be sufficient as well, transiting between two equilibrium configurations along some path with piecewise negative compressibility would not be an unthinkable operation...

## References

Callen, H. B. (1960): *Thermodynamics* (John Wiley & Sons)

Courant, R., Hilbert, D. (1962): *Methods of Mathematical Physics, Vol II* (Interscience, New York)

Hadamard, J. (1964): *La Théorie des Equations aux Dérivées Partielles* (Editions Scientifiques, Pékin)

# Part III

# Non-Linear Waves

# Hysteresis and interaction of standing waves with Faraday excitation


A.D.D. Craik[1], J.G.M. Armitage[2]

[1]School of Mathematical & Computational Sciences,
[2]School of Physics & Astronomy,
University of St.Andrews,
St.Andrews, Fife, KY16 9SS
Scotland, U.K.



**Abstract**: Water waves in tanks subjected to small-amplitude periodic vertical vibration display a variety of phenomena. Asymptotic theories for the linear onset of resonantly-excited standing waves, their nonlinear equilibration and stability to neighbouring modes are reviewed, and results are compared with experiments.


## 1. Introduction

A number of interesting problems in asymptotic analysis arise in the study of weakly nonlinear water waves. Here, a few of these are discussed, that have arisen in the study of surface waves in closed containers subjected to small-amplitude vertical vibrations: this is commonly called 'Faraday excitation', the first such experiments being those of Faraday (1831). Although asymptotic analysis has had real success in modelling observed phenomena, some features seen in recent experiments are not captured by the customary scalings, and more elaborate models must be sought. Here, some of these phenomena, and the proposed models for them, are reviewed and some forthcoming work is previewed.

In inviscid theories of weakly nonlinear water waves, the small parameter, say $\varepsilon$, upon which asymptotic expansions are usually based is a measure of the wave slope. Since the time of Stokes (1847), expansion in ascending powers of waveslope has been a common and successful procedure. However, real fluids are viscous and real waves are damped by viscosity unless some forcing is supplied. In order to incorporate damping and forcing into a consistent asymptotic theory, it is necessary to make the apparently innocuous assumption that such terms are formally $O(\varepsilon^2)$: they then do not affect the 'zero-order' linear inviscid approximation that yields the dispersion relation, but appear at the same level as the leading-order, conservative nonlinear term. The latter, for a single dominant wavemode, simply corresponds to a slight frequency shift, often called the 'leading-order Stokes correction'. But, as will be seen, this model is not always good enough. In particular, higher-order damping and forcing terms must be included in models of hysteresis: see section 2.

It has also been observed that the first excited standing wavemode may be unstable to other neighbouring modes. There has been much study of this and related phenomena, particularly the various attempts to model the fascinating experiment of Ciliberto & Gollub (1985), which first revealed clear evidence of periodic and chaotic modulations as two neghbouring modes 'compete'. But two-mode competition is sometimes less important than a three-mode scenario, involving the first excited mode and a pair of 'sidebands', much as in nonlinear Eckhaus (1963) and Benjamin-Feir (1967) instability. The only theoretical study of such three-mode models with Faraday excitation is a very recent paper by Craik (1994a). This is briefly summarised in section 3, where the results are compared with some unpublished photographs taken from a videotape recording of the experiments by Craik & Armitage (1994).

## 2. Hysteresis

Commonly, in a closed tank subject to Faraday excitation, a single dominant standing wavemode emerges from the discrete spectrum of available modes. This has a frequency precisely half that of the imposed oscillations, and the preferred mode is that with natural linear frequency closest to this value. The linear inviscid stability problem was shown by Benjamin & Ursell (1954) to be governed by a Mathieu equation. The subharmonic instability band, which has strongest growth, is that usually observed, since weak damping affects it least. It is for this reason that one may restrict attention to modes with frequencies close to this subharmonic resonance. Then, the resultant slow temporal evolution of the wave, on a timescale $\tau = \varepsilon^2 t$, is governed by a complex amplitude equation of the form

$$-i(d/d\tau + \mu)A = \Omega A + FA^* + p|A|^2A + O(\varepsilon^2) \qquad (1)$$

(see, for example, the review article of Miles & Henderson, 1990). Here, $\varepsilon^2\mu > 0$ denotes weak viscous damping, $\varepsilon^2\Omega$ measures a small frequency mismatch from the precisely resonant forcing, $\varepsilon^2F$ measures the amplitude of the imposed tank oscillations and p is a real nonlinear coefficient that depends on the tank configuration and the particular wavemode selected. Without loss of generality, $\mu$ and $|p|$ may be set to unity by means of a rescaling, but it is preferable to retain the present form in order that the inviscid limit is easily recovered.

The model equation (1), with error terms ignored, yields linear instability when $F^2 > \mu^2 + \Omega^2$, and finite-amplitude equilibria $A = a\exp(i\theta)$ where

$$a = \{p^{-1}[-\Omega \pm (F^2 - \mu^2)^{1/2}]\}^{1/2}, \quad \sin 2\theta = \mu/F \le 1. \qquad (2)$$

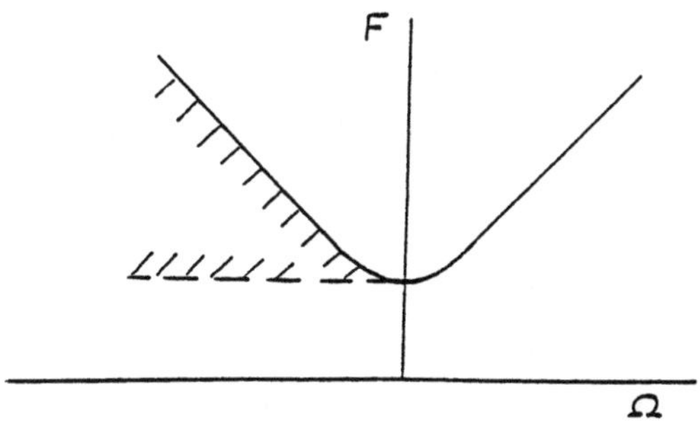


**Fig.1.** Hyperbolic linear stability boundary and hatched hysteresis region for p > 0, from (1).

The smaller solution a exists only outside the region of linear instability and corresponds to a saddle point. The larger denotes a stable focus of (1) and so corresponds to a realisable equilibrium state. A region of hysteresis therefore exists, between the linear stability boundary and the line $F = \mu$, having $\Omega < 0$ when $p > 0$ and $\Omega > 0$ when $p < 0$: the former case is shown in Figure 1. In this region, an undisturbed surface ($A = 0$) and the larger solution (2) are both stable to small disturbances of the complex amplitude A (though not necessarily to other combinations of modes: see later). This hysteresis may be understood as due to the nonlinear detuning of the natural wave frequency by the p-term.

However, the observed hysteresis regions in experiments of Simonelli & Gollub (1989) and Craik & Armitage (1994) for waves in rectangular containers take various shapes that differ from that just described, and Douady (1990) found no hysteresis at all near the minimum of the linear neutral stability boundary. The reason for this discrepancy was studied theoretically by Craik & Armitage (1994), who show that a variety of shapes for the lower hysteresis boundary is recovered if certain extra terms are retained in the model evolution equation. In particular, this boundary bifurcates from the stability boundary at some point away from the minimum at $F = \mu$, $\Omega = 0$; it may then extend to F-values substantially less than $F = \mu$, and may, or may not, curve upwards again as $|\Omega|$ increases. The extra terms are the same in form as those recently retained by Miles (1993), and derive from nonlinear modifications of the forcing and damping that depend on terms that are cubic in wave amplitude. Such a nonlinear damping term was earlier proposed by Milner (1991) and Douady (1990); but Miles (1993) was the first to derive the extra forcing terms, using a Hamiltonian formulation and retaining all non-viscous terms that appear at cubic order in wave amplitude.

The revised model equation is then

$$-i(d/d\tau + \mu)A = \Omega A + (F + \gamma|A|^2)A^* + \delta A^3 + (p + i\nu)|A|^2 A. \tag{3}$$

The new term in $\nu$ augments the linear damping when $\nu > 0$; but the value of this parameter is hard to determine, whether experimentally or theoretically, and there is no a priori reason to suppose that it must be positive. The terms in $\gamma$ and $\delta$ describe a nonlinear modification to the forcing and expressions for these parameters were derived by Miles (1993), using his truncated Hamiltonian formulation. Both $\gamma$ and $\delta$ are proportional to F, and $\delta = \gamma/3$ for Miles' configuration. Craik & Armitage (1994) consider the hysteresis region associated with equation (3). They obtain analyic solutions for the lower hysteresis boundary when $\delta = 0$ and they argue that the hysteresis boundary will be rather similar when $\delta = \gamma/3$, though it must then be found numerically. They show that the shape of the hysteresis boundary in the $\Omega$-F plane depends sensitively on the values of $\nu$ and $\Gamma = \gamma/F$: see Figure 2, which plots hysteresis regions in the plane (X, Y) where $X = \Omega|\Gamma|$ and $Y = (F\Gamma)^2$ for various cases (with $p = 1$ and $\delta = 0$).

Efforts are presently underway (Decent & Craik, in preparation) both to provide a more precise comparison of these theoretical results with experiment, and to give a more rigorous derivation of equation (3), or an extension thereof. The difficulty with (3), from the viewpoint of rational asymptotics, is that the scalings outlined above relegate the cubic damping and forcing terms to a higher power of $\varepsilon$ than the terms that also appear in (1). This comes about because linear forcing and damping are assumed to be $O(\varepsilon^2)$, and so the next-order corrections, that are cubic in wave amplitude, must normally be smaller by an $O(\varepsilon^2)$ factor than the cubic conservative term in p.

Three separate analyical approaches appear to be fruitful, each valid in a particular locality. The first employs equation (1) as a first approximation, giving the equilibrium solution (2) on the lower hysteresis boundary $F = \mu$, $\Omega < 0$ (for $p > 0$). A higher approximation, involving small modifications to both F and $\Omega$, and evolution on a slower timescale $\tau_1 = \varepsilon^4 t$, then yields an evolution equation that contains the terms in $\nu, \gamma, \delta$ that appear in (3), but with the addition of a further term in $|A|^4 A$ that comes from the next-order conservative frequency shift. Examination of this equation then leads to a small correction of the first-order lower hysteresis boundary $F = \mu$. Though this scaling constrains this deviation to be small, results are in line with expectations from (3).

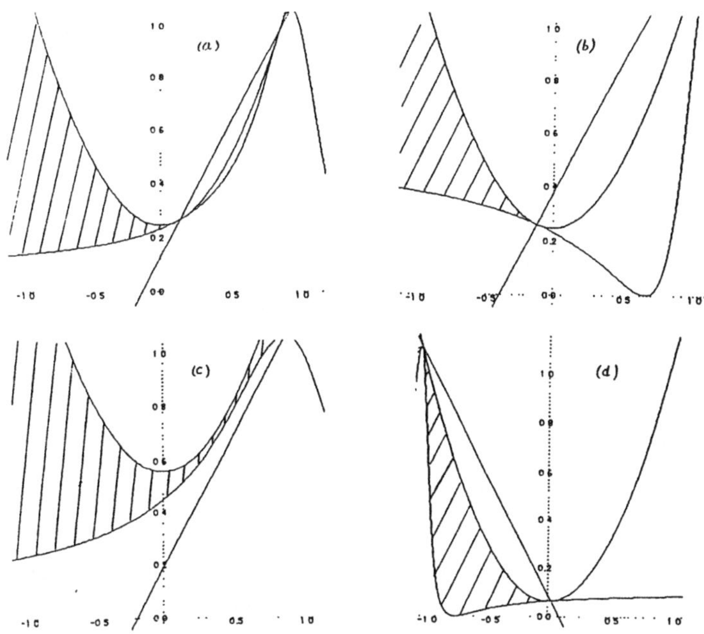


**Fig.2 a-d.** Modified (hatched) hysteresis regions for equation (2) with scaled variables (X,Y) for $p = 1$, $\delta = 0$. Cases (a) to (d) have $(\mu\Gamma, \nu)$ = (0.5, 0.25), (0.5, 0.75), (0.75, 0.25), (-0.25, -0.333) respectively.

But this representation breaks down very close to the linear neutral curve. Near there, an alternative expansion is applicable. This has, as a first approximation, the linear equation

$$-i(d/d\tau + \mu)A = \Omega A + FA^* \tag{4}$$

with the equilibrium solution $F = (\Omega^2 + \mu^2)^{1/2}$, $2ph(A) = \sin^{-1}(\mu/F)$ and $|A|$ as yet arbitrary. At next order, small $O(\varepsilon^2)$ modifications to F and $\Omega$ may be introduced. If p is then formally taken to be $O(\varepsilon^2)$, so that it appears at the same order as the nonlinear damping and forcing terms (even though it will normally be numerically much larger than $\nu$ or $\gamma$), a consistent scaling is achieved that gives the bifurcation point of the lower hysteresis boundary from the linear neutral curve.

Lastly, a global analysis may be carried out of the hysteresis properties of an equation close to (3), but with the addition of a quintic conservative term in $|A|^4A$. Together, these complementary analyses reinforce the general conclusions of Craik & Armitage (1994) based on the model equation (3), though with some modification of detail. Also, estimation of the various parameter values appropriate to their, and some new, experimental results enables detailed comparisons to be made. Full details are in course of preparation (Decent & Craik, 1994).

### 3. Stability and mode-competition of standing waves

The mutual interaction of two modes of different spatial forms but closely similar natural frequencies has been studied by a number of authors, for example Meron & Procaccia (1986), Feng & Sethna (1989), Nagata (1989, 1991), Umeki & Kambe (1989), Kambe & Umeki (1990), Umeki (1991) The initial impetus for these studies was undoubtedly the now well-known experiment of Ciliberto & Gollub (1985), who found periodic and chaotic modulations in distinct regions of the plane of forcing-amplitude versus forcing frequency. Other experiments of note are those of Simonelli & Gollub (1989), Douady & Fauve (1988), Douady (1990), Douady, Fauve & Thual (1989), Craik & Armitage (1994) in containers of various shapes and dimensions. Also noteworthy are the studies of Funakoshi & Inoue (1988, 1992) with containers subject to horizontal rather than vertical vibration.

This work on two-mode interaction is reviewed by Craik (1994a), who also describes the detailed properties of the coupled system

$$\begin{aligned} -i(A_\tau + \mu A) &= (\Omega - 1)A + FA^* + A(-|A|^2 + \nu_1|B|^2) \\ -i(B_\tau + \mu B) &= (\Omega + 1)B + FB^* + B(\nu_2|A|^2 - |B|^2) \end{aligned} \tag{5}$$

where the subscript $\tau$ denotes $d/d\tau$. These equations are a simplified representation of two modes, A and B with different, but similar, natural (linear) frequencies at $\Omega = 1$ and $-1$ respectively, each with nonlinear 'p-coefficients' taken as -1, and with interaction through the terms in $\nu_1$ and $\nu_2$. The coefficients $\nu_1$ and $\nu_2$ are taken to be real, so representing conservative interactions. For simplicity, any resonant terms of form $A^2B^*$ or $B^2A^*$ are ignored, though these are certainly important in some contexts. The respective hysteresis regions of each mode, taken separately, then both lie to the right of the minima of their neutral stability boundaries, at $(F, \Omega) = (\mu, 1)$ and $(\mu, -1)$, and so overlap for $\Omega > 1$. (This is not the case in Ciliberto & Gollub's experiment, but it is in those of Simonelli & Gollub and Craik & Armitage). Four families of mixed-mode equilibria also exist, in certain regions of the $(F,\Omega)$ plane, and these, like the single-mode equilibria, may be locally stable or unstable to small perturbations of A and B. These stability properties are fully described in Craik (1994a) and compared with related studies of Feng & Sethna (1989) and Kambe & Umeki (1990). Computer solutions of (5) show a range of behaviour. Particularly noteworthy is the fact that, with $\nu_1 = \nu_2 = -1$, all cases studied evolved to a stable single-mode equilibrium or flat surface whenever $\mu > 0$; even though evolution with $\mu = 0$ revealed a rich structure, with instances of period-doubling and homoclinic chaos. Such 'collapse' of homoclinic chaos was also noted by Umeki (1991).

The corresponding model equations when the sign of the 'p-coefficient' of mode B is changed from -1 to +1 have very different behaviour. Then, the respective single-mode hysteresis regions do not overlap (and so, in this respect at least, conform to the Ciliberto & Gollub experiment) and the structure of equilibrium states is totaly altered. So too is the temporal evolution of solutions, which can now admit a rich variety of limit-cycle and chaotic behaviour, even with $\nu_1 = \nu_2 = -1$, $\mu > 0$ and when resonance terms are neglected.

However, there are good reasons to believe that a two-mode model is inappropriate for those configurations that admit modes of nearly similar spatial structure with similar frequencies. One such example is the experiment of Craik & Armitage (1994), who investigate two-dimensional standing waves in a rectangular tank that is about eleven times longer than the wavelengths studied. Then, wave modes with 21, 22 and 23 half-wavelengths within the tank length certainly have similar structures and frequencies. Craik (1994a) postulates that a single mode - say that with 22 half-wavelengths - will first be selected and achieve its equilibrium state, according to (1) and (2) (ignoring, for simplicity, the further complexities of nonlinear forcing and damping described above). But this mode may trigger a parametric instability of the pair of neighbouring modes, say B and C. With the dominate A-mode represented by $A_0$ and taken as constant, the linearised B and C equations have the form

$$-i(B_\tau + \mu B) = (\Omega - 1)B + FB^* + \nu_1|A_0|^2B + sA_0^2C^*$$

$$-i(C_\tau + \mu C) = (\Omega + 1)C + FC^* + \nu_2|A_0|^2C + sA_0^2B^*. \tag{6}$$

This scenario resembles Eckhaus (1963) and Benjamin-Feir (1967) instability, with the B and C modes in the role of sidebands. A standard stability analysis of this 4-dimensional system, with $A_0$ given by (2), leads to an eigenvalue relation for exponential solutions proportional to $\exp(\lambda\tau)$. With $\mu = 0$, the inviscid approximation to this is

$$\begin{aligned}\lambda^2 = (s\gamma)^2 - 1 - \gamma[\Omega(2+\nu_1+\nu_2) + \nu_2 - \nu_1] - (1/2)\gamma^2(\nu_1{}^2+\nu_2{}^2-2)\\ \pm\{(s\gamma)^2[4(\Omega-\gamma)^2 - 4 - \gamma^2(\nu_1-\nu_2)^2]\\ +[2 + \gamma(\nu_2-\nu_1)^2][\Omega + (\gamma/2)(\nu_1+\nu_2)]^2\}^{1/2},\end{aligned} \tag{7}$$

where $\gamma = |A_0|^2 = \Omega+F$. Instability occurs when the real part of $\lambda$ can take positive values.

The particular case $\nu_1 = \nu_2 = -1$ gives the following instability criteria:

$$\Omega > F[-1 + s^{-1}(1-F^2)^{-1/2}] \quad (F^2 < 1) \qquad \text{(growing oscillations)}, \tag{8a}$$

$$\text{either} \quad \Omega+F > s^{-1} \quad (F^2 > 1/2) \quad \text{or} \quad 0 < \Omega+F < s^{-1}(4F^2 - 1)^{1/2} \quad (F^2 > 1/4)$$

$$\text{(exponential growth)}. \tag{8b}$$

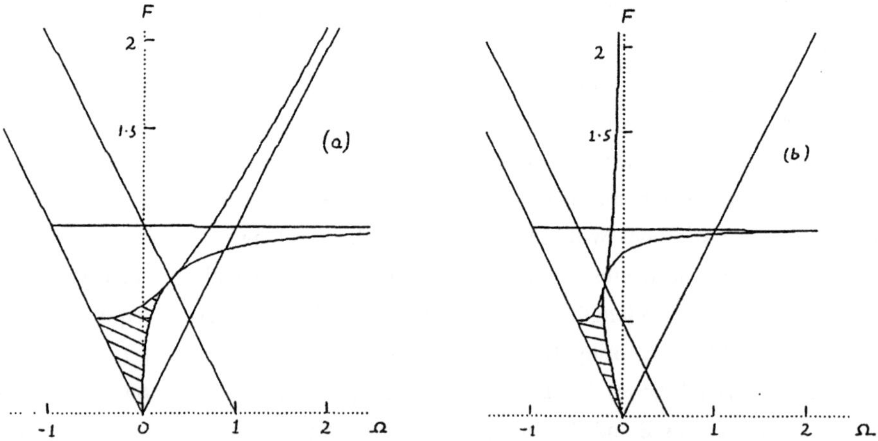


**Fig. 3 a,b.** Regions of Ω-F plane (hatched) where mode $A_0$ is stable to sidebands B and C, from equations (6) with $\nu_1 = \nu_2 = -1$ and with s = 1 and 2 respectively in (a) and (b).

The region of the F-$\Omega$ plane within which the $A_0$ wave is stable to small perturbations B and C is found to extend to small F-values. Figure 3, from Craik (1994a), shows the results from (8a,b) in the cases $s = 1$ and $s = 2$, only the shaded region giving a stable A-wave: note that the linear stability boundary for this mode is now given by $F = \pm\Omega$, since $\mu = 0$.

Work is presently underway to extend this analysis to compute the various parameter values that are characteristic of the experiment of Craik & Armitage (Decent & Craik, in preparation). Only when this is achieved can a precise comparison of theory and experiment be made. But, meanwhile, it is useful to compare the typical appearance of three-mode motion with that seen in videotape recordings made by Craik & Armitage and not previously published.

First, we show in Figure 4a three stills from the videotape that demonstrate onset of wave motion due to Faraday excitation. There, the liquid depth is 11.5mm, the forcing period is 108.2ms and the vertical periodic tank displacement between maximum and minimum is 0.46mm. The top still shows a flat surface (at say t = 0 secs.), the middle shows onset of wave motion at t = 90secs, and the bottom shows fully-developed standing waves at t = 110secs. Under these conditions, the standing waves were stable and persisted indefinitely.

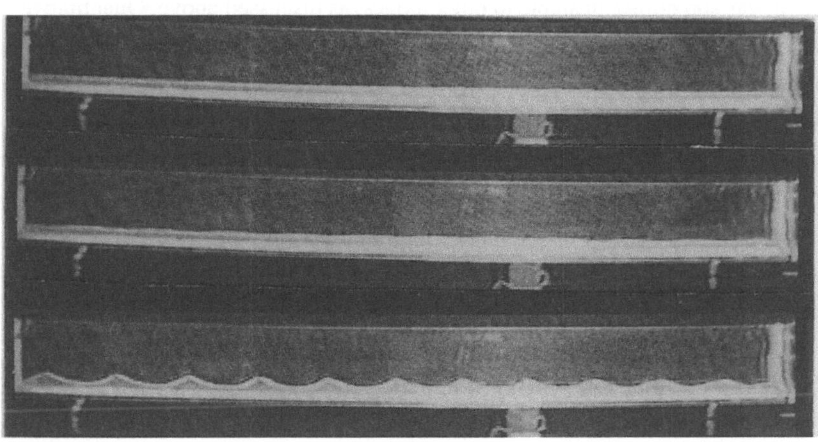

**Fig. 4a**

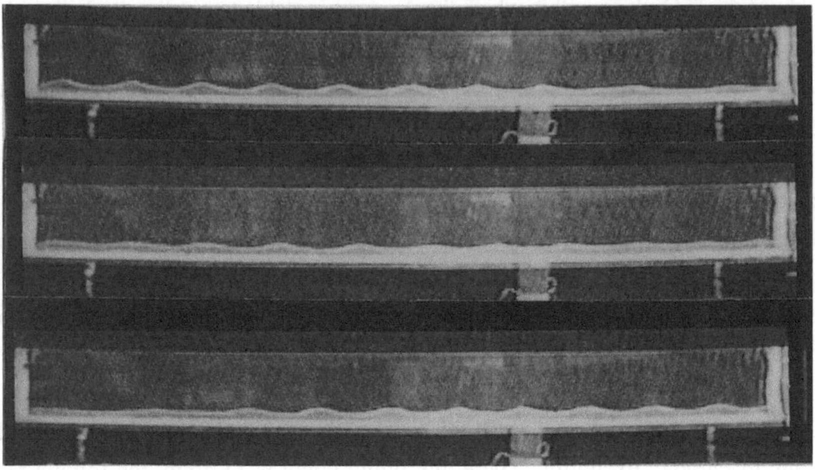

(b)

**Fig. 4 a,b.** Videotape stills from experiment by Craik & Armitage (1994): (a) onset of regular stable standing waves from flat surface, (b) periodic modulations.

At other forcing frequencies and amplitudes, although a regular standing-wave pattern first emerged, it was sometimes unstable to other modes, as discussed above. Quantitative data are provided by Craik & Armitage (1994); see also Craik (1994b). Sometimes, a modulating state was then established, with a slow period of several seconds: the waves then gave the appearance of a packet propagating to and fro from end to end of the tank. One such case is shown in Figure 4b, for forcing period of 110ms and tank displacement of 0.58mm. The sideband instability outlined above, which gives rise to growing oscillations, is believed to be responsible. Supposing that quasi-steady modulations among three participating modes becomes established, a demonstration of typical behaviour is modelled in Figure 5 by the function

$$a \cos kx + \cos(\omega t)\cos[(k+1)x] + \sin(\omega t)\cos[(k-1)x] \qquad (0 < x < \pi)$$

with $k = 22$, $a = 1$ and $\omega t = \pi N/6$ ($N = 1,2,3,.....12$), where, to separate the graphs, each is displaced vertically by 4N, for each N. Rather similar results are obtained with other values of a. This simple model certainly exhibits the same qualitative behaviour as is shown in Figure 4b and gives confidence in the likely success of a more precise three-mode analysis, yet to be carried out.

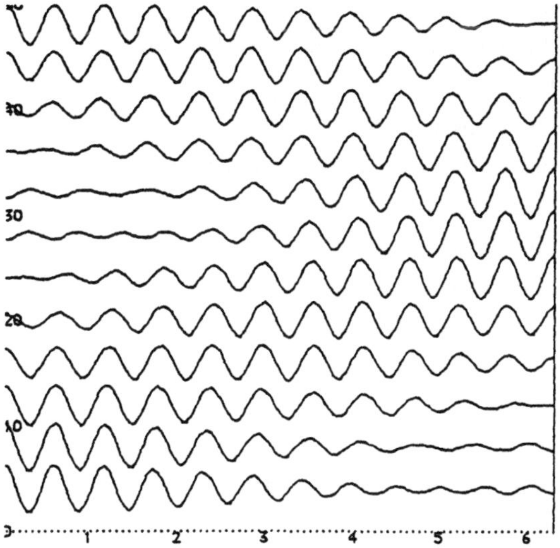

**Fig.5.** Periodic state resulting from superposition of three modes.

However, the modulations may be more complex than just described. Experimental cases revealed that a nearly-periodic modulation sometimes died out, leaving an almost flat surface, which then, being unstable, gradually gave rise to regular standing waves. These waves, themselves unstable to sidebands, were again replaced by the nearly-periodic modulations in a cycle of recurrence of several minutes' duration. Clearly, theoretical challenges still remain!

**Acknowledgment:** ADDC is grateful to the organisers for their kind invitation to attend this Colloquium.

## References

Benjamin, T.B. & Ursell, F. 1954 *Proc. Roy. Soc. Lond.* A225, 505-515

Benjamin, T.B. & Feir, J.E. 1967 J. *Fluid Mech.* 27, 417-430

Craik, A.D.D. 1994a Nonlinear interaction of standing waves with Faraday excitation. *Submitted*

Craik, A.D.D. 1994b to appear in *Proc. IUTAM Symp*. July 1993, Clarkson Univ. NY (ed. W.R.C. Phillips & S.P. Lin), Springer

Craik, A.D.D. & Armitage, J. 1994 Faraday excitation, hysteresis and wave instability in a narrow rectangular wave tank. *Submitted*

Ciliberto, S. & Gollub, J.P. 1985 *J. Fluid Mech.* 158, 381-398.

Decent, S.P. & Craik A.D.D. 1994 Hysteresis in Faraday resonance. *Preprint*

Douady, S. 1990 *J. Fluid Mech.* 221, 383-409

Douady, S. & Fauve, S. 1988 *Europhys. Lett.* 6, 221-226.

Douady, S. Fauve, S. & Thual, O. 1989 *Europhys. Lett.* 10, 309-315.

Eckhaus, W. 1963 *J. Mecanique* 2, 153-172.

Faraday, M. 1831 *Phil. Trans. Roy. Soc. Lond.* 121, 319-346.

Feng, Z.C. & Sethna, P.R. 1989 *J. Fluid Mech.* 199, 495-518.

Funakoshi, M. & Inoue, S. 1988 *J. Fluid Mech.* 192, 219-247.

Funakoshi, M. & Inoue, S. 1992 Nonlinear Dispersive Wave Systems 635-666 (ed. L. Debnath), *World Scientific*

Kambe, T. & Umeki, M. 1990 *J. Fluid Mech.* 212, 373-393

Meron, E. & Procaccia, I. 1986 *Phys. Rev.* A, 34, 3221-3237

Miles, J.W. 1993 *J. Fluid Mech.* 248, 671-683

Miles, J.W. & Henderson, D. 1990 *Ann. Rev. Fluid Mech.* 22, 143-165

Milner, S.T.1991 *J. Fluid Mech.* 225, 81-100

Nagata, M. 1989 *J. Fluid Mech.* 209, 265-284

Nagata, M. 1991 *Euro. J. Mech. B/Fluids* 10(2), Suppl. 61-63

Simonelli, F. & Gollub, J.P. 1989 J. *Fluid Mech.* 199, 471-494

Stokes, G.G. 1847 *Trans. Camb. Phil. Soc.* 8, 441-455. (Reprinted in *Math. Phys. Papers* 1, 197-219)

Umeki, M. & Kambe, T. 1989 *J. Phys. Soc. Japan* 58, 140-154

Umeki, M. 1991 *J. Fluid Mech.* 227, 161-192

# About the radiation diagram of an underwater acoustic source in the presence of gravity waves

Daniel Euvrard †

Groupe Simulation et Modélisation des phénomènes de Propagation
Equipe associée au CNRS et à l'Université Pierre et Marie Curie
Laboratoire de Mécanique, ENSTA, Centre de l'Yvette,
chemin de la Hunière, 91 120 – Palaiseau, France

**Abstract:** A time-harmonic acoustic source of amplitude $\varepsilon$ being located at a point $A$ of depth a, the corresponding acoustic field in the absence of gravity waves is well-known: it can be obtained by introducing a sink of amplitude $-\varepsilon$ at the point symmetrical of $A$ with respect to the free surface. If there are also gravity waves of small amplitude $\mathcal{A}$ and if the ratio $\ell$ of the acoustic to the gravity wave lengths is of order 1, the reflexion of the acoustic waves by the free-surface generates an acoustic perturbation of order $\varepsilon \mathcal{A}$. In a previous report (Euvrard and Mechiche Alami 1992) an explicit expression of this second-order acoustic pressure has been found, and then justified via a limiting-amplitude procedure: it is a Helmholtz double layer distribution on the mean free-surface. Here the asymptotic behaviour of the second-order acoustic pressure for large values of the horizontal distance is exhibited with help of matched asymptotic expansions, and then proved using Lebesgue's theorem together with stationary phase upper bounds. Finally this asymptotic behaviour can be written in a closed form as a function of $\ell$; it does not satisfy the standard radiation condition but exhibits some interesting features.

## 1 Introduction

### 1.1 Motivation

A linearized underwater monochromatic acoustic wave is described by a pressure field $\varepsilon\, p_1(x,\, y,\, z\,;\, t)$, $\varepsilon$ being a small and dimensionless parameter, and defined for instance by

$$p_1(x,\, y,\, z\,; t) = \mathcal{I}\mathrm{m}\left\{P_1(x,\, y,\, z)\, e^{-i\,\omega\, t}\right\}. \tag{1}$$

† Daniel Euvrard was not only an outstanding scientist and a warm man, much appreciated by his colleagues, but he was a, long experienced, licensed pilot. He died accidently, on July 10, 1994, while flying.

The complex function $P_1$ must satisfy the Helmholtz equation

$$\Delta P_1 + k^2 P_1 = 0\,, \quad k = \omega / c\,, \tag{2}$$

where $c$ (of the order of 1 400m/s) is the sound speed. If for instance this acoustic wave is radiated by a submarine, the normal velocity of the fluid, whence the normal derivative of $P_1$, will be prescribed on the mean position $\Sigma$ of the hull: this is a non-homogeneous Neumann condition. If we assume the ocean to be infinitely deep, and calm, $P_1$ must vanish on the free-surface $z = 0$ ($z$ being the elevation), which is a homogeneous Dirichlet condition. To close the problem we shall add the standard "Radiation Condition" at infinity in all horizontal and downward directions. Such a problem is well-posed and can be solved using a variety of numerical methods, for instance Boundary Element Methods.

To get a more realistic physical pattern we can introduce a stratification of the ocean, a finite and variable depth... or other complicated phenomena. Instead of that we shall consider here the presence of monochromatic gravity waves of small reduced amplitude $\mathcal{A}$ (see figure 1), and focus our attention on the case when the acoustic and gravity waves are of the same order, whence a maximum perturbation of acoustics by gravity waves can be expected. For usual gravity waves this corresponds to very low acoustic frequencies, say a few $Hz$ (Pot 1989). Conversely the submarine will be assumed to be deeply immersed, whence the diffraction of the gravity waves by the submarine will be disregarded.

### 1.2 Physical pattern

The corresponding physical pattern has been studied in Mechiche Alami 1992, assuming an isentropic compressibility law $p = f(\varrho)$ where $\varrho$ is the water density. In fact up to the orders considered here it was shown that the sound speed $c$ could be considered as a constant. A third small dimensionless parameter appeared naturally and characterizes the water compressibility, namely

$$\delta = \frac{g\,\Lambda}{c^2}\,,$$

where $g$ is the gravity acceleration and $\Lambda$ the acoustic wave length. The parameter $\varepsilon$ is very small, $\mathcal{A}$ of the order of a few per cent and $\delta$ of the order of $10^{-3}$. All physical variables have been scaled with respect to the acoustic wave length, the acoustic period and the water density at $z = 0$ for an ocean at rest. As a consequence the sound speed will be 1. The gravity wave length will be $1/\ell$ – or the gravity wave number will be $\ell$; $\ell$ will represent a parameter of our study. The asymptotic expansion of the acoustic pressure was shown to be

$$p = \varepsilon\, p_1 + \varepsilon\, \mathcal{A} p_2 + \varepsilon\, \mathcal{A} \sqrt{\delta}\, p_3 + \cdots, \tag{3}$$

where $p_1$ is the acoustic pressure in the absence of gravity waves. If the gravity waves travel in the positive $x$ direction, the corresponding free-surface elevation being

$$z_{FS} = \mathcal{A} \cos\left[2\pi\left(\ell x - \tilde{t} + \tilde{t}_0\right)\right], \quad \tilde{t} = \sqrt{\frac{\delta \ell}{2\pi}}\, t, \text{ where } \tilde{t}_0 \text{ is a constant phase}, \tag{4}$$

then it turns out that $p_2$ must satisfy the same wave equation as $p_1$, together with the following non-homogeneous Dirichlet condition

$$p_{2|z=0} = -\cos\left[2\pi\left(\ell x - \tilde{t} + \tilde{t}_0\right)\right] \frac{\partial p_1}{\partial z}_{|z=0}, \tag{5}$$

which means that the acoustic pressure should vanish on the true free surface rather than on the plane $z = 0$. Clearly $p_2$ corresponds to the diffraction of the acoustic waves by the free surface (see figure 1). The time variable $\tilde{t}$ is slow in comparison with $t$: there are two time scales, a fast one for acoustics and a slow one for gravity waves. It must be noticed that the condition (5) implies the assumption that the acoustic and gravity wave lengths are of the same order. The second-order pressure $p_2$ can be written as

$$p_2 = \mathcal{I}\mathrm{m}\left\{\exp\left[-2i\pi\left(t + \tilde{t} - \tilde{t}_0\right)\right] P_2^{(1)}(x, y, z) + \exp\left[-2i\pi\left(t - \tilde{t} + \tilde{t}_0\right)\right] P_2^{(2)}(x, y, z)\right\}. \tag{6}$$

In (6) the frequency is either the sum or the difference of the acoustic and gravity frequencies. $P_2^{(1)}$ and $P_2^{(2)}$ must satisfy the same Helmholtz equation (2) as $P_1$, together with the non-homogeneous Dirichlet condition

$$P_2^{(j)}{}_{|z=0} = Q^{(j)} = -\frac{1}{2}\exp\left[(-1)^{j+1}\, 2i\pi\ell x\right] \frac{\partial P_1}{\partial z}_{|z=0}, \quad j = 1, 2. \tag{7}$$

As we have $P_2^{(2)}(x, y, z) = P_2^{(1)}(-x, y, z)$, we can restrict our attention to $P_2^{(1)}$, which we shall denote $P_2$, while $Q^{(1)}$ will be denoted $Q$.

The pressure distribution $p_3$ must satisfy a non-homogeneous wave equation which right-hand side depends on the fluid velocity associated with the gravity waves, together with a homogeneous Dirichlet condition on $z = 0$. Such a term, which is not typical of underwater acoustics, will be disregarded here.

### 1.3 Case of a pinpoint source

For a given vibrating submarine, the evaluation of $P_2$ exhibits two difficulties:
– a technical one, namely the presence of a boundary condition (7) up to infinity,
– and a more basic one, namely "what to impose to $P_2$ at infinity to get a unique solution?". The standard radiation condition is surely valid for the whole expansion (3), and also for $P_1$, but it is not obvious that it should also apply to $P_2$.

In order to get an insight into this problem, let us replace the submarine by a pinpoint source of intensity 1 at the point $A$ of coordinates $(0, 0, -a)$, with $a > 0$ (see figure 1).

The corresponding first-order pressure $P_1$ can be easily calculated: it is the sum of the source of intensity 1 at point $A$ and of a sink of intensity $-1$ at point $B$, where $B$ is symmetrical of $A$ with respect to $z = 0$ (its coordinates are $0, 0,\ a$):

$$
\begin{aligned}
&P_1 = -\frac{1}{4\pi}\left\{\frac{\exp(2\,i\,\pi\,R_A)}{R_A} - \frac{\exp(2\,i\,\pi\,R_B)}{R_B}\right\},\\
&R_A = \sqrt{x^2+y^2+(z+a)^2}\,,\ R_B = \sqrt{x^2+y^2+(z-a)^2}\,.
\end{aligned}
\tag{8}
$$

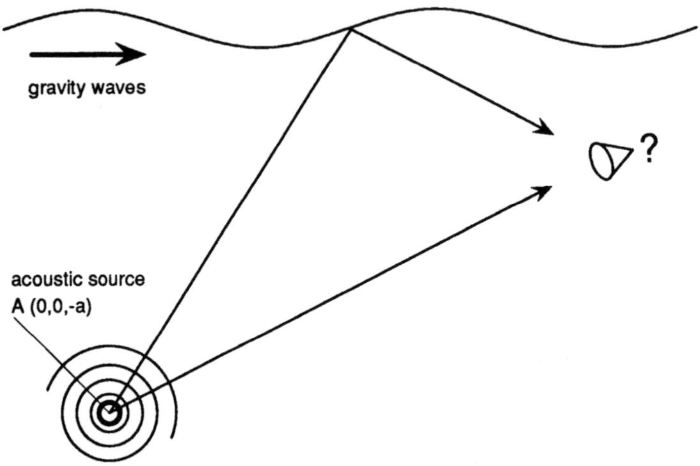


**Fig. 1.** Pinpoint acoustic source and gravity waves

The second-order pressure $P_2$ must satisfy

$$
\begin{aligned}
&(a)\quad (\nabla^2 + 4\,\pi^2)\,P_2 = 0 \quad \text{for } z < 0\,,\\
&(b)\quad P_2 = Q \qquad\qquad\quad\ \ \text{for } z = 0\,,
\end{aligned}
\tag{9}
$$

together with some unknown condition at infinity, $Q$ being given in (7) for $j = 1$ [see the comment following (7)]. In (Pot 1989) a solution has been found for **(9)**, namely

$$
\begin{cases}
P_2 = -2\,Q\,(x,\,y) \overset{x,\,y}{*} \frac{\partial G}{\partial z}(x,\,y,\,z) = -2\,\iint_{\mathbb{R}^2} Q\,(\xi,\,\eta)\,\frac{\partial G}{\partial z}\,(x-\xi,\,y-\eta,\,z)\,d\xi d\eta\,,\\
\text{where } G = \dfrac{\exp(2\,i\,\pi\,r)}{-4\,\pi\,r}\,.
\end{cases}
\tag{10}
$$

Its numerical computation exhibits a peak of $P_2$ just over the source $A$. In Euvrard and Mechiche Alami 1992 it was proved via a limiting-amplitude procedure that (10) is actually the physical solution. Having thus calculated and confirmed the second-order solution $P_2$, we now aim to exhibit its behaviour when the horizontal distance $r = \sqrt{x^2 + y^2}$ tends to infinity, in any fixed direction $\theta$ and for any fixed elevation $z < 0$.

# 2 Formal asymptotic behaviour at infinity

Let $r = \sqrt{x^2 + y^2}$ be the horizontal distance between the source A and a point $M$ $(x,\ y,\ z)$. We aim to exhibit the asymptotic behaviour of $P_2$ for $r \to +\infty$.

## 2.1 The behaviour of $\mathbf{P_1}$ and Q

It is easy to expand the explicit functions (8) and (7). It is found

$$P_1 = -i\,a\,z\,\frac{\exp(2\,i\,\pi\,r)}{r^2} + O\left(\frac{1}{r^3}\right) \tag{11}$$

and

$$P_{2|z=0} = Q = i\,\frac{a}{2}\exp(2\,i\,\pi\,\ell\,x)\,\frac{\exp(2\,i\,\pi\,r)}{r^2} + O\left(\frac{1}{r^3}\right), \tag{12}$$

so that the expansion (3) seems to be uniformly valid, even at infinity.

## 2.2 Straight-in limit of $\mathbf{P_2}$

Let $x = r\cos\theta$, $y = r\sin\theta$. $\theta$ and $z$ will remain fixed while $r$ will tend to infinity, and

$$\begin{cases} \xi = \varrho\cos\varphi\,,\ \eta = \varrho\sin\varphi\,;\ \xi - x = \tilde{\varrho}\cos\tilde{\varphi}\,,\ \eta - y = \tilde{\varrho}\sin\tilde{\varphi} \\ \Rightarrow\ \tilde{\varrho}^2 = r^2 + \varrho^2 - 2\,r\,\varrho\cos(\varphi-\theta)\,,\quad \dfrac{\sin(\varphi-\theta)}{\tilde{\varrho}} = \dfrac{\sin(\tilde{\varphi}-\theta)}{\varrho}\,. \end{cases} \tag{13}$$

With these notations (10) can be written as

$$\begin{cases} P_2 = \dfrac{a\,z}{8\,\pi^2}\displaystyle\iint_{\mathbb{R}^2}\exp(2\,i\,\pi\,\ell\,\varrho\cos\varphi)\,F(\varrho,\ a)\,F(\tilde{\varrho},\ |z|)\,\varrho\,d\varrho\,d\varphi\,, \\ \text{with } F(\varrho,\ a) = \left[\dfrac{2\,i\,\pi}{\varrho^2 + a^2} - \dfrac{1}{(\varrho^2+a^2)^{3/2}}\right]\exp\left(2\,i\,\pi\sqrt{\varrho^2+a^2}\right). \end{cases} \tag{14}$$

For fixed values of $r$, $\theta$ and $z$, the integrand in (14) behaves like $\varrho^{-4}$ at infinity and thence is Lebesgue-integrable. Let us abruptly introduce the limiting procedure $r \to +\infty$ in $F(\tilde{\varrho},\ |z|)$, with $\tilde{\varrho} = r - \varrho\cos(\varphi - \theta) + O(1/r)$ due to (13). Then (14) becomes

$$P_2 \sim \frac{i\,a\,z}{4\,\pi}\,\frac{\exp(2\,i\,\pi\,r)}{r^2}\int_{\varrho=0}^{+\infty}\int_{\varphi=-\pi}^{+\pi}\exp\{2\,i\,\pi\,\varrho\,[\ell\cos\varphi - \cos(\varphi-\theta)]\}\,F(\varrho,\ a)\,\varrho\,d\varrho\,d\varphi\,. \tag{15}$$

In fact such a straight-in approach ... is surely wrong! First for $z = 0$ the behaviour (15) does not match with $Q$ as given by (12). And, more essentially, assuming $r \to +\infty$ for fixed $\varrho$ in the integrand of (14) implicitely excludes the integration points $(\xi,\ \eta)$ wich remain close to the point $(x,\ y\ )$ as $M$ goes to infinity. So we need a more sophisticated limiting procedure. We shall use a standard method described in Darrozès 1978.

## 2.3 A method involving matched asymptotic expansions

Let

$$I_\varepsilon = \int_0^1 f_\varepsilon(x)\,dx$$

be the integral on a fixed domain (0, 1) of a function $f_\varepsilon$ depending on a small positive parameter $\varepsilon$. $f_\varepsilon$ is assumed to admit an outer expansion for fixed $x \in ]0,\ \ell]$ and $\varepsilon \to 0$, and also an inner expansion for fixed $\tilde{x} = x/\varepsilon$ and $\varepsilon \to 0$. We want to expand $I_\varepsilon$ with respect to $\varepsilon$.

First it must be checked that both expansions actually match together. Then an intermediate boundary $\ell = \varepsilon\,\tilde{\ell} \in ]0,\ 1[$ is introduced.

– The integral on $(\ell,\ 1)$ is expanded using the outer expansion of $f_\varepsilon$, and later on the inner limiting procedure is applied which implies $\ell \to 0$.

– The integral on $(0,\ \tilde{\ell})$ is expanded using the inner expansion of $f_\varepsilon$, and later on the outer limiting procedure is applied which implies $\tilde{\ell} \to +\infty$.

– Finally both expansions are superimposed but it must be checked that their sum is actually independant of $\ell$ (or $\tilde{\ell}$).

## 2.4 Application of the previous method to our problem

**(a)** Let $P(\xi,\ \eta)$ be the integration point, $M(x,\ y,\ z)$ the observation point and $m$ the point $(x,\ y)$. We have $r = \sqrt{x^2+y^2} \to +\infty$: $r^{-1}$ plays the role of $\varepsilon$ in the general procedure of § 2.3. Moreover (14) involves a double integral on an unbounded domain ... When the distance $\varrho$ of $P$ to the origin $O$ remains finite, the limiting procedure will be referred to as being "outer"; when the distance $\tilde{\varrho}$ of $P$ to $m$ remains finite, the limiting procedure will be referred to as being "inner". Namely (see figure 2):

$$\begin{cases} \text{Outer limiting procedure : } \varrho \text{ and } \varphi \text{ fixed}, \theta \text{ fixed and } r \to +\infty\,; \\ \text{Inner limiting procedure : } \tilde{\varrho} \text{ and } \tilde{\varphi} \text{ fixed}, \theta \text{ fixed and } r \to +\infty\,. \end{cases} \tag{16}$$

**(b)** Let us expand the integrand in (14), using "O E" for its "outer expansion" and "I E" for its "inner expansion". It is easily found

$$OE \sim F(\varrho,\ a)\exp\{2\,i\,\pi\,\varrho\,[\ell\cos\varphi - \cos(\varphi-\theta)]\}\,\frac{2\,i\,\pi\exp(2\,i\,\pi\,r)}{r^2}\,, \tag{17}$$

$$IE \sim F(\tilde{\varrho},\ |z|)\,\exp\{2\,i\,\pi\,\tilde{\varrho}\,[\ell\cos\tilde{\varphi} + \cos(\tilde{\varphi}-\theta)]\}\;\frac{2\,i\,\pi\exp[2\,i\,\pi\,(1+\ell\cos\theta)\,r]}{r^2}\,. \tag{18}$$

Both expansions (17) and (18) can be matched according to Van Dyke 1964, and taking into account the obvious relation $\varrho\cos\varphi = r\cos\theta + \tilde{\varrho}\cos\tilde{\varphi}$:

$$OE\,(IE) \sim IE\,(OE) \sim -4\,\pi^2\exp(2\,i\,\pi\,\varrho\,\ell\cos\varphi)\,\frac{\exp(2\,i\,\pi\,r)}{r^4}\,.\ \square \tag{19}$$

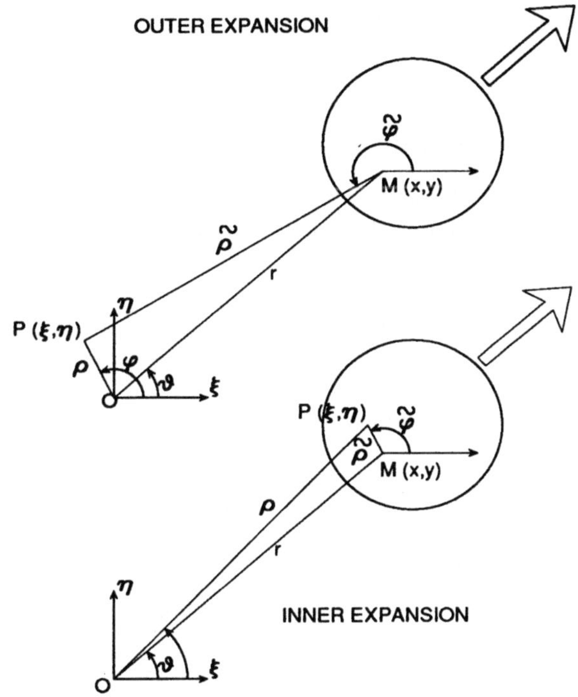


**Fig. 2.** Outer and inner limiting procedures

**(c)** The further steps of Darrozès' procedure consist in
– restricting the double integral in (14) to a disk of center $O$ and radius $\mathcal{R}$, passing to the outer limit for fixed $\mathcal{R}$, then passing to the inner limit which implies $\mathcal{R} \to +\infty$;
– also restricting the double integral in (14) to a disk of center $m$ and radius $\tilde{\mathcal{R}}$, passing to the inner limit for fixed $\tilde{\mathcal{R}}$, then passing to the outer limit which implies $\tilde{\mathcal{R}} \to +\infty$;
– and finally add both results.

The first result is nothing but (15). The latter is

$$\begin{aligned} &\frac{i\,a\,z}{4\,\pi}\;\frac{\exp\left[2\,i\,\pi\,(1+\ell\cos\theta)\,r\right]}{r^2} \\ &\times\int_{\tilde{\varrho}=0}^{+\infty}\int_{\tilde{\varphi}=-\pi}^{+\pi}\exp\left\{2\,i\,\pi\,\tilde{\varrho}\left[\ell\cos\tilde{\varphi}+\cos\left(\tilde{\varphi}-\theta\right)\right]\right\}F\left(\tilde{\varrho},\,|z|\right)\tilde{\varrho}\,d\tilde{\varrho}\,d\tilde{\varphi}\,. \end{aligned} \tag{20}$$

$\mathcal{R}$ and $\tilde{\mathcal{R}}$ are the counterparts of $\ell$ and $\tilde{\ell}$ in the general procedure of § 2.4. And both terms (15) and (20), whence their sum, being pure limits, are independant of $\mathcal{R}$ and $\tilde{\mathcal{R}}$. □

### 2.5 Simplification of the result

If we introduce

$$\mathcal{C} = \sqrt{(\ell - \cos\theta)^2 + \sin^2\theta} = \sqrt{\ell^2 - 2\,\ell\cos\theta + 1}\,, \tag{21}$$

then $\ell\cos\varphi - \cos(\varphi - \theta)$ can be written as $\mathcal{C}\cos(\varphi^* - \varphi)$, where $\cos\varphi^* = (\ell - \cos\theta)/\mathcal{C}$ and $\sin\varphi^* = -\sin\theta/\mathcal{C}$. Whence, thanks to Abramowitz and Stegun 1970, formula 9.1.18 p. 360:

$$\begin{aligned} \int_{\varphi=-\pi}^{+\pi} \exp\{2\,i\,\pi\,\varrho\,[\ell\cos\varphi - \cos(\varphi-\theta)]\}\,d\varphi &= 2\int_0^\pi \cos\{2\,\pi\,\mathcal{C}\,\varrho\cos\varphi\}\,d\varphi \\ &= 2\,\pi\,\mathrm{J}_0\,(2\,\pi\,\mathcal{C}\,\varrho)\,. \end{aligned} \tag{22}$$

The same procedure can be applied to (20), replacing $\mathcal{C}$ by

$$\mathcal{D} = \sqrt{(\ell + \cos\theta)^2 + \sin^2\theta} = \sqrt{\ell^2 + 2\,\ell\cos\theta + 1}\,. \tag{23}$$

Taking into account the fact that the derivative of $\mathrm{J}_0$ is $-\mathrm{J}_1$ and integrating by parts, the final result turns out to be

$$\begin{cases} P_2 \sim \dfrac{i\,a\,z}{2}\left\{I_1\,\dfrac{\exp(2\,i\,\pi\,r)}{r^2} + I_2\,\dfrac{\exp[2\,i\,\pi\,(1+\ell\cos\theta)\,r]}{r^2}\right\}\,, \quad \text{where} \\ I_1 = -\dfrac{\exp(2\,i\,\pi\,a)}{a} + 2\,\pi\,\mathcal{C}\displaystyle\int_0^{+\infty} \frac{\exp\left(2\,i\,\pi\,\sqrt{\varrho^2+a^2}\right)}{\sqrt{\varrho^2+a^2}}\,\mathrm{J}_1\,(2\,\pi\,\mathcal{C}\,\varrho)\,d\varrho\,, \\ \text{and } I_1 \to I_2 \text{ by changing } \mathcal{C} \text{ into } \mathcal{D} \text{ and } a \text{ into } |z|\,. \end{cases} \tag{24}$$

Using Lebesgue's theorem of dominated convergence, it is now easy to check that, due to the presence of $I_2$, the behaviour (24) now matches with that of $Q$ in (12). □

# 3 Full justification of the asymptotic behaviour at infinity

Let us introduce the disks $\mathcal{D}$ and $\tilde{\mathcal{D}}$, respectively centered in $O$ and $M$ and of common radius $\mathcal{R} = \tilde{\mathcal{R}} = r^{1/3}$(in fact, any power strictly included between 0 and 1 would be suitable). It must be pointed out that their radius grows indefinitely but that the distance $r$ of their centres grows still faster. Let $\mathcal{M}$ be the mid-perpendicular to $OM$, $\Delta$ the half plane limited by $\mathcal{M}$ and containing $O$ but deprived of the $\mathcal{D}$ disk, and

$\tilde{\Delta}$ the complementary half-plane deprived of the $\tilde{\mathcal{D}}$ disk (see figure 3). The integral in (14) will be split into 3 parts, namely the integrals on $\mathcal{D}$, $\tilde{\mathcal{D}}$ and $\Delta \cup \tilde{\Delta}$.

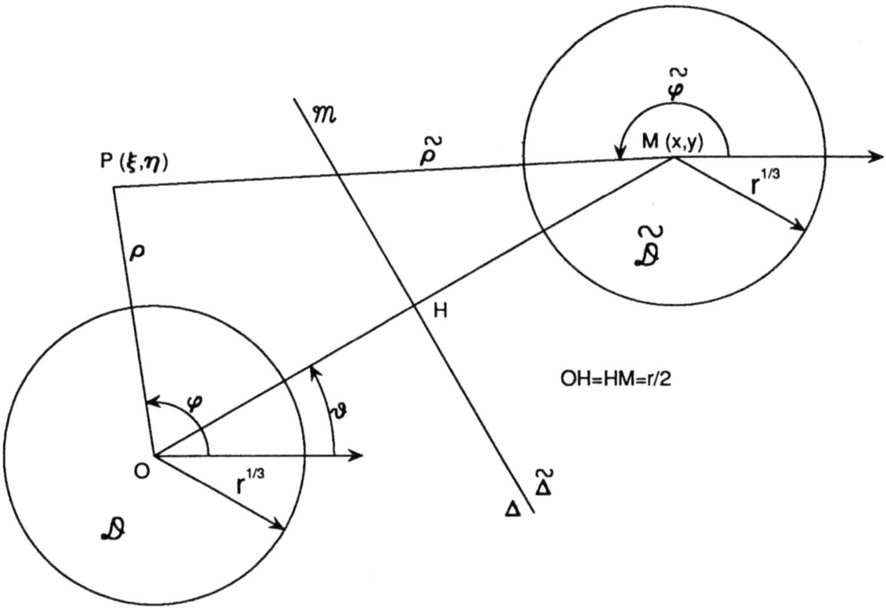


**Fig. 3.** The regions $\mathcal{D}$, $\tilde{\mathcal{D}}$, $\Delta$ and $\tilde{\Delta}$

### 3.1 Integrals on the disks $\mathcal{D}$ and $\tilde{\mathcal{D}}$

– On $\mathcal{D}$ we have $\tilde{\varrho} = r - \varrho\cos(\varphi - \theta) + O\left(r^{-1/3}\right)$, $\varrho$ being bounded by $O\left(r^{1/3}\right)$. Whence it can be easily shown that the integral on $\mathcal{D}$ behaves like (15), with a complementary term of order $r^{-7/3}$.
– Similarly the integral on $\tilde{\mathcal{D}}$ behaves like (20).

### 3.2 The remaining problem

The problem is now to prove that the integral on $\Delta \cup \tilde{\Delta}$ vanishes when $r \to +\infty$ faster than $r^{-2}$. The problem in $\tilde{\Delta}$ can be reduced to that on $\Delta$ by exchanging $\tilde{\varrho}$ and $\varrho$, $\tilde{\varphi}$ and $\varphi$, a and $|z|$. Moreover a symmetry with respect to the line $OM$ can be found, and the whole figure can be rotated for the sake of convenience.

Finally we can restrict our attention to the quarter of a plane $OH\mathcal{M}$ deprived of a half-disk of center $O$ and radius $r^{1/3}$ (see figure 4). This domain will then be cut into two parts:
– the half-ring $\Delta_1$, defined by $r^{1/3} \le \varrho \le \bar{t}\, r$,

– and the unbounded region $\Delta_2$, corresponding to $\varrho \geq \bar{t}\,r$, where $\bar{t}$ is a positive but fixed number to be determined later on.

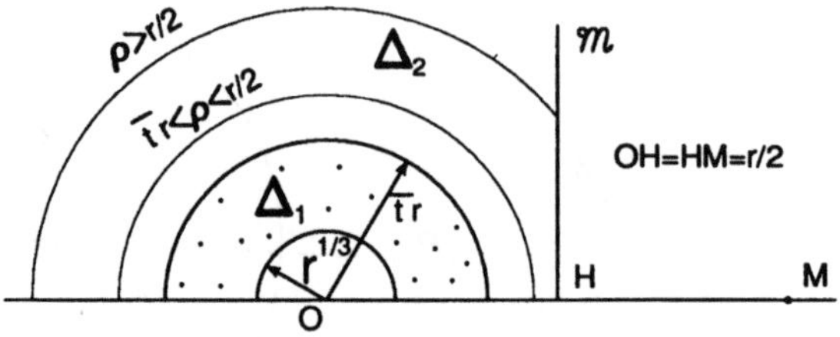


**Fig. 4.** The regions $\Delta_1$ and $\Delta_2$

Now on $\Delta_i$, $i = 1, 2$, from $\varrho \geq r^{1/3}$ it can easily be seen that the integral to evaluate, apart from a non-important constant factor, is

$$\begin{cases} I(\Delta_i) = \displaystyle\int\!\!\int_{\Delta_i} \exp\{2\,i\,\pi\,[\varrho\,\ell\cos(\varphi+\theta)+\varrho+\tilde{\varrho}]\}\,\frac{d\varrho\,d\varphi}{\varrho\,\tilde{\varrho}^2} + O\left(r^{-7/3}\right), \\ \text{with } \tilde{\varrho} = \left(\varrho^2 - 2\,\varrho\,r\cos\varphi + r^2\right)^{1/2}. \end{cases} \tag{25}$$

### 3.3 A theorem of stationary phase

Let us remind one of the standard theorems of stationary phase. Let $(a,\, b)$ be a bounded interval of $\mathbb{R}$, and $c \in ]a,\, b[$. Let $g$ and $h$ be real functions, belonging respectively to $C^1([a,b])$ and $C^2([a,\,b])$, and respectively of classes $C^2$ and $C^4$ in the vicinity of $c$.

Let us consider the following cases:

**(i)** $h'(t)$ negative for $a < t < c$ and positive for $c < t < b$, with $h'(c) = 0$ and $h''(c) > 0$,

**(ii)** $h'(t)$ positive for $a < t < c$ and negative for $c < t < b$, with $h'(c) = 0$ and $h''(c) < 0$.

Then, when $r$ tends to $+\infty$, we have

$$\int_a^b e^{i\,r\,h(\varphi)} g(\varphi)\,d\varphi = \frac{\exp\{i\,[r\,h(c) + \varepsilon\,\pi/4]\}}{\sqrt{r}}\, g(c)\sqrt{\frac{2\,\pi}{|h''(c)|}} + O\left(\frac{1}{r}\right), \tag{26}$$

with $\varepsilon = +1$ in the case (i) and $\varepsilon = -1$ in the case (ii).

### 3.4 Behaviour of $\mathbf{I}(\Delta_1)$ for $\ell = 1$, $\theta \neq 0$ and $\theta \neq \pi$

Introducing $t = \varrho/r$, $I(\Delta_1)$ can be written as

$$\begin{cases} I(\Delta_1) = \dfrac{1}{r^2}\displaystyle\int_{r^{-2/3}}^{\bar{t}} \frac{\exp(2\,i\,\pi\,r\,t)}{t}\,\mathcal{L}_r(t)\,dt\,, \text{ with } \mathcal{L}_r(t) = \int_0^{\pi} \frac{\exp\{2\,i\,\pi\,r\,H\}}{\mathcal{T}}\,d\varphi\,, \\ \mathcal{T} = 1 - 2\,t\cos\varphi + t^2\,,\ H = t\,\ell\cos(\varphi+\theta) + \sqrt{\mathcal{T}}\,. \end{cases} \tag{27}$$

Let us first apply (26) to $\mathcal{L}_r(t)$, with $a = 0$, $b = \pi$, $h(\varphi) = 2\,\pi\,H$ and $g(\varphi) = 1/\mathcal{T}$. The first derivatives of $H$ are

$$\begin{cases} \dfrac{\partial H}{\partial\varphi} = t\left[-\ell\sin(\varphi+\theta) + \dfrac{\sin\varphi}{\sqrt{\mathcal{T}}}\right], \\ \dfrac{\partial^2 H}{\partial\varphi^2} = t\left[-\ell\cos(\varphi+\theta) + \dfrac{\cos\varphi}{\sqrt{\mathcal{T}}} - t\,\dfrac{\sin^2\varphi}{\mathcal{T}^{3/2}}\right], \end{cases} \tag{28}$$

with of course $\ell = 1$. For $t = 0$, $1/t\,\partial H/\partial\varphi$ vanishes when $\varphi$ equals the unique determination of $(\pi - \theta)/2$ belonging to $]0,\,\pi[$. For small and positive $t$, $\partial H/\partial\varphi$ has a unique root in $]0,\,\pi[$, namely

$$\varphi_0 = \frac{\pi}{2} - \frac{\theta}{2} - \frac{1}{2}\cos\left(\frac{\theta}{2}\right)t + O\left(t^2\right), \tag{29}$$

and $\varphi_0$ depends continuously of $t$. For $\varphi = \varphi_0$ we find

$$\frac{\partial^2 H}{\partial\varphi^2}(\varphi_0) = 2\sin\left(\frac{\theta}{2}\right)t - \cos\theta\,t^2 + O\left(t^3\right), \tag{30}$$

which does not vanish. The result (26) can be applied for $c = \varphi_0$. For small $t$, $g(c)$ is bounded and $|h''(c)|$ is équivalent to $2\sin(\theta/2)\,t$. Finally $|\mathcal{L}_r(t)|$ can be bounded by

$$C\,\frac{1}{\sqrt{r}}\,\frac{1}{\sqrt{t}}\,, \text{ with } C = \frac{K}{\sqrt{2\,|\sin(\theta/2)|}}\,,\ K = \text{constant}\,,$$

provided that $\bar{t}$ should be small enough. Whence

$$|I(\Delta_1)| \le \frac{C}{r^{5/2}}\int_{r^{-2/3}}^{\bar{t}} \frac{dt}{t^{3/2}} = \frac{2\,C}{r^{5/2}}\left(r^{1/3} - \frac{1}{\sqrt{\bar{t}}}\right) = O\left(\frac{1}{r^{13/6}}\right) \ll \frac{1}{r^2}\,.\ \square \tag{31}$$

### 3.5 Behaviour of $\mathbf{I}(\Delta_1)$ for $\ell \neq 1$, $\theta \neq 0$ and $\theta \neq \pi$

For $t = 0$, the roots of $1/t\,\partial H/\partial\varphi$ are the intersections of two sine curves of different amplitudes and phases, so they form a finite set. But $\partial H/\partial\varphi$ and $\partial^2 H/\partial\varphi^2$ cannot vanish simultaneously since we have $\operatorname{tg}\varphi \neq \operatorname{tg}(\varphi+\theta)$. Moreover for small and positive $\bar{t}$ the same conclusions are valid and the roots of $\partial H/\partial\varphi$ depend continuously on $t$, even if it cannot be found explicit results like (29) and (30). In (26) we have several terms instead of one, but the result (31) remains valid. $\square$

**3.6 Behaviour of $(\Delta_1)$ for $\ell > 0$ and $\theta = \pi$**

(28) becomes

$$\begin{cases} \dfrac{\partial H}{\partial \varphi} = t \sin\varphi \left(\ell + \dfrac{1}{\sqrt{T}}\right), \\ \dfrac{\partial^2 H}{\partial \varphi^2} = t \left[\cos\varphi \left(\ell + \dfrac{1}{\sqrt{T}}\right) - t\, \dfrac{\sin^2\varphi}{T^{3/2}}\right]. \end{cases} \tag{32}$$

$1/t\,\partial H/\partial\varphi$ vanishes for $\varphi = 0$ and for $\varphi = \pi$, while $1/t\,\partial^2 H/\partial\varphi^2$ does not vanish for $\varphi = 0$ nor for $\varphi = \pi$. In (26) we must take into account two terms, one for $\varphi = 0$ and one for $\varphi = \pi$ but with coefficients twice smaller. The result (31) remains valid. □

**3.7 Behaviour of $\mathbf{I}(\Delta_1)$ for $\ell \neq 1$and $\theta = 0$**

(28) becomes

$$\begin{cases} \dfrac{\partial H}{\partial \varphi} = t \sin\varphi \left(-\ell + \dfrac{1}{\sqrt{T}}\right), \\ \dfrac{\partial^2 H}{\partial \varphi^2} = t \left[\cos\varphi \left(-\ell + \dfrac{1}{\sqrt{T}}\right) - t\, \dfrac{\sin^2\varphi}{T^{3/2}}\right]. \end{cases} \tag{33}$$

$1/t\,\partial H/\partial\varphi$ vanishes for $\varphi = 0$ and for $\varphi = \pi$, while $1/t\,\partial^2 H/\partial\varphi^2$ does not vanish for $\varphi = 0$ nor for $\varphi = \pi$. Like in the previous subsection the result (31) remains valid. □

**3.8 Behaviour of $\mathbf{I}(\Delta_1)$ for $\ell = 1$ and $\theta = 0$**

Here we need more refined evaluations.

For $\varphi = 0$ and for $\varphi = \pi$, $\partial^2 H/\partial\varphi^2$ vanishes like $t^2$. Moreover $1/t\,\partial H/\partial\varphi$ also vanishes for $\varphi = \arccos(t/2) \sim \pi/2$, while $\partial^2 H/\partial\varphi^2$ vanishes like $t^2$ too. In all three cases, we have

$$h(\varphi) = 2\pi\,[1 + \beta(t)], \text{ with} \begin{cases} \beta = 0 \text{ for } \varphi = 0 \text{ or } \varphi = \pi, \\ \beta = t^2/2 \text{ for } \varphi = \arccos(t/2). \end{cases}$$

For small $t$ and apart from a non-important constant factor, $\mathcal{L}_r(t)$ behaves like

$$\mathcal{L}_r(t) \sim \frac{1}{\sqrt{r}}\, \frac{\exp\{2 i \pi r\,[1 + \beta(t)]\}}{t}.$$

**(a)** If $\beta = 0$, $I(\Delta_1)$ behaves like

$$I(\Delta_1) \sim \frac{\exp(2 i \pi r)}{r^2}\, \frac{1}{\sqrt{r}}\, \mathcal{I},\ \mathcal{I} = \int_{r^{-2/3}}^{\bar{t}} \exp(2 i \pi r t)\, \frac{dt}{t^2}. \tag{34}$$

Going back to the variable $\varrho = r\,t$, $\mathcal{I}$ can be written as

$$\begin{cases} \mathcal{I} = r \displaystyle\int_{r^{1/3}}^{\bar{t}\,r} \exp(2\,i\,\pi\,\varrho)\,\frac{d\varrho}{\varrho^2} = r\,(\mathcal{I}_1 - \mathcal{I}_2)\,, \\ \mathcal{I}_1 = \displaystyle\int_{r^{1/3}}^{+\infty} \exp(2\,i\,\pi\,\varrho)\,\frac{d\varrho}{\varrho^2}\,,\; \mathcal{I}_2 = \displaystyle\int_{\bar{t}r}^{+\infty} \exp(2\,i\,\pi\,\varrho)\,\frac{d\varrho}{\varrho^2}\,. \end{cases}$$

Due to Abel's theorem, $|\mathcal{I}_1|$ and $|\mathcal{I}_2|$ are respectively bounded by $r^{-2/3}/(2\,\pi)$ and $(\bar{t}\,r)^{-2}/(2\,\pi)$, so $|\mathcal{I}|$ is of order $r^{1/3}$ and $I\,(\Delta_1)$ of order $r^{-13/6} \ll r^{-2}$. □

**(b)** If $\beta = t^2/2$, let us change the integration variable in $\mathcal{I}$, introducing:

$$\tau = t + \beta\,(t) = t + t^2/2 \Leftrightarrow t = -1 + \sqrt{1 + 2\,\tau}\,, \text{ whence } d\tau/dt = 1 + t\,.$$

In (34), $\mathcal{I}$ then becomes

$$\begin{aligned} \int_{r^{-2/3}}^{\bar{t}} \frac{\exp\{2\,i\,\pi\,r\,[t + \beta\,(t)]\}}{t^2}\,dt &= \int_{r^{-2/3}+r^{-4/3}/2}^{\bar{t}+\bar{t}^2/2} \frac{\exp(2\,i\,\pi\,r\,\tau)}{\sqrt{1+2\,\tau}\,\left[2\,(1 + \tau - \sqrt{1 + 2\,\tau})\right]}\,d\tau \\ &= \int_{r^{-2/3}(1+r^{-2/3}/2)}^{\bar{t}\,(1+\bar{t}/2)} \exp(2\,i\,\pi\,r\,\tau)\left[\frac{1}{\tau^2} + O\left(\frac{1}{\tau}\right)\right] d\tau\,. \end{aligned}$$

The modulus of the integral of $O\,(1/\tau)$ is bounded by something in $\ln r$: the corresponding term in $I\,(\Delta_1)$ is the of order $r^{-5/2}\ln r$, and is therefore negligible with respect to $r^{-2}$. □

And with regard to the term in $1/\tau^2$ in $\mathcal{I}$, the result is the same as in (a). □

### 3.9 Behaviour of $\mathbf{I}\,(\Delta_2)$

For $I\,(\Delta_2)$ we have integrals similar to those in (27), except that $\mathcal{L}_r(t)$ is an integral from $\alpha\,(t)$ to $\pi$, with $\alpha\,(t) = 0$ for $t \le 1/2$ and $\alpha\,(t) = \arccos\,[1/(2\,t)]$ for $t > 1/2$, and that $t$ varies between $\bar{t}$ and $+\infty$.

For a fixed value of $r$, several theorems of stationary phase must be applied, according to the number and the type of roots of $\partial H/\partial\varphi$ on the interval $[\alpha\,(t), \pi]$. For instance if there is no root $\mathcal{L}_r(t)$ is of order $r^{-1}$, if all the roots are simple $\mathcal{L}_r(t)$ is of order $r^{-1/2}$, if there is a double root $\mathcal{L}_r(t)$ is of order $r^{-1/3}$, etc... In any case however $L_r(t)$ vanishes when $r \to +\infty$ for nearly any $t$.

For $\bar{t} \le t \le 1$, the modulus of the integrand of the double integral in $\varphi$ and $t$ is bounded by a constant. For $t \ge 1$, we have $\alpha\,(t) \ge \arccos(1/2)$ whence $\cos\varphi \le 1/2$ and $\mathcal{T} \ge 1 - t + t^2$, and the modulus of the integrand is bounded by $[t\,(1 - t + t^2)]^{-1}$, which is a Lebesgue-integrable function independant of $r$.

This justifies Fubini's theorem (namely the integration with respect to $\varphi$ and then to $t$). Moreover Lebesgue's theorem of dominated convergence shows that the double integral vanishes when $r \to +\infty$. Whence (27) vanishes faster than $r^{-2}$. □

It must be pointed out that due to the denominator $t$ in the first formula (27) it is impossible to apply Lebesgue's theorem in $\Delta_1 \cup \Delta_2$. But conversely for small $\bar{t}$ we have a precise knowlage of the roots of $H$ in $\Delta_1$, which would not be the case in $\Delta_2$.

### 3.10 Last comments on § 3

Because of a lack of room we have only given the outlines of the demonstrations. In § 3.4 to 3.7 as we have to integrate with respect to $t$ after having used the stationary phase concept we need a bit more than (26), namely upper bounds: this can be found in the appendices of Euvrard and Mechiche Alami 1992.

With regard to § 3.8, we acknowledge that the demonstration given in Euvrard and Mechiche Alami 1992 was wrong; it is why we specially detailed here the new proof...

# 4 Final result in closed form

Combining the formulae (18) and (21) from Bateman 1954, page 20 and extending the result via Lebesgue's theorem to the limiting case $y = b$, it is easily found

$$\int_0^{+\infty} \frac{\exp\left(i\,b\sqrt{x^2+a^2}\right)}{\sqrt{x^2+a^2}}\, \mathrm{J}_1(x\,y)\,dx = \begin{cases} \dfrac{\exp(i\,a\,b) - \exp\left(i\,a\sqrt{b^2-y^2}\right)}{a\,y} & \text{if } 0 < y \le b\,, \\ \dfrac{\exp(i\,a\,b) - \exp\left(-a\sqrt{y^2-b^2}\right)}{a\,y} & \text{if } b \le y\,. \end{cases}$$

Then changing $x$ into $\varrho$, and taking $b = 2\pi$ and $y = 2\pi C$, it is possible to express $I_1$ from (24) in a closed form – and of course $I_2$ too. Finally the equation (24) can be written in the form:

$$\begin{aligned} &P_2 \sim \frac{i}{2}\left\{|z|\,\mathcal{I}_1\,\frac{\exp(2\,i\,\pi\,r)}{r^2} + a\,\mathcal{I}_2\,\frac{\exp[2\,i\,\pi\,(1+\ell\cos\theta)\,r]}{r^2}\right\} \text{ for } r\to\infty\text{, with} \\ &\mathcal{I}_1 = \begin{cases} \exp\left(2\,i\,\pi\,a\sqrt{1-\mathcal{C}^2}\right) & \text{if } \mathcal{C} = \sqrt{\ell^2 - 2\,\ell\cos\theta + 1} \le 1\,, \\ \exp\left(-2\,\pi\,a\sqrt{\mathcal{C}^2-1}\right) & \text{if } \mathcal{C} = \sqrt{\ell^2 - 2\,\ell\cos\theta + 1} \ge 1\,, \end{cases} \\ &\mathcal{I}_2 = \begin{cases} \exp\left(2\,i\,\pi\,|z|\sqrt{1-\mathcal{D}^2}\right) & \text{if } \mathcal{D} = \sqrt{\ell^2 + 2\,\ell\cos\theta + 1} \le 1\,, \\ \exp\left(-2\,\pi\,|z|\sqrt{\mathcal{D}^2-1}\right) & \text{if } \mathcal{D} = \sqrt{\ell^2 + 2\,\ell\cos\theta + 1} \ge 1\,. \end{cases} \end{aligned} \tag{35}$$

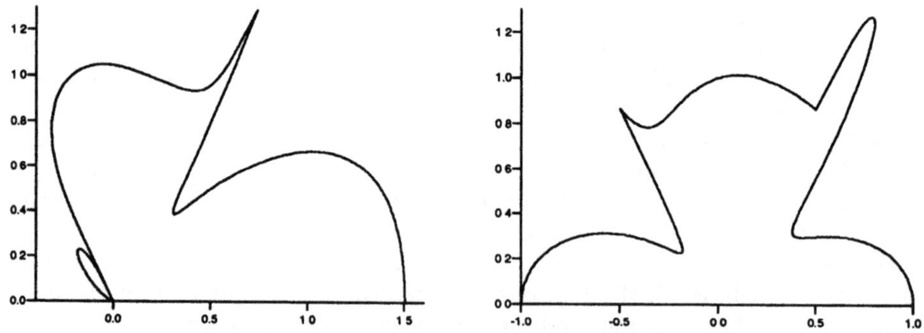


**Fig. 5.** Re $\{P_2\}$ (left diagram) and Im $\{P_2\}$ (right diagram) for $r = 1$

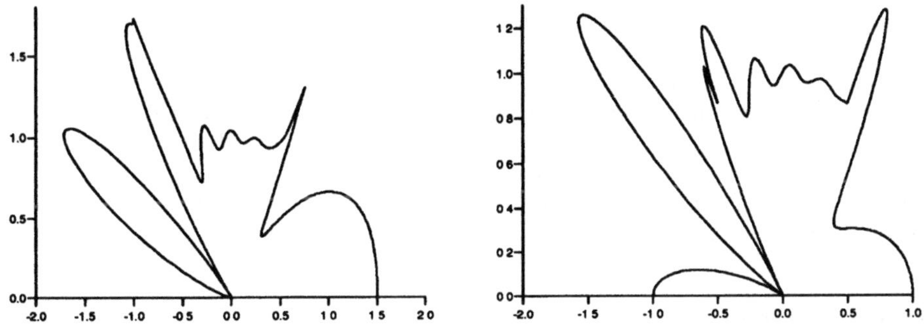


**Fig. 6.** Re $\{P_2\}$ (left diagram) and Im $\{P_2\}$ (right diagram) for $r = 4$

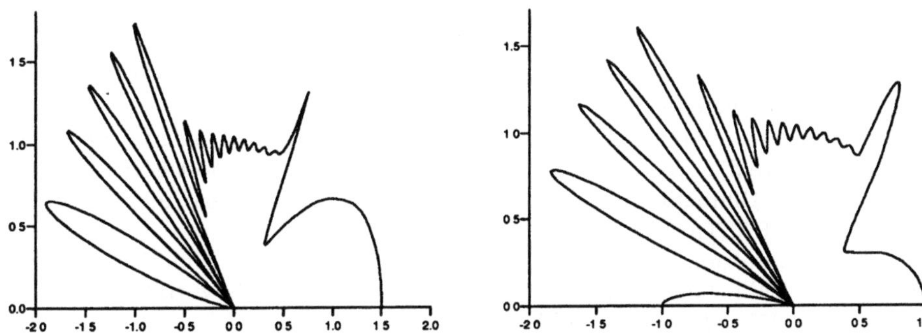


**Fig. 7.** Re $\{P_2\}$ (left diagram) and Im $\{P_2\}$ (right diagram) for $r = 10$

The asymptotic behaviour (35) of $P_2$ obviously matches, when $z$ vanishes, with the asymptotic behaviour of $Q$ as given by (12). □

It is also easy to check that the right-hand side of (35) asymptotically satisfies the Helmholtz equation:

$$\forall \ell,\ \forall z,\ \forall \theta,\ (\nabla^2 + 4\pi^2) P_2 = o(r^{-2}),\ P_2 \text{ being given by equ. (35)}. \ \square$$

*The conventional radiation condition* for a time-harmonic pressure $P$ satisfying the Helmholtz equation together with homogeneous boundary conditions (this is in particular the case of $P_1$) can be written as $\partial P/\partial r \sim 2i\pi P$ when $r \to \infty$. This condition *is* obviously *satisfied by the first term of equ. (35), but not by the second one.* For $\ell > 1$, the ratio of $\partial P_2/\partial r$ to $2i\pi P_2$ may even become negative. This result is obviously due to the non-homogeneous boundary conditions (9) (b). It is not so strange if we remind that the "radiation" condition means a positive radiation of energy; this physical condition is surely satisfied by the whole pressure field, and also by $P_1$; but there is no reason for the same condition to be satisfied by the leading term of their difference, namely $P_2$...

The asymptotic behaviour of a pressure field is usually the product of a gauge function of $r$ by a function $f(\theta)$. But *here* this function $f$ also depends on $r$, so that *we shall find a different radiation diagram for each value of* $r$! In the figures 5, 6 and 7 we present three samples of radiation diagrams, for $r = 1$ (i.e. at a distance of one wave-length), $r = 4$ and $r = 10$. In each figure we show $1 + \mathrm{Re}\, P_2$ or $1 + \mathrm{Im}\, P_2$ ($\geq 0$) versus $\theta$ for $0 \leq \theta \leq \pi$ (because of a symmetry).

In conclusion: *the further we are, the more $P_2$ oscillates with respect to $\theta$ and the more complicated the radiation diagram is...*

## References

Darrozès, J. S. (1978): "Exposé d'une méthode de calcul approché d'intégrales contenant un petit paramètre", note de recherche ENSTA N° 039.

Euvrard, D., Mechiche Alami, O. (1992): "Propagation acoustique sous-marine tridimensionnelle en présence de la houle", rapport de recherche ENSTA N° 264 (including an extended list of references).

Mechiche Alami, O. (1992): "Influence de la houle sur le rayonnement acoustique à très basse fréquence d'un corps sous-marin", Doctorat UPMC, rapport ENSTA N° 259.

Pot, G. (1989): "Diffraction d'une onde acoustique sous-marine par une houle plane simple", rapport ENSTA N° 233.

Abramowitz, M., Stegun, I. A.(1970): *Handbook of Mathematical Functions* (Dover).

Bateman, H. (1954): *Tables of Integral Transforms*, vol 2 (Mc Graw Hill).

Van Dyke, M. (1975): *Perturbation Methods in Fluid Mechanics* (The Parabolic Press, Stanford, California).

# FINITE-AMPLITUDE, PULSED, ULTRASONIC BEAMS

François Coulouvrat[1], Kjell-Eivind Frøysa[2]

[1] Laboratoire de Modélisation en Mécanique, Université Pierre et Marie Curie & C.N.R.S., case 162, 4, place Jussieu, F75252 Paris Cedex 05, France

[2] Matematisk Institutt, Universitetet i Bergen, Allégaten 55, N5007 Bergen, Norway (Present Address: Christian Michelsen Research, Fantoftvn.38, N5036 Fantoft, Norway)

**Abstract** An analytical, approximate solution of the inviscid KZK equation for a nonlinear pulsed sound beam radiated by an acoustic source with a Gaussian velocity distribution, is obtained by means of the renormalization method. This method involves two steps. First, the transient, weakly nonlinear field is computed. However, because of cumulative nonlinear effects, that expansion is non-uniform and breaks down at some distance away from the source. So, in order to extend its validity, it is re-written in a new frame of co-ordinates, better suited to following the nonlinear distorsion of the wave profile. Basically, the nonlinear co-ordinate transform introduces additional terms in the expansion, which are chosen so as to counterbalance the non-uniform ones. Special care is devoted to the treatment of shock waves. Finally, comparisons with the results of a finite-difference scheme turn out favorable, and show the efficiency of the method for a rather large range of parameters.

## 1. Introduction

With the development of non-invasive therapy technics such as lithotripsy or hyperthermia, the use of very intense ultrasonic fields for biomedical applications became more and more common during the recent years. The signals used for these applications are generally short pulses generated by high frequency transducers (of order 1MHz). Such acoustic fields are described by the so-called KZK equation (Zabolotskaya and Khokhlov 1969, Kuznetsov 1970). This scalar equation can be viewed as a parabolic approximation of the nonlinear wave equation. It takes into account consistently the three different mechanisms involved: acoustic propagation, diffraction by the source edges, and non-linearities.

Numerical solutions of the KZK equation have been developped for several years (Aanonsen *et al.* 1984, Bakhvalov *et al.* 1987) for periodic signals. Recently, a new and more efficient numerical scheme for pulsed signals has been proposed (Frøysa *et al.* 1993, Frøysa *et al.* 1994). However, the computation costs remain rather high. Therefore, analytical approximations of finite-amplitude sound fields may be useful, in order to quickly evaluate the influence of nonlinear effects on the signal duration, amplitude and spectrum as it propagates away from the source. With this in view, an asympotic method was proposed recently for a monochromatic input signal (Coulouvrat 1991). This paper presents a generalization of the method for nonlinear transient ultrasonic beams.

## 2. The KZK equation

The ultrasonic field radiated by a plane, axisymmetric transducer into an inviscid and compressible fluid (with ambient sound velocity $c$ and ambient density $\rho$) involves three characteristic lengths: (1) the wavelength $\lambda = 2\pi\ c/\omega_0 = 2\pi/k$, related to the mean frequency $\omega_0$ of the input signal, (2) the Rayleigh distance $R_0 = ka^2/2$, where $a$ is the source radius, and (3) the plane shock-wave distance $z_s = c^2/(1+B/2A)\omega_0^2 v_0$, where $B/A$ is the nonlinear parameter (Beyer, 1974) and $v_0$ is the input signal velocity amplitude. These three lengths measure respectively the characteristic distances over which the effects of acoustic propagation, diffraction and non-linearities take place. In acoustics, the Mach number $M = v_0/c$ is always very small, so that the shock wave distance is very large compared to the wavelength. Similarly, for a rather high frequency input signal, the Rayleigh distance is large compared to the wavelength. Consequently, over a few wavelengths, the two mechanisms of diffraction and non-linearities are negligible, so that the ultrasonic signal propagates almost as a plane wave. It is only over much larger distances that non-linearities and diffraction significantly alter the signal. Consequently, using the formalism of multiple scales (Naze Tjøtta and Tjøtta 1981, Coulouvrat 1992), the radiated field can be shown to be governed by the so-called KZK equation:

$$\frac{\partial^2 p}{\partial z \partial \tau} - \frac{1}{4}\left(\frac{\partial^2 p}{\partial x^2} + \frac{1}{x}\frac{\partial p}{\partial x}\right) = \frac{1}{2N}\frac{\partial^2 p^2}{\partial \tau^2}. \tag{1}$$

In this equation, the acoustic pressure $p$ is normalized to the quantity $cv_0\rho$. The axial co-ordinate $z$ is normalized to the Rayleigh distance, while the radial co-ordinate $x$ is normalized to the source radius. The dimensionless retarded time $\tau$ is defined by $\tau = \omega_0(t - R_0 z/c)$, and the parameter $N$ is the ratio of the shock distance to the Rayleigh distance. A large value of this parameter means that the shock wave distance is large compared to the Rayleigh distance, so that nonlinear effects take place only in the farfield. On the contrary, for a small value, nonlinear effects are prevalent even in the nearfield.

That equation is the simplest one taking into account consistently the mechanisms of acoustic propagation, diffraction and non-linearities. When nonlinear effects are neglected $(N >> 1)$, the linearized KZK equation is equivalent to the parabolic approximation of the wave equation. The ability of that approximation to describe the nearfield of an ultrasonic source has been investigated precisely (Kunitsyn and Rudenko 1978, Naze Tjøtta and Tjøtta 1980, Coulouvrat 1992). When nonlinear effects are prevalent $(N << 1)$, the KZK equation reduces to the well-known Burgers's equation (Burgers 1948). Its solutions can be obtained exactly (Hopf 1950, Cole 1951), which gives a deep understanding of the mechanisms of nonlinear wave propagation (Lighthill 1956, Blackstock 1964, Whitham 1974, Rudenko and Soluyan 1977).

The aim of this study is to obtain analytical approximations of the solutions of Eq.(1). With this in view, we proceed in two steps. First, a quasilinear approximation (QLA) is obtained as a straightforward expansion in powers of $1/N$. Of course, that weakly nonlinear expansion is limited to large $N$ values. It is also valid only up to a certain distance, as it is well-known that nonlinear effects are cumulative, and, with no absorption, lead to the distorsion of the wave profile, energy transfers among the frequency components and eventually the formation of weak shock waves. These strongly nonlinear effects cannot be described through the QLA. In order to extend the validity of this one, a new expansion is then obtained. The principle is to re-write the QLA in a new frame of co-ordinates, which is better suited to following the nonlinear distorsion of the wave profile. The choice of these new co-ordinates is guided by the form of the exact solutions of Burgers' equation.

## 3. The quasilinear expansion

The quasilinear expansion is calculated for a pulsed signal radiated by an emitter with a Gaussian velocity distribution:

$$p(z=0,x,\tau)=f(\tau)\exp(-x^2). \tag{2}$$

Then, the exact analytical form of the QLA:

$$p(z,x,\tau,N)=p_1(z,x,\tau)+\frac{1}{N}p_2(z,x,\tau)+O(1/N^2), \tag{3}$$

is obtained by means of Fourier transforms (Frøysa *et al.* 1993, Frøysa 1994):

$$p_1=\frac{1}{\sqrt{2\pi}}\int_{-\infty}^{+\infty}\frac{F(\omega)}{1+iz/\omega}\exp\left(\frac{-x^2}{1+iz/\omega}-i\omega\tau\right)d\omega, \tag{4}$$

$$p_2=\frac{1}{4\pi}\iint_{-\infty}^{+\infty}\frac{\omega^2 s(\omega-s)F(s)F(\omega-s)}{2s(\omega-s)-\omega(\omega+iz)}\exp\left(\frac{x^2\omega^2}{2s(\omega-s)-\omega(\omega+iz)}-i\omega\tau\right)I(\omega,s)\,ds\,d\omega. \tag{5}$$

In Eq.(5), $F(\omega)$ is the Fourier transform of the input signal $f(\tau)$ and $I(\omega,s)$ is given by:

$$I=Ei\left(\frac{ix^2\omega(4s(\omega-s)-\omega^2)}{(2s(\omega-s)-\omega(\omega+iz))(i\omega-2z)}\right)-Ei\left(\frac{ix^2 s(\omega-s)(4s(\omega-s)-\omega^2)}{(2s(\omega-s)-\omega(\omega+iz))(is(\omega-s)-z(\omega+iz))}\right), \tag{6}$$

with $Ei(.)$ the exponential integral function. As the KZK equation is basically a high-frequency approximation of the nonlinear wave equation, its derivation requires in particular that the zero-frequency component of the input-signal vanishes $(F(0)=0)$ (Bakhvalov *et al.* 1987). Then it is easy to check that the zero-frequency component of the QLA (3-6) also vanishes. Equivalents of the QLA can be obtained along the acoustical axis in the near- and farfields:

$$\begin{array}{ll} p_1(z,0,\tau)\underset{z\to 0}{\approx} f(\tau) & p_2(z,0,\tau)\underset{z\to 0}{\approx} zf'(\tau)f(\tau) \\ p_1(z,0,\tau)\underset{z\to\infty}{\approx} f'(\tau)/z & p_2(z,0,\tau)\underset{z\to\infty}{\approx} \ln(z)f'(\tau)f''(\tau)/z \end{array} \tag{7}$$

It is easy to recognize in Eq.(7) the quasilinear approximations of the solutions of the plane (in the nearfield) and spherical (in the farfield) Burgers' equations. With equivalents (7), the validity limits of the QLA are determined more precisely: it is necessary that the second-order term of Eq.(3) be much smaller than the first-order one, i.e. $z \ll N$ in the nearfield and

$ln(z) << N$ in the farfield. It must be pointed out that, whatever the value of $N$, the QLA is non-uniform (the second-order becomes of the same magnitude as the first-order one for sufficiently large distances): nonlinear effects are cumulative. However, if $N$ is rather large, nonlinear effects grow very slowly because of the geometric spreading. In this case, the actual distance of shock formation is very large (of order $exp(N)$), and the QLA is valid very far from the source.

## 4. Renormalization

In order to extend the validity of the QLA when nonlinear effects become significant, expansion (3) is re-written within a new frame of co-ordinates. According to the exact solutions of Burgers' equation, the nonlinear co-ordinate transform:

$$\tau = \varphi - \frac{1}{N} r(z, x, \varphi), \tag{8}$$

is introduced directly into the QLA (3). The renormalized approximation (RNA) is obtained, by making formally a new expansion up to the same order, but now in terms of variable $\varphi$ instead of variable $\tau$:

$$p(z, x, \tau, N) = p_1(z, x, \varphi) + \frac{1}{N}\left( p_2(z, x, \varphi) - r(z, x, \varphi)\frac{\partial p_1}{\partial \varphi}(z, x, \varphi)\right) \Big/ \left( 1 - \frac{1}{N}\frac{\partial r}{\partial \varphi}(z, x, \varphi)\right) + O(\frac{1}{N^2}) \tag{9}$$

Formally, the denominator $(1 - 1/N\partial r/\partial\varphi)$ does not change the order of the whole expansion. Its importance will be outlined later on. Transform (8) introduces additional terms in expansion (9). The idea is to choose the (yet) unspecified renormalization function $r$ so as to counterbalance the non-uniform terms appearing in the QLA. Ideally, $r$ should be chosen equal to $p_2/\partial p_1/\partial\varphi$ . This would cancel the whole second-order term, and the RNA would not be non-uniform at all, at least up to the considered order. However, this choice is not possible here, because, generally, the second-order pressure $\partial p_1/\partial\varphi$ might vanish while $p_2$ does not. Then, the renormalization function would be unbounded, which is unphysical. This is due to diffraction effects. For instance, for a monochromatic case (Coulouvrat 1991), diffraction produces a phase shift between the fundamental and the first harmonic components. Only in the case of Burgers' equation is it possible to cancel entirely the second-order term of the RNA. Then, the resulting solution is an *exact* solution of the fully nonlinear Burgers' equation. In this particular case, renormalization is dramatically efficient: a fundamentally asymptotic method, based on a limited expansion up to only the second order, yields the exact solution! This result gives us some confidence in the method, though we cannot expect here anymore an exact solution, neither an approximation valid for all ranges of parameters.

Here, the renormalization function is chosen so as to match continuously the exact solutions of the plane and spherical Burgers' equations along the axis in the near- and farfields:

$$r(z, x, \varphi) = r_0(z, x, \varphi) + \psi_F(z, x) + \psi_N(z, x), \tag{10}$$

with:

$$r_0(z,x,\varphi)=\frac{1}{\sqrt{2\pi}}\int_{-\infty}^{+\infty}-i\omega\ F(\omega)\,ln(1+iz/\omega)\,exp\left(\frac{-x^2}{1+iz/\omega}\right)exp(-i\omega\varphi\ )d\omega. \tag{11}$$

It is easily verified that:

$$r_0(z,0,\varphi)\underset{z\to 0}{\approx} zf(\varphi)\quad r_0(z,0,\varphi)\underset{z\to\infty}{\approx} ln(z)f'(\varphi), \tag{12}$$

so that the exact solutions of Burgers' equations in the near- and farfield are recovered.
The two constants $\psi_F$ and $\psi_N$ remain to be determined. This will be done later on, according to considerations about the formation of shock waves.

## 5. Shock waves

As a consequence of the high-frequency assumption underlying the derivation of the KZK equation, the zero-frequency component of the pressure field must vanish:

$$B_0(z,x)=\int_{-\infty}^{+\infty}p(z,x,\tau)d\tau=0. \tag{13}$$

Calculating that quantity for the asymptotic expression (9), three cases arise.
(i) If the nonlinear co-ordinate transform (8) can be inverted at all points, it is easy to check that condition (13) is always fulfilled. This results from the introduction of the denominator in Eq.(9), which compensates for the term $\partial\tau\ /\partial\varphi$ appearing in performing transform (8) under integral (13). In this case, there is no shock wave: the wave profile remains continuous.
(ii) If (8) can be inverted (at least locally) in the vicinity of a point $\varphi_0$ but at that point:

$$\frac{\partial\tau}{\partial\varphi}(\varphi\neq\varphi_0)>0,\quad \frac{\partial\tau}{\partial\varphi}(\varphi_0)=0, \tag{14}$$

the renormalized expansion would be unbounded. In order to avoid this, the constant $\psi_F$ is chosen so as to cancel the numerator of the second-order term in Eq.(9):

$$\psi_F\left(\sum_i\frac{\frac{\partial p_1}{\partial\varphi}(\varphi_0^{(i)})}{\left(1-\frac{1}{N}\frac{\partial r_0}{\partial\varphi}(\varphi_0^{(i)})\right)^2}\right)=\left(\sum_i\frac{(p_2-r_0\frac{\partial p_1}{\partial\varphi})(\varphi_0^{(i)})}{\left(1-\frac{1}{N}\frac{\partial r_0}{\partial\varphi}(\varphi_0^{(i)})\right)^2}\right), \tag{15}$$

where points $\varphi_0^{(i)}$ denote all the local maxima of $\partial\ r_0/\partial\varphi$ . This case means that one shock wave is about to form.
(iii) There are (one or) several intervals over which transform (8) cannot be inverted anymore. In order that pressure remains single-valued in terms of the physical variable $\tau$, it is necessary

to jump from one invertible part of the curve $\tau(\varphi)$ to the other. Each jump corresponds to one shock wave. The shock position of the jth shock is determined by the lefthand $\varphi_1^{(j)}$ and righthand $\varphi_2^{(j)}$ values of variable $\varphi$ on each side of the shock. They must verify:

$$\tau^{(j)} = \varphi_1^{(j)} - \frac{1}{N} r(\varphi_1^{(j)}) = \varphi_2^{(j)} - \frac{1}{N} r(\varphi_2^{(j)}) \tag{16}$$

with:

$$1 - \frac{1}{N}\frac{\partial r}{\partial \varphi}(\varphi_i^{(j)}) > 0 \quad (i = 1, 2) \quad and \quad \varphi_1^{(j)} < \varphi_2^{(j)}. \tag{17}$$

Then, the zero-frequency component is:

$$B_0 = \frac{1}{N}\psi_N \sum_j \left(p_1(\varphi_2^{(j)}) - p_1(\varphi_1^{(j)})\right) - \sum_j b_0^{(j)}, \tag{18}$$

with:

$$b_0^{(j)} = \int_{\varphi_1^{(j)}}^{\varphi_2^{(j)}} \left( p_1 + \frac{1}{N}(p_2 - (r_0 + \psi_F)\frac{\partial p_1}{\partial \varphi} - p_1 \frac{\partial r_0}{\partial \varphi}) \right) d\varphi. \tag{19}$$

The condition $b_0^{(j)} = 0$ generalizes the law of equal area (Whitham 1974, Rudenko and Soluyan 1977). It is solved numerically by means of a Newton algorithm. However, that equation has not always a solution. Then, the position of the shock position is fixed at some given position. The quantity $b_0^{(j)}$ has then some definite value, and the quantity $\psi_N$ is calculated according to Eq.(18), so that the global zero-frequency component be equal to zero. More details on the determination of the shock positions can be found in Frøysa and Coulouvrat 1994.

## 6. Comparisons with a finite-difference scheme

The validity of the method is estimated by comparing the analytical wave profiles with the results of a finite-difference computation. The numerical code solves the KZK equation by a finite-difference procedure, advancing plane by plane away from the source. The dependance on time variable is dealt with, partly in the time domain (for non-linearities), partly in the frequency domain (for diffraction). Details are given in Frøysa *et al.* (1993, 1994).
On Fig.1 are compared the on-axis wave profiles at half the Rayleigh distance $(z = 0.5)$ for $N = 0.5$, resulting from the QLA (long-short dashed line), the renormalization procedure (continuous line), or the finite-differences code (dotted line).
The input pulse is a Gaussian modulated sine:

$$f(\tau) = \sin\tau \exp(-\tau^2/9). \tag{20}$$

It clearly appears that, while the QLA is invalid, the wave profile predicted by the renormalization method remains extremely close to the numerical solution. In this case, the nonlinear effects are strong enough to significantly alter the wave profile and make the QLA inappropriate, but not enough to produce a shock wave. Renormalization is especially well suited to this moderately nonlinear situation.

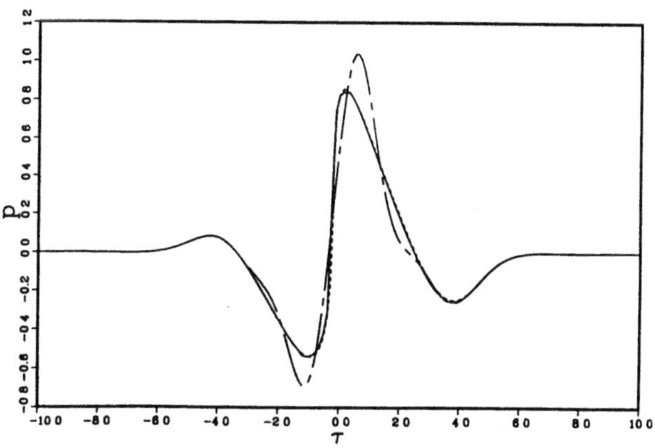

Fig.1: Wave profiles ($N$=0.5, $x = 0.$, $z = 0.5$) - renormalization (————), quasilinear approximation (— - — - —) and finite-differences (- - - - - - - - -)

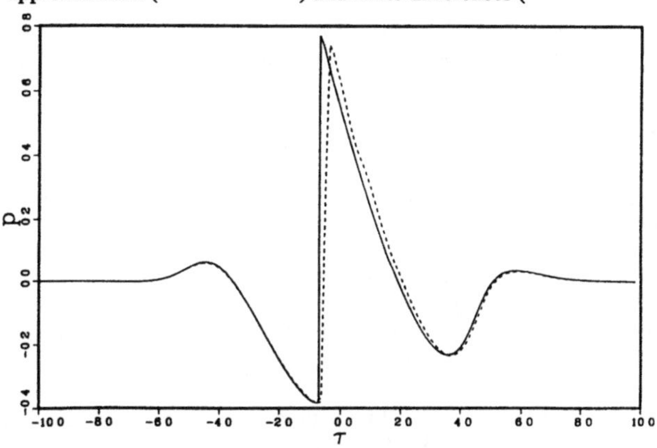

Fig.2: Wave profiles ($N$=0.5, $x = 0.$, $z = 1.$) - renormalization (————) and finite-differences (- - - - - - - - -)

On Fig.2 are compared the renormalized (continuous line) and numerical (dotted line) solutions for the same parameters, except the distance now equal to one Rayleigh distance $(z = 1)$. Here a shock wave has formed. Once again, the two methods are in remarkable agreement. In particular, the shock position and amplitude are predicted by the renormalization procedure with a good precision.

Other comparisons are presented in Coulouvrat and Frøysa 1993, and Frøysa and Coulouvrat 1994. They show the ability of the method to describe precisely the nonlinear distorsion of the wave profiles, the formation of shock waves and their early evolution. However, as the method is only an approximate one, it does not provide solutions valid for all ranges of parameters. For very strong nonlinearities, the renormalization approximation breaks down. A quantitative parameter has been proposed for classifying the validity of renormalized solutions. Nevertheless, the method considerably extends the validity of the QLA. To our knowledge, it is the only analytical approximation dealing with pulsed signals and taking into account shock waves. Generalizations to weakly focused sources, or other velocity distributions on the emitter surface, are possible.

## References


Aanonsen, S.I., Barkve, T., Naze Tjøtta, J., Tjøtta, S. (1984): "Distorsion and harmonic generation in the nearfield of a finite amplitude sound beam", *J. Acoust. Soc. Am.*, Vol.75, pp. 749-768

Bakhvalov, N.S., Zhileikin, Ya.M., Zabolotskaya, E.A. (1987): *Nonlinear theory of sound beams* (American Institue of Physics, New York)

Beyer, R.T. (1974): *Nonlinear acoustics* (Naval Ship Systems Command, Department of the Navy)

Blackstock, D.T. (1964): "Thermoviscous attenuation of plane, periodic, finite amplitude sound waves", *J. Acoust. Soc. Am.*, Vol.36, pp. 534-542

Burgers, J.M. (1948): "A mathematical model illustrating the theory of turbulence", *Adv. Appl. Mech.*, Vol.1, pp. 171-199

Cole, J.D. (1951): "On a quasilinear parabolic equation occuring in aerodynamics", *Quat. Appl. Math.*, Vol.3, pp. 225-236

Coulouvrat, F. (1991): "An analytical approximation of strong nonlinear effects in bounded sound beams", *J. Acoust. Soc. Am.*, Vol.90, pp. 1592-1600

Coulouvrat, F. (1992): "On the equations of nonlinear acoustics", *J. Acoustique*, Vol.5, pp. 321-359

Coulouvrat, F., Frøysa, K.E. (1993) "An approximate solution for the propagation of finite amplitude pulsed sound beams", in *Advances in nonlinear acoustics*, Proc. 13th Int. Symp. Nonlinear Acoustics, ed. by H. Hobæk (World Scientific, Singapore), pp. 239-244

Frøysa, K.E. (1994): "Weakly nonlinear propagation of a pulsed sound beam", *J. Acoust. Soc. Am.*, Vol.95, pp. 123-130

Frøysa, K.E., Coulouvrat, F. (1994): "A renormalization method for nonlinear pulsed sound beams", submitted to *J. Acoust. Soc. Am.*

Frøysa, K.E., Naze Tjøtta, J., Berntsen, J. (1993): "Finite amplitude effects in sound beams. Pure tone and pulsed excitation", in *Advances in nonlinear acoustics*, Proc. 13th Int. Symp. Nonlinear Acoustics, ed. by H. Hobæk (World Scientific, Singapore), pp. 233-238

Frøysa, K.E., Naze Tjøtta, J., Berntsen, J., Tjøtta, S. (1994): "Propagation of finite-amplitude sound beams, monochromatic and pulsed excitation, numerical and physical considerations", submitted to *J. Comp. Acoust.*

Frøysa, K.E., Naze Tjøtta, J., Tjøtta, S. (1993): "Linear propagation of a pulsed sound beam from a plane or focusing source", *J. Acoust. Soc. Am.*, Vol.93, pp. 80-92

Hopf, E. (1950): "The partial differential equation $u_t + uu_x = \mu\ u_{xx}$", *Comm. Pure Appl. Math.*, Vol.3, pp. 201-230, 1950

Kunitsyn, V.E., Rudenko, O.V. (1978): "Second-harmonic genration in the field of a piston radiator", *Sov. Phys. Acoust.*, Vol.24, pp. 310-313

Kuznetsov, V.P. (1970): "Equations of nonlinear acoustics", *Sov. Phys. Acoust.*, Vol.16, pp. 467-470

Lighthill, M.J. (1956): "Viscosity effects in sound waves of finite amplitude", in *Surveys in Mechanics*, ed. by G.K. Batchelor and R.M. Davies (Cambridge University Press, Cambridge), pp. 250-351

Naze Tjøtta, J., Tjøtta S. (1981): "Nonlinear equations of acoustics, with application to parametric arrays", *J. Acoust. Soc. Am.*, Vol.69, pp. 1644-1652

Rudenko, O.V., Soluyan, S.I. (1977): *Theoretical foundations of nonlinear acoustics* (Plenum, New York)

Whitham, G.B. (1974), *Linear and nonlinear waves* (Wiley, New York)

Zabolotskaya, E.A., Khokhlov, R.V. (1969): "Quasi-plane waves in the nonlinear acoustics of confined beams", *Sov. Phys. Acoust.*, Vol.15, pp. 35-40

# Long - Waves on Thin Viscous Liquid Film : Derivation of Model Equations

Radyadour Kh. Zeytounian

Université de Lille I , L.M.L C.N.R.S - URA 1441 ,
59655 Villeneuve d'Ascq, Cédex, France

**Abstract :** The present paper is devoted to a consistent asymptotic derivation of model equations for two-dimensional nonlinear (2DNL) waves on a viscous dilatable liquid film. Various approximate equations describing the NL evolution in time - space of the thickness of the film are obtained . More precisely, for small amplitudes in the long wave approximation, we consider three cases : high Reynolds number boundary layer regime, moderate Reynolds number of Poiseuille type one, and low Reynolds number second Poiseuille type regime.These various situations arise from the smallness of the thin film aspect ratio parameter $\varepsilon$ and are distinguished from each other by selecting, as far as order of magnitude are concerned, some dimensionless parameters with respect to $\varepsilon$ . They are Reynolds, Froude, Marangoni, Weber, Grashof numbers as well as a Biot one which is assumed, throughout, to be of order unity. For the high Reynolds number case we obtain an equation for the thickness h(t,x) of the film, which is similar to equation previously derived by Oron and Rosenau 1992 . From this last evolution equation for h(t,x) a new limiting process, for small amplitude (long) waves, leads to a modified (by Boussinesq effects ) Kuramoto- Sivashinsky (KS) equation. The two other regimes do not lead to a unique consistent equation for h(t,x) unless we focus on small amplitude (long) waves; one finds a KS or a KS with Korteweg-de-Vries (KdV) dispersive term equations. The role of buoyancy is quite different in the various regimes considered; for the low Reynolds number regime the buoyancy coupling effect is "effectively" significant in the KS - KdV model equation.

## 1 . Introduction

At present time a great number of studies of the hydrodynamics of a wavy thin falling film are present in the literature (a review and further references can be found in Lin and Wang 1985). For a non dilatable , incompressible, viscous liquid and two- dimensional wave flow, a discussion of basic results is given in recent papers by Trifonov and Tsvelodub 1991 and Prokopiou, Cheng and Chang 1991. For the vertical film , a recent paper by Chang 1994, gives an excellent review mostly concerning the various transition regimes on a free-falling vertical film. Concerning the derivation of equations describing the evolution of a film interface in the case of the combined thermocapillary and thermogravitational flow on a plane warmer than the ambient air, see a by Garazo and Verlade 1991 and Oron and Rosenau 1992.

In the present short paper, we consider three limiting regimes for the high, moderate and low Reynolds numbers, in the long wave approximation, and derive systematically various model equations for the film thickness. A new interesting result is the derivation of a consistent KdV - KS model equation for small amplitude wave in the long-wave approximation for the low Reynolds number regime.

## 2 . Governing Equations and Boundary Conditions

Using dimensionless variables, within the framework of the *Boussinesq approximation* (the dilatability of the liquid film is taken into account only when it is *coupled* with the gravity, it generates the *buoyancy* effect- parameter $\lambda$), the governing equations and boundary conditions for the liquid film problem are as follows (for a dimensional form of these equations and conditions, see Garazo and Velarde 1991 and also Oron and Rosenau 1992) :

$$\partial u/\partial x + \partial w/\partial z = 0 \ ; \tag{1}$$

$$Du/Dt + \partial p/\partial x - \sin\beta /\varepsilon Fr^2 + \sin\beta \ (\lambda /\varepsilon)\, \theta$$

$$= (1/\varepsilon Re)[\partial^2 u/\partial z^2 + \varepsilon^2 \partial^2 u/\partial x^2]; \tag{2}$$

$$\varepsilon^2 Dw/Dt + \partial p/\partial z + \cos\beta / Fr^2 - \cos\beta \ \lambda\, \theta$$

$$= (\varepsilon /Re)[\partial^2 w/\partial z^2 + \varepsilon^2 \partial^2 w/\partial x^2]; \tag{3}$$

$$Pr\, D\theta /Dt = (1/\varepsilon Re)[\partial^2\theta/\partial z^2 + \varepsilon^2 \partial^2\theta/\partial x^2]: \tag{4}$$

*on the solid surface* $z = 0: \ u = 0 ; w = 0$ and $\theta = 1$ ; (5)

*on the free surface* $z = 1 + \delta\, \eta\, (t,x) \equiv h(t,x)$ :

$$w = \partial h/\partial t + u\, \partial h/\partial x \ ; \tag{6a}$$

$$\partial u/\partial z + \varepsilon^2 \partial w/\partial x + 4\, (\varepsilon^2 / N\text{-})\partial h/\partial x\, \partial w/\partial z$$

$$+ \varepsilon\, Ma\, (N^{1/2} / N\text{-})[\partial\theta /\partial x + \partial h/\partial x\, \partial\theta/\partial z\, ]= 0 \ ; \tag{6b}$$

$$p - p_a = (\varepsilon^2 / N^{3/2}\ )\ [\ (Ma/ Re\ )\, \theta - \ We\ ]\ \partial^2 h/\partial x^2$$

$$+ 2\, (\varepsilon /Re)(1/N)\ [\ (N\text{-})\ \partial w/\partial z - \partial h/\partial x\ (\ \partial u/\partial z + \varepsilon^2 \partial w/\partial x)]\ \ ; \tag{6c}$$

$$\partial\theta /\partial z - \ \varepsilon^2 \partial h/\partial x\, \partial\theta/\partial x + \ N^{1/2}\ (1 + Bi\, \theta\, ) = 0\ , \tag{6d}$$

where

$$D/Dt \ = \ \partial/ \partial t + u\, \partial/\partial x + w\, \partial/\partial z\ , \ \text{ and } \ N = 1 + \varepsilon^2 (\partial h/\partial x)^2\ , \quad N\text{-} = 1 - \varepsilon^2 (\partial h/\partial x)^2\ .$$

In eqs. (1)- (4) , Re is the classical Reynolds number (viscosity) , Fr is the Froude number (gravity), $\beta$ is the inclination angle of the solid plane from the horizontal, $\lambda = Gr / (Re)^2$ is the *buoyancy* parameter, where Gr is the Grashof number (dilatable liquid), Pr is the Prandtl number (conduction) and $\varepsilon = h_o / l_o$ is the "long- wave" main small parameter, where $h_o$ is the average film thickness and $l_o$ is a characteristic value for the horizontal length . For the derivation of consistent equations for the (perturbed) thickness $\eta$ (t,x) it is necessary to assume that the *amplitude* parameter $\delta = a_o/h_o$ is also small and to write similarity relations between $\varepsilon$ and $\delta$. In eq. (6b), Ma is the Marangoni number (temperature variation of the interfacial tension); in Eq. (6c), We is the classical Weber number (effect of the constant of tension) and in Eq. (6d), Bi is the so-called Biot number which takes into account the effect of the *air-liquid transfer* characteristic. We note that the Strouhal number $S = l_o/t_o u_c \approx 1$. In the dimensionless equations (1)- (4) and conditions (5), (6) the units are based on: $h_o$ for the vertical co-ordinate, $l_o$ for the horizontal co-ordinate, $t_o = l_o/u_c$ for the time, $u_c$ for the horizontal velocity component ,$\varepsilon u_c$ for the vertical velocity component and $\rho_o (u_c)^2$ for the pressure variation, where $\rho_o$ is a constant value of the density. For a *Boussinesq dilatable fluid* we have (with dimensionless variables ) the following equation of state:

$$\rho = 1 - \alpha \Delta To\, \theta , \tag{7a}$$

where $\alpha$ is the coefficient of volumetric expansion for the liquid while:

$$\theta = (T - T_o) / \Delta To,$$

with T the (absolute) temperature. We assume that the temperature of the solid plane (z = 0) is $T_o + \Delta To$ while $T_o$ is a reference value for the temperature taken at the undeformed liquid - air interface (z = 1) in the quiescent state. Concerning the air- liquid interfacial tension $\sigma$ (function of the temperature) we assume also that:

$$\sigma = 1 - s\, \Delta To\, \theta, \tag{7b}$$

where s = Const takes into account the Marangoni effect. In fact, we have for the dimensionless parameters (Ma , We , Bi , Gr , Re , Pr and Fr) the following expressions:

$$Ma = s\, \Delta To / \rho_o \nu_o u_c\ , We = \sigma_o / \rho_o h_o (u_c)^2\ , \ Bi = q_o h_o, \ \ Gr = \alpha\, g\, \Delta To\, ( h_o)^3 / (\nu_o)^2\ ,$$

$$Re = h_o u_c / \nu_o\ , Pr = \nu_o / \kappa_o\ , Fr = u_c / (g h_o)^{1/2}\ , \tag{8}$$

where $\sigma_o$ is a constant value of the interfacial tension, $q_o$ is the constant ratio between the heat transfer rates by convection and conduction, $\kappa_o$ is the constant thermal diffusivity, g is the acceleration due to gravity and $\nu_o$ is the constant kinematic viscosity.

From the continuity equation (1) and the kinematic condition on the free surface (6a), with (5), we derive the following *averaged* continuity equation:

$$\partial h/\partial t + \partial q/\partial x = 0\ , \qquad \text{where} \qquad q = \int_o^{h(t,x)} u(t,x,z)\, dz, \tag{9}$$

The function q(t,x) is the average over the film thickness of the (parallel to the solid wall z = 0 ) component u of the velocity.

Due to the fact that q(t,x), as a general rule, cannot be expressed in terms of h(t,x) or some of its spatial derivatives, equation (9) is not a closed form evolution equation for the thickness h(t,x). Our main goal is to examine some situations, arising mainly from distinguished limiting processes, from which a closed equation may be obtained for h(t,x). We shall examine also how some limiting equations may be derived for the amplitude $\eta$ (t,x) (see just before (6a)), concerning long waves.

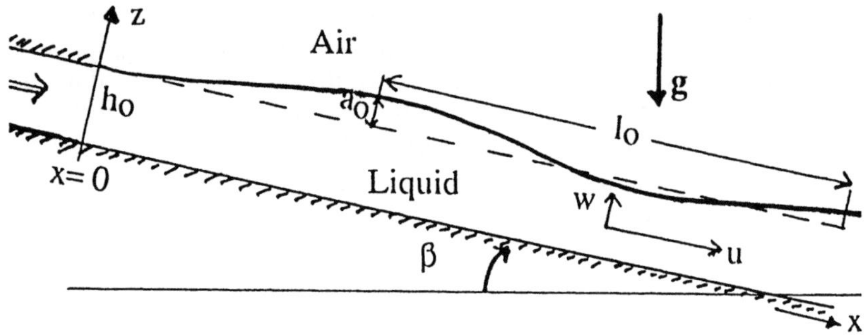


**Fig. 1** . Schematic geometry of the film problem

### 3 . High Reynolds Number Boundary Layer Regime - Re >>1

In this section we consider the following situation :

$$\varepsilon << 1 \quad \text{and} \quad Re >> 1 \quad \text{but} \quad \varepsilon\, Re = Re^* = O(1)\ , \tag{10}$$

with:

$$\varepsilon\, Ma = Ma^* = O(1)\ ,\ \varepsilon^2 We = We^* = O(1)\ ,\ \lambda/\varepsilon = \lambda^* = O(1) \tag{11a}$$

$$Bi = O(1) \quad \text{and} \quad Fr^2 = Re\ . \tag{11b}$$

Setting $\varepsilon \to 0$ in (1)- (6), integrating (3) over the film and taking (6c) into account, we get, to the limit:

$$\partial u/\partial x + \partial w/\partial z = 0\ ; \tag{12a}$$

$$Du/Dt - \sin\beta\,[\,1/Re^* - \lambda^*\,\theta] - We^*\,\partial^3 h/\partial x^3 = (1/Re^*)\,\partial^2 u/\partial z^2; \tag{12b}$$

$$Pr\, D\theta/Dt = (1/Re^*)\,\partial^2\theta/\partial z^2\ ; \tag{12c}$$

*on the solid wall* ( z = 0):

$$u = 0,\ w = 0 \quad \text{and} \quad \theta = 1; \tag{12d}$$

*on the free surface* ($z = 1 + \delta\,\eta\,(t,x) \equiv h(t,x)$):

$$w = \partial h/\partial t + u\,\partial h/\partial x\ , \tag{12e}$$

$$\partial u/\partial z = - Ma^*[\partial\theta /\partial x + \partial h/\partial x\, \partial\theta/\partial z], \qquad (12f)$$

$$\partial\theta /\partial z + 1 + Bi\, \theta = 0, \qquad (12g)$$

and we observe that according to the form of (12b) we may refer to a *boundary layer regime.*

We have, obviously, got some simplifications but they are not sufficient for obtaining, from (9), an evolution equation for h(t,x), in closed form, as was obtained , for example, by Oron and Rosenau 1992. A way of achieving this goal is to perform a new limiting process $Re^* \to 0$ , by assuming that $Re^*\, We^*$ remains O(1), while $Re^*\, \lambda^* = G\, r /Re \equiv \alpha\, \Delta T_0 \to 0$ (which allows to remain within the framework of the Boussinesq approximation ).It is straightforward, in particular , to integrate out (12b) and to derive the desired evolution equation for the thickness of the film h(t,x):

$$\partial h/\partial t + (1/3)\, \partial/\partial x \left\{ h^3 [\, \sin\beta + Re^*\, We^*\, \partial^3 h/\partial x^3] \right.$$

$$\left. + (3/2)\, Ma^*\, (1+Bi)\;\; h^2\, \partial h/\partial x\;\; [1/ (1+Bi\, h)^2] \right\}= 0. \qquad (13)$$

When starting from (12), one has dropped two time derivatives by the limiting process and one may ask what is the order of magnitude of time necessary for establishing the velocity u and temperature $\theta$ which are associated with the solution for h(t,x). The answer is that this time for adjustment is O(Re*) and that the rate is exponential. The only interesting issue with this point is that one has not to be anxious about any oscillations which might persist without attenuation after the O(Re*) period.

Up to now a smallness of $\delta$ in $h(t,x)\equiv 1 + \delta\, \eta\, (t,x)$ has not been taken into account so that (13) is a *full* evolution equation with this respect. If one wishes to investigate what comes out of the assumption $\delta << 1$, a significant result comes in with the study of long times $O(1/\delta)$ but the form of (13) suggests that convection with a speed close to $\sin\beta$ occurs and one finds appropriate to rescale the main variables according to:

$$\tau = \delta\, t \qquad \text{and} \qquad \xi = x - \sin\beta\;\, t\, ; \qquad \eta\, (t,x)= \eta^*(\tau,\xi). \qquad (14)$$

If we assume, further, that:

$$We^* = \varepsilon\;\, We^{**}\, ,\; Ma^* = \varepsilon\;\, Ma^{**} = \varepsilon\;\, Ma\;\; , \delta = \varepsilon, \qquad (15)$$

with $We^{**} = O(1)$, and $Ma = O(1)$ , one gets to the limit $\varepsilon \to 0$ , the following Kuramoto-Sivashinsky (KS) equation:

$$\partial\, \eta^* /\partial\tau + 2\;\, \sin\beta\;\; \eta^*\, \partial\, \eta^* /\partial\xi + [\, (1/2)\, M\, a / (1+Bi\, )]\;\; \partial^{\,2}\eta^* /\partial\xi^2$$

$$+\;\; (1/3) Re^*\, We^{**}\, \partial^{\,4}\eta^* /\partial\xi^{\,4} = 0\, . \qquad (16)$$

This last KS equation (16) for $\eta^*(\tau,\xi)$ is valid to order $O(\delta\, )$.

Finally, referring to (13) we may remember that we have used at this time, the Boussinesq approximation issued from $\alpha\ \Delta T_o \ll 1$ . But curiously , it is possible to take care of a "non Boussinesq " correction to (16) by assuming :

$$\alpha\ \Delta To = \alpha_* \delta = \alpha_* \varepsilon\ ,\quad \alpha_* = O(1), \tag{17}$$

which allows to derive, from $\delta \to 0$ , the following *modified* KS equation:

$$\partial \eta_* /\partial \tau - \chi\ \partial \eta_* /\partial \xi + 2\ \sin\beta\ \eta_* \partial \eta_* /\partial \xi + [\,(1/2)\,Ma\,/\,(1+Bi\,)]\ \partial^2 \eta_* /\partial \xi^2$$

$$+ (1/3) Re^* We^{**} \partial^4 \eta_* /\partial \xi^4 = 0\ , \tag{18}$$

where:

$$\chi = (1/6)\sin\beta\ \alpha_*[\,1 + (5/4)\,Bi\,/\,(1 + Bi)\,]\ . \tag{19}$$

The evolution equation (18) is still valid to order $O(\delta\,)$.

## 4. Moderate Reynolds Number, Poiseuille Regime - $Re = O(1)$

Assuming that $Fr^2 = Re = O(1)$ and that $\alpha\ \Delta To = \alpha_* \varepsilon$ , $\alpha_* = O(1)$ ,which, according to $\lambda\,Re = \alpha_* \varepsilon$ , involves $\lambda = O(\varepsilon)$, we get that $\varepsilon$ is the main small parameter occurring within equations (1)- (4), for the full problem. Concerning the boundary conditions, on the free surface, we assume:

$$Ma = O(1)\ ,\quad \varepsilon^2\,We = We^* = O(1)\ ,\quad \text{and}\quad Bi = O(1)\ , \tag{20}$$

and we look for an expansion, for the solution of (1)- (6) , in the form proposed by Benney 1966, namely:

$$U = [\,u, w, p, \theta\,]^T = U_o + \varepsilon\ U_1 + O(\varepsilon^2), \text{when } \varepsilon \to 0. \tag{21}$$

but we do not expand h(t,x) for the moment .

The solution $U_o$ is got in a straightforward way, ( assuming $\beta = \pi / 2$ for simplicity) we obtain:

$$u_o = -\,(1/2)\,z^2 + h\,z\ ,\quad wo = -\,(1/2)\,(\partial h/\partial x)\,z^2\ , \tag{22a}$$

$$p_o = p_a - We^*\,\partial^2 h/\partial x^2\ , \tag{22b}$$

$$\theta o = 1 - [\,(1+Bi\,)\,/\,(1+Bi\,h)]\,z\ , \tag{22c}$$

and from (9) we get

$$\partial h/\partial t + h^2\,\partial h/\partial x = O(\,\varepsilon\,). \tag{23}$$

Writing out the set of equations and boundary conditions for $U_1$, and, again, assuming that h(t,x) is not expanded, it is easy to get an analytic expression for $u_1$, as a function of z. Using $u_o$ and $u_1$, we may compute $q_o$ and $q_1$ within the expansion of:

$$q = \int_o^{h(t,x)} u(t,x,z)\, dz = q_o + \varepsilon q_1 + O(\varepsilon^2).$$

Concerning $q_o = (1/3)\, h^3$ , this has already been taken into account with (23), while we get for $q_1$ the following expression :

$$q_1(t,x) = (1/3)\, Re\, We^*\, h^3\, \partial^3 h/\partial x^3 + (1/2)\, Ma\, (1+Bi)\, h^2\, \partial h/\partial x\, [1/(1+Bi\, h)^2]$$

$$+ (2/15)\, Re\, h^6\, \partial h/\partial x$$

$$+ (1/3)\, \alpha_*\, h^3 \{(5/8)\, [(1+Bi)/(1+Bi\, h)]\, h - 1\}. \qquad (24)$$

From equation (9), we may get an equation for the thickness of the film h(t,x), involving the O (ε ) term occurring in (23) :

$$\partial h/\partial t + h^2\, \partial h/\partial x + \varepsilon\, \partial/\partial x \Big\{ (1/3)\, Re\, We^*\, h^3\, \partial^3 h/\partial x^3$$

$$+ (1/2)\, Ma\, (1+Bi)\, h^2\, \partial h/\partial x\, [1/(1+Bi\, h)^2] + (2/15)\, Re\, h^6\, \partial h/\partial x$$

$$+ (1/3)\, \alpha_*\, h^3 \{(5/8)\, [(1+Bi)/(1+Bi\, h)]\, h - 1\} \Big\} = 0. \qquad (25)$$

This evolution equation of the “Benney type” contains the small parameter ε , a fact which is due to that h(t,x) has not been expanded as it should in a fully consistent asymptotic approach through an expansion with respect to ε. Of course we may expand h(t,x) in different ways and we shall investigate, here, only the same kind of phenomenon that the one which led to the KS equation (16). In order to achieve, this we put:

$$\tau = \varepsilon\, t, \quad \xi = x - t; \quad h(t,x) = 1 + \varepsilon\, \eta_*(\tau,\xi) + \ldots, \quad \varepsilon = \delta, \qquad (26)$$

and we let $\varepsilon \to 0$ within the transformed version of equation (25), getting:

$$\partial\eta_*/\partial\tau - \chi^o\, \partial\eta_*/\partial\xi + 2\, \eta_*\, \partial\eta_*/\partial\xi + A\, \partial^2\eta_*/\partial\xi^2$$

$$+ B\, \partial^4\eta_*/\partial\xi^4 = 0, \qquad (27)$$

with:

$$\chi^o = (1/6)\, \alpha_*\, [1 + (5/4)\, Bi/(1+Bi)], \qquad (28a)$$

$$A = (2/15)\, Re + (1/2)\, Ma/(1+Bi), \quad B = (1/3)\, Re\, We^*. \qquad (28b)$$

We observe that (27) is, again, a modified KS equation looking like (18). The only difference occurs in the coefficients A and B (remember that sinβ = 1, according to our hypothesis ). However, it is important to note that the full equation (13), for the thickness of the film $h(t,x)$ (valid for the high Reynolds number boundary layer regime), is more significant than the "Benney type" equation (25) for the moderate Reynolds number, Poiseuille regime.

## 5. Low Reynolds Number , Second Poiseuille Regime - $Re \ll 1$

The problem considered in this section 5 is interesting when we look at (25). As a matter of fact this is a singular perturbation of the hyperbolic equation : $\partial h/\partial t + h^2\, \partial h/\partial x = 0$ .

Curiously, we get again a singular perturbation of this same equation, of an another type, if we make the assumption ( again for a vertical plane : $\beta = \pi / 2$) :

$$Re / \varepsilon = \mathcal{R} \quad , Ma / \varepsilon = \mathcal{M} \quad , \lambda / \varepsilon = \lambda* , \quad Fr^2 = Re \quad \text{and} \quad \varepsilon^2\, We = We, \tag{29}$$

with $\mathcal{R}$ , $\mathcal{M}$ , $\lambda *$, $We*$ and $Bi$ all $O(1)$ .

The formal expansion (21) has, then, to be modified into:

$$U = [\, u, w, p, \theta\, ]^T = U_0 + \varepsilon^2\, U_2 + \ldots , \quad \text{when } \varepsilon \to 0, \tag{30}$$

and the only modification to (22) occurs in the expression for $p_0$ (see, (22b) ) :

$$p_0 = p_a - We^*\, \partial^2 h/\partial x^2 - (1 / \mathcal{R})\, [z + h]\, \partial h/\partial x\, . \tag{31}$$

From (9) we get, this time, in place of (23) :

$$\partial h/\partial t + h^2\, \partial h/\partial x = O(\varepsilon^2) . \tag{32}$$

Writing out the set of equations and boundary conditions for $U_2$, and assuming, yet, that $h(t,x)$ is not expanded, we may get an awkward expression for $u_2$ that may be integrated with respect to z in order to obtain the following expression for $q_2$ ( in place of (24)) :

$$\begin{aligned} q_2 = {} & (1/3)\, h^3\, [\, \mathcal{R}\, We^*\, \partial^3 h/\partial x^3 + 7\, (\partial h/\partial x)^2 - \mathcal{R}\, \lambda_*\, ] + (2/15)\, \mathcal{R}\, h^6\, \partial h/\partial x \\ & + (1/2)\, \mathcal{M}\, (1+Bi)\, h^2\, \partial h/\partial x\, [1/ (1+Bi\, h)^2] \\ & + h^4\, \{\, \partial^2 h/\partial x^2 + \mathcal{R}\, \lambda*\, (5/24)\, [\, (1+Bi\, )/ (1+Bi\, h)]\, \}. \end{aligned} \tag{33}$$

The final result is analogous to (25) (but with some additional terms ) which reads:

$$\partial h/\partial t + h^2\, \partial h/\partial x + \varepsilon^2\, \partial/\partial x \Big\{ (1/3)\, h^3\, [\, \mathcal{R}\, We^*\, \partial^3 h/\partial x^3 + 7\, (\partial h/\partial x)^2 - \mathcal{R}\, \lambda_*]$$

$$+ (2/15)\, \mathcal{R}\, h^6\, \partial h/\partial x + (1/2)\, \mathcal{M}\, (1+Bi)\, h^2\, \partial h/\partial x\, [1/ (1+Bi\, h)^2]$$

$$+ h^4 \{ \partial^2 h/\partial x^2 + R\,\lambda_* \;\; (5/24)\, [\, (1+\text{Bi}\,)/\,(1+\text{Bi}\, h)]\, \}\} = 0. \tag{34}$$

We intend to play, with this last evolution equation (34), the same game as the one we considered in Section 4 . We use (Cf. (26)), namely:

$$\tau = \delta\ t, \quad \xi = x - t; \quad h(t,x) = 1 + (1/\varphi)\ \varepsilon^2\ \eta_*(\tau,\xi) + ..., \qquad \delta = (1/\varphi)\,\varepsilon^2, \tag{35}$$

where $\varphi$ is the *dispersive* similarity parameter.

Carrying out, again, the limiting process $\varepsilon \to 0$ , we find an equation which cumulates the features of KdV on the one hand and KS on the other hand:

$$\partial\eta_*/\partial\tau - \nu\, \partial\eta_*/\partial\xi + 2\, \eta_*\, \partial\eta_*/\partial\xi + \sigma\, \partial^2\eta_*/\partial\xi^2$$
$$+ \phi\, \partial^3\eta_*/\partial\xi^3 + \mu\, \partial^4\eta_*/\partial\xi^4 = 0, \tag{36}$$

where:

$$\nu = (1/6)\, \phi\, R\, \lambda_*[\, 1 + (5/4)\, \text{Bi} / (1 + \text{Bi})\, ]\, , \tag{37a}$$

$$\sigma = \phi\, [\, (2/15)\, R + (1/2)\, M / (1 + \text{Bi}\,)\, ]\, , \tag{37b}$$

$$\mu = (1/3)\, \phi\, R\, \text{We}_*\ . \tag{37c}$$

The evolution equation (36) is again a significant model equation valid for large time with an error of $O(\delta\,)$ . The coefficients $\sigma$, $\mu$ and $\phi$ are all positive constants characterising *instability*, *dissipation* and *dispersion* respectively.

## 6. Conclusion

Since the new KdV - KS evolution equation (36) for the "perturbed thickness" of the film $\eta_*(\tau,\xi)$ , valid to order $O(\delta\,)$, includes the relevant physical factors (through the Weber, Marangoni, Biot, Reynolds and Grashof numbers ) it can be studied as a consistent asymptotic model for the problem of the stability and the *rupture of the free surface film* for time *large* enough.

However, for the stability problem, it seems preferable to use an *ad hoc* but convenient simplification of (12) , arbitrarily assuming a self- similar parabolic flow profile beneath the film (Shkadov 1967). For Pr tending to zero, this reduces (12) to the simplified form of the "integral boundary- layer equations" (IBL) or "averaged" equations, for h(t,x) and q(t,x), which generalize the Shkadov 1967 averaged equations (Zeytounian 1993). In a recent paper (Zeytounian 1994) we have investigated the possibility of a bifurcation of the solution of these averaged equations and we have obtained the value of the cut- off wave number, which is a function of $\text{We}^*$ , $\text{Ma}^*$ and $\text{Re}^*$ . For the particular case ,when $\text{Re}^*$ and $\text{Ma}^*$ tend to zero (but $\text{Ma}^*/\text{Re}^* = \Lambda = O(1)$ ), we obtain for the cut- off wave number, $k^*$, the following relation:

$$k^* = [\ (1 + (9/2)\ \Lambda\ )/3\ We^*\ ]^{1/2}\ .$$

When $\Lambda = 0$ we derive again the classical result corresponding to Ma = 0 ( Trifonov and Tsvelodub 1991). Finally, we note that a weakly nonlinear analysis of the averaged equations for h(t,x) and q(t,x) , for small amplitude $\delta$ , may be considered for the investigation of the rupture phenomenon of the film thickness, which is based on a calculation of the "blowup time".

**References**


Benney, D.J. (1966): Long - Waves on Liquid Films , *J. Math. and Phys.* **45** ; 150 - 155

Chang , H.- Ch . (1994): Wave Evolution on a Falling film, *Ann. Rev. Fluid Mech.* **26** ; 103 - 136

Garazo, A.N. and Velarde, M.G. (1991): Dissipative KdV Description of Marangoni - Bénard Oscillatory Convection, *Phys. Fluids*, **A3** (10) ; 2295 - 2300

Lin, S.P. and Wang, C.Y. (1985): Modeling wavy film flows " , in *Encyclopedia of Fluid Mechanics*, Gulf, Houston , Vol. **1** ; p. 93

Oron, A. and Rosenau, Ph. (1992): Formation of Patterns Induced by Thermocapillarity. *J.Phys. II France* **2** ; 131- 146

Prokopiuo, Th. , Cheng,M. and Chang, H.-Ch. (1991) : Long waves on inclined films at High Reynolds number, *J. Fluid Mech.*, **222** ; 665 - 691

Shkadov, V. Ya. (1967) : Wave conditions in the flow of thin layer of a viscous liquid under the action of gravity, *Izv. Akad. Nauk SSSR, Mekh. Zhidk. i Gaza,* 1 ; 43-50

Trifonov, Yu. Ya. and Tsvelodub, O. Yu. (1991) : *J. Fluid Mech.*, **229** ; 531 - 554

Zeytounian, R . Kh. (1993) : Nonlinear Dynamics of Waves on the Free Surface of a Falling Liquid Film Revisited. *Preprint*

Zeytounian, R . Kh. (1994) : Equations d'évolution pour les films dilatables, visqueux, minces. *Projet de Note aux CRAS*

# Part IV

# Aerodynamics and Rarefied Gas Flows

# On global existence theorems for the initial-boundary value problem for the Boltzmann equation

Carlo Cercignani

Dipartimento di Matematica
Politecnico di Milano
Milano, Italy

**Abstract:** Existence theorems have been recently proved by several authors for the initial-boundary value problem for the nonlinear Boltzmann equation. Here we wish to survey and extend these results and show the basic role that an inequality first introduced by Darrozès and Guiraud plays in this matter.

## 1. Introduction

In this paper, I shall deal with the initial-boundary value problem which arises when we consider the time evolution of a rarefied gas in a vessel $\Omega$ ($\Omega$ is a bounded open set of $\Re^3$ with a sufficiently smooth boundary $\partial\Omega$), endowed with unit normal $n(x)$ (pointing into $\Omega$) or outside a solid body (see Sect. 7). Different points of the boundaries will be assumed to possess, generally speaking, different velocities and temperatures.

Although I shall occasionally mention perturbation methods, this is not, strictly speaking, a paper devoted to asymptotic techniques. This is not due to the fact that perturbation problems are lacking in kinetic theory; they are plenty, starting with the rigorous derivation of the Boltzmann equation from Hamiltonian dynamics. The motivation of this paper is rather to illustrate the role that the work of Jean-Pierre Guiraud played in the development of rigorous kinetic theory and, in particular, to underline the importance of an inequality, first introduced by Jean-Pierre Guiraud in a paper in collaboration with Jean Sylvestre Darrozès, for the most recent developments of the subject.

The evolution equation for the distribution function $f(x, \xi, t)$ is the Boltzmann equation (Cercignani 1988, 1990a):

$$\Lambda f = Q(f, f) \qquad \text{in } \mathcal{D} \quad (\mathcal{D} = \Omega \times \Re^3 \times (0, T)) \tag{1.1}$$

with the boundary condition

$$\gamma^+ f(x,\xi,t) = K\gamma^- f \qquad \text{on } E^+ \tag{1.2}$$

(where, as explained below, $K$ is a linear integral operator and $\gamma^\pm f$ denote the traces of $f$ on $E^\pm = \{(t,x,\xi) \in \partial\Omega \times \Re^3 \times (0,T) \mid \pm\xi \cdot n(x) > 0\}$) and the initial condition

$$f(x,\xi,0) = f_0(x,\xi) \tag{1.3}$$

In Eq. (1.1) $\Lambda f$ is the free streaming operator defined by

$$\Lambda f = \frac{\partial f}{\partial t} + \xi \cdot \frac{\partial f}{\partial x} \tag{1.4}$$

and $Q(f,f)$ the collision term

$$Q(f,f)(x,\xi,t) = \int_{\Re^3}\int_{\mathcal{B}^+} (f'f'_* - ff_*)B(V,n)d\xi_* dn \quad (V = \xi - \xi_*) \tag{1.5}$$

Here $B(V,n)$ is a kernel containing the details of the molecular interaction, $f', f'_*, f_*$ are the same as $f$, except for the fact that the velocity argument $\xi$ ($\in \Re^3$) is replaced by $\xi', \xi'_*, \xi_*$, respectively, $\xi_*$ being an integration variable (having the meaning of the velocity of a molecule colliding with the molecule of velocity $\xi$, whose path we are following). $\xi'$ and $\xi'_*$ are the velocities of two molecules entering a collision that will bring them to have velocities $\xi$ and $\xi_*$, whereas the unit vector $n$ describes a hemisphere $\mathcal{B}^+$ of the unit sphere $\mathcal{B}$. The relations between $\xi'$, $\xi'_*$, on one hand, and $\xi$, $\xi_*$, on the other hand, read as follows:

$$\begin{aligned} \xi' &= \xi - n[(\xi - \xi_*)\cdot n] \\ \xi'_* &= \xi_* + n[(\xi - \xi_*)\cdot n] \end{aligned} \tag{1.6}$$

We shall assume that the gain and loss parts of $Q(f,f)$, denoted by $Q^\pm$ are separately meaningful (a.e.). Please remark that the loss term $Q^-(f,f)$ equals $f$ times an expression linear in $f$ that we shall denote by $\nu(f)$. Occasionally we simply write $Q$ or $Q^\pm$ to denote $Q(f,f)$ or $Q^\pm(f,f)$.

If we work in spaces of sufficiently regular solutions, we can introduce the semigroup $U(t)$ associated with the collisionless evolution ($\Lambda f = 0$). Then the above initial-boundary value problem reduces to solving the following integral equation

$$f(t) = U(t)f_0 + \int_0^t U(t-s)Q(f(s),f(s))ds \tag{1.7}$$

It is easy to prove local existence of this equation in several spaces, typically:

$$X_{\alpha,\beta} = \{f \mid (1+|\xi|^2)^{\alpha/2}\exp(-\beta|\xi|^2)f \in L^\infty(\Omega \times \Re^3)\} \tag{1.8}$$

with norm

$$\| f \|_{\alpha,\beta} = \| (1+|\xi|^2)^{\alpha/2}\exp(-\beta|\xi|^2)f \|_{L^\infty(\Omega\times\Re^3)} \tag{1.9}$$

A case in which one can say a lot more about the solutions of initial boundary value problems is the case when the data are compatible with a solution close to a uniform Maxwellian distribution $M$, given by

$$M = A\exp(-\beta|\xi - u|^2) \tag{1.10}$$

where $\beta$ is the inverse temperature and $u$ the bulk velocity (frequently assumed to be zero). Then techniques from rigorous perturbation theory (Kato, 1966) can be used. To this end, let us introduce the perturbation $h$ such that

$$f = M + M^{1/2}h \tag{1.11}$$

and assume (for simplicity) that $M$ coincides with the wall Maxwellian, so that Eq. (1.2) is satisfied by the restrictions of $M$ to $E^{\pm}$. Eqs. (1.1-3) can then be rewritten in the following way:

$$\Lambda h = Lh + \Gamma(h,h) \qquad \text{in } \mathcal{D} \tag{1.12}$$

$$\gamma^+ h(x,\xi,t) = \hat{K}\gamma^- h \qquad \text{on } E^+ \tag{1.13}$$

$$h(x,\xi,0) = h_0(x,\xi) \tag{1.14}$$

where $L$ and $\Gamma$ are two suitable operators, linear and quadratic, respectively, whereas

$$\hat{K} = M^{-1/2}KM^{1/2} \tag{1.15}$$

Now the global solution can be found by perturbation techniques, provided the linearized operator

$$B = -\xi \cdot \partial/\partial x + L \tag{1.16}$$

with the boundary conditions (1.13) generates a semigroup $T(t)$ with a nice decay.

One finds two types of result: the first of them with a decay like $t^{-\alpha}$ ($\alpha > 0$) applies to unbounded domains, whereas the second, with an exponential decay, applies to bounded domains.

There are several papers dealing with the proofs of the behaviors indicated above. The case of a bounded domain has been considered in the case of diffuse reflection (Guiraud, 1975) and in the case of specular reflection ( Shizuta and Asano, 1977). Both authors assumed that $\Omega$ is convex. The case of unbounded domains exterior to a bounded convex obstacle was treated by several Japanese authors ( Ukai, 1986; Ukai and Asano, 1980).

Here we want to deal with the more difficult case of arbitrarily large data. In this case we can hope for some significant results only if we use the $L^1$ framework and the techniques which have been used to deal with the Cauchy problem (DiPerna and Lions, 1989), as discussed with more detail in the next few sections.

## 2. An inequality introduced by Darrozès and Guiraud

On $\partial\Omega$ we impose a linear boundary condition of a rather standard form (Cercignani 1988, 1990a):

$$\gamma^+ f(t,x,\xi) = \int_{\xi'\cdot n<0} K(\xi' \to \xi; x,t)\gamma^- f(t,x,\xi')d\xi' \equiv K\gamma^- f \qquad (x \in \partial\Omega,\ \xi\cdot n > 0) \tag{2.1}$$

where $K(\xi' \to \xi; x, t)$ is a kernel (the boundary scattering kernel) such that:

$$K(\xi' \to \xi; x,t) \geq 0 \tag{2.2}$$

$$\int_{\xi\cdot n>0} K(\xi' \to \xi; x,t)|\xi\cdot n|d\xi = |\xi'\cdot n| \tag{2.3}$$

$$M_w(\xi) = \int_{\xi'\cdot n<0} K(\xi' \to \xi; x,t) M_w(\xi')d\xi' \tag{2.4}$$

where $M_w$ is the wall Maxwellian given by Eq. (1.10) with $u = 0$ and $\beta = \beta_w$ ($\beta_w$ being the inverse temperature of the wall), whereas $\gamma^{\pm}$ are the trace operators introduced in Sect. 1.

The case of an isothermal boundary was the first to be considered (Hamdache, 1992) . In the case of non-isothermal data along $\partial\Omega$ the initial-boundary value problem possesses boundary data which are compatible, not with a Maxwellian, but rather with one of those steady solutions, whose theory is still in its infancy (Arkeryd *et al.*, 1991); thus one cannot expect the solution to tend toward a Maxwellian when $t \to \infty$ as has been recently shown for other kinds of boundary conditions (Desvillettes, 1990; Cercignani, 1990b; Arkeryd, 1992; Lions, 1993). The main difficulties in tackling this problem seem to lie with large velocities. For this reason, a modified Boltzmann equation was recently introduced (Arkeryd and Cercignani, 1993), in which all the collisions such that the sum of the squares of two colliding molecules is larger than $m^2$ are cut off (here $m$ is an assigned positive constant). The only place where the cutoff was used was in the entropy estimate and the need for the cutoff disappears when the temperature is constant. Thus the abovementioned paper (Arkeryd and Cercignani, 1993) contains also a slightly different proof of Hamdache's theorem, with an extension to more general boundary conditions, to a more detailed study of the boundary behavior, and for the full class of collision operators of the DiPerna-Lions existence context.

A central observation for these proofs was an inequality introduced long time ago (Darrozès and Guiraud, 1966) and subsequently discussed by several authors (Cercignani and Lampis, 1971; Cercignani, 1972; Truesdell and Muncaster, 1980; Cercignani, 1988). We shall state this inequality in the form of a lemma:

**Lemma 0**: If Eqs. (2.1), (2.2), (2.3) and (2.4) hold, then:

$$\int \xi\cdot n\gamma f \log\gamma f d\xi \leq -\beta_w \int \xi\cdot n|\xi|^2 \gamma f d\xi \qquad \text{(a.e. in } t \text{ and } x \in \partial\Omega) \tag{2.5}$$

where $\beta_w$ is the inverse temperature of the wall, evaluated at the point $x \in \partial\Omega$. Unless the kernel in Eq. (2.1) is a delta function, equality holds in Eq. (2.9) if and only if the trace $\gamma f$ of $f$ on $\partial\Omega$ coincides with $M_w$ (the wall Maxwellian).

For a proof see Refs. 18 and 19.

If the wall is moving, the above relations hold in the reference frame of the wall. Then, since the Maxwellian $M_w$ has a drift velocity $u_w$, if we want to adopt a reference frame, with respect to which the wall moves, then $\xi$ must be replaced by $\xi - u_w$ .

In a recent paper (Arkeryd and Maslova, 1994) the previous results (Arkeryd and Cercignani 1993) were extended to the noncutoff case at the price of introducing some restrictions on the kernel $K(\xi' \to \xi; x, t)$. One of the aims of this paper is to show that one of the conditions can be replaced by a more natural one thanks to the above inequality.

## 3. A historical remark

The discovery of the inequality (2.5) (Darrozès and Guiraud, 1966) is a very important turning point in the study of the boundary conditions for the Boltzmann equation. More controversial is the fact whether the paper by Darrozès and Guiraud contains a proof of the above Lemma without invoking properties extraneous to those of the kernel $K(\xi' \to \xi; x, t)$.

In fact our authors say (the notation is changed in order to agree with the notation of the present paper):

"Des considération physiques évidentes impose sur $K$ deux restrictions:

(i) il existe une distribution maxwellienne $M_w$ (et une seule si le niveau de densité est fixé) telle que $M_w^+ = KM_w^-$;

(ii) si la paroi est non poreuse, il en resulte pour $K$ une relation qu'il est facile d'expliciter. "

Thus they seem to claim that the relation mentioned under (ii) is a consequence of obvious physical considerations. When reading their paper I understood that (ii) was inequality (2.5) and the obvious physical considerations were based on the Second Law of Thermodynamics. A few years later I tried to prove Lemma 0 as a mathematical consequence of (2.2), (2.3) and (2.4). The proof was initially obtained (Cercignani and Lampis, 1971) by first showing that $K$ has a contraction property in a weighted $L^p$ and then letting $p \to 1$ in a suitable way. Subsequently I obtained a direct proof based on convexity (Cercignani, 1972).

Then it occurred to me that there was a reference to convexity in the paper under discussion (Darrozès and Guiraud, 1966). In fact a few lines after the passage quoted above the authors say (again, the notation is changed):

"Introduisons la fonction convexe définie pour $f \geq 0$, égale á

$$I(f) = f \log(f/M_w) - f + M_w$$

et posons

$$\Sigma(f) = \tilde{K}[I(f)] - I(\tilde{K}f)$$

oú $\tilde{\psi}^+$ est une fonction du type +, obtenue à partir de $\psi^-$ par rèflection à partir de $\xi \cdot n = 0$. On a $\Sigma \geq 0$ avec égalité si et seulement si $f/M_w$ est une constante."

When reading this again, after obtaining my proof, I had the impression that reference was made to a mathematical property rather than to physical considerations. But was this a re-interpretation or not? I was still puzzzled by the previous passage that I have quoted. More recently, I understood that in the above passage the "relation easy to make explicit" must be mass conservation (as suggested by the fact that the wall is asumed to be non-porous) and not (2.5). Thus we must credit Darrozès and Guiraud with the idea of the proof although their presentation is not so explicit.

## 4. The results of Arkeryd and Cercignani

Before discussing the existence theorem one needs to recall some trace results giving the $L^1$ regularity of the trace of $f$ on the boundary and study the semigroup generated by the free streaming operator, including a sort of Green's formula, that will be used in Sect. 6. This is done in detail in another paper (Arkeryd and Cercignani, 1993) and will not be repeated here.

In order to deal with the existence theorem in a vessel at rest, with a temperature that varies from one point to another, it is convenient to remark that there is a Maxwellian naturally associated with the problem at each point of the boundary, *i. e.* the wall Maxwellian $M_w$; an exception is offered by specularly reflecting boundaries, which will not be considered here because they have no temperature associated with the boundary. Eq. (2.5) gives (for smooth solutions):

$$\int \xi \cdot n\gamma f \log \gamma f d\xi + \beta_w \int \xi \cdot n|\xi|^2 \gamma f d\xi \leq 0 \quad \text{(a.e. in } t \text{ and } x \in \partial\Omega) \tag{4.1}$$

For this reason Arkeryd and Cercignani consider an inverse temperature $\beta(x)$ with $\inf \beta(x) > 0$ which reduces to $\beta_w$ at each point of the boundary and otherwise depends smoothly on $x$ and the modified $H$-functional:

$$H = \int f \log f d\xi dx + \int \beta(x)|\xi|^2 f d\xi dx \tag{4.2}$$

In general $H$ will not decrease in time, as a consequence of the Boltzmann equation and inequality (4.1), because a simple calculation shows that

$$\frac{dH}{dt} \leq -\int \xi \cdot \frac{\partial \beta}{\partial x} |\xi|^2 f d\xi dx \tag{4.3}$$

Thanks to the truncation for large speeds (Arkeryd and Cercignani, 1993) $|\xi| \leq m$ and the right hand side of Eq. (4.3) is bounded by a constant $C$ given by

$$C = m^3 \parallel \frac{\partial \beta}{\partial x} \parallel_{L^\infty} \int f_0 d\xi dx \tag{4.4}$$

Thus $H$ is bounded by $H_0 + CT$ on $[0, T]$ if bounded initially by $H_0$.

This is the only point where the truncation for large speeds is needed; further, the truncation can be dispensed with if $\beta$ is constant because the left hand side of Eq. (4.3) is bounded by 0.

Eq. (4.2) implies (Arkeryd and Cercignani, 1933) that both $\int f|\log f| d\xi dx$ and $\int |\xi|^2 f d\xi dx$ are separately bounded in terms of the initial data.

I shall use the following notation:

$$< f, g >= \int_{\mathcal{D}} fg dx d\xi dt \tag{4.5}$$

$$< f, g >_{\pm} = \int_{\partial E^{\pm}} f g d\sigma^{\pm} \tag{4.6}$$

$$< f, g >_{t} = \int_{\Omega \times R^3} f(t, x, \xi) g(t, x, \xi) dx d\xi \tag{4.7}$$

$$(f, g)_{\pm} = \int_{\pm \xi \cdot n > 0} f g |\xi \cdot n(x)| d\xi. \tag{4.8}$$

I also define the backward and forward stay times as

$$t^{+} = t^{+}(x, \xi, t) = \inf\{s > 0; x - s\xi \in \partial\Omega\} \tag{4.9}$$

$$t^{-} = t^{-}(x, \xi, t) = \inf\{s > 0; x + s\xi \in \partial\Omega\} \tag{4.10}$$

with the related quantities

$$s^{+}(x, \xi, t) = \min\{t, t^{+}(x, \xi, t)\}, \tag{4.11}$$

$$s^{-}(x, \xi, t) = \min\{T - t, t^{-}(x, \xi, t)\}. \tag{4.12}$$

It is convenient to use the mappings

$$R^{s} : \overline{\mathcal{D}} \to \overline{\mathcal{D}}$$

with

$$R^{s}(x, \xi, t) = (t + s, x + s\xi, \xi) \tag{4.13}$$

to define

$$f^{\#}(s, x, \xi, t) = f^{\#}(s) = f(R^{s}((x, \xi, t))). \tag{4.14}$$

As hinted at in Sect. 1, throughout this paper I will use the equivalent concepts of exponential, mild, and renormalized solutions as defined by Di Perna and Lions and such solutions will be found as limits of functions solving truncated equations.

The definitions of these solutions require some comment because of the boundary conditions, which are not satisfied exactly but only in the form of an inequality, as Hamdache first pointed out in 1991. This aspect of the matter has been recently discussed in some detail (Arkeryd and Maslova, 1994). The basic point is that, when approximating a solution with a sequence, we partially lose control upon the traces, that can only be shown to tend to measures $\mu^{\pm}$ of the spaces $\mathcal{M}^{\pm}$ of $\sigma$-finite measures defined on the $\sigma$-algebras $\mathcal{B}^{\pm}$ of Borel sets from $E^{\pm}$ . Each of these measures can be decomposed into a part completely continuous $\mu_c^{\pm}$ with density $f^{\pm}$ with respect to the Lebesgue measure $d\sigma^{\pm}$, and a singular part $\mu_s^{\pm}$. The measures $\mu_s^{\pm}$ satisfy the boundary conditions:

$$\mu^{+} = K\mu^{-} \tag{4.15}$$

where $K$ can be defined for measures via

$$< \varphi, K\mu^{-} >_{+} = < K^{*}\varphi, \mu_{-} >_{-} \tag{4.16}$$

where the adjoint operator $K^*$ is defined by:

$$K^*\varphi(x,\xi',t) = \int_{\xi\cdot n(x)>0} \varphi(x,\xi,t)K(\xi' \to \xi;t,x)|\xi\cdot n(x)|/|\xi'\cdot n(x)|d\xi. \tag{4.17}$$

and assumed to carry $C_0(E^+)$ into $C_b(E^-)$.

The traces of a solution $\gamma^\pm f$ will only satisfy

$$\frac{d\mu^\pm}{d\sigma^\pm} \geq \gamma^\pm f \tag{4.18}$$

Then we can introduce the following definitions:

**Definition 4.1.** $f$ is a mild solution of (1.1-3) if

$$f \in L^1(\mathcal{D}), \quad f \geq 0, \quad (Q^\pm)^\# \in L^1([0, s_-(x,\xi,t)]) \tag{4.19}$$

$$f^\#(s,x,\xi,t) = f^\#(\tau,x,\xi,t) + \int_\tau^s Q^\#(z,x,\xi,t)dz, \quad 0 \leq s \leq \tau \leq s^-(x,\xi,t) \tag{4.20}$$

$$f = f_0(x,\xi) \quad \text{in } E^0 \qquad (E^0 = (x,\xi,t \in \overline{\mathcal{D}}; t=0) \tag{4.21}$$

for a.e. $(x,\xi,t) \in E^\pm \cup E^0$, and there are $\mu^\pm \in \mathcal{M}^\pm$ satisfying (4.15) and (4.18).

**Definition 4.2** $f$ is a solution in exponential multiplier form (or exponential solution for short) of (1.1-3) if

$$f \in L^1(\mathcal{D}), \quad f \geq 0, \quad \nu(f) \in L^1_{\text{loc}}(\mathcal{D}) \tag{4.22}$$

$$f^\#(s,x,\xi,t) = [f_0(x,\xi)\chi^0 + \gamma_+ f\chi^+]\exp(-\int_0^s (\nu(f))^\#(z,x,\xi,t)dz) +$$

$$\int_0^s (Q^+)^\#(z,x,\xi,t)\exp(-\int_z^s (\nu(f))^\#(z',x,\xi,t)dz')dz, \quad 0 \leq s \leq s^-(x,\xi,t) \tag{4.23}$$

for a.e. $(x,\xi,t) \in E^\pm \cup E^0$, and there are $\mu^\pm \in \mathcal{M}^\pm$ satisfying (4.15) and (4.18). Here $\chi^+$ and $\chi^0$ denote the characteristic functions of $E^+$ and $E^0$.

**Definition 4.3** $f$ is a renormalized solution of (1.1-3) if

$$f \in L^1(\mathcal{D}), \quad f \geq 0, \quad \nu(f) \in L^1_{\text{loc}}(\mathcal{D}) \tag{4.24}$$

and $f$ is a weak solution (with test functions vanishing in the neighborhood of $E^\pm$) of

$$\Lambda \log(1+f) = \frac{Q(f,f)}{1+f} \qquad \text{in } \mathcal{D} \tag{4.25}$$

and there are $\mu^\pm \in \mathcal{M}^\pm$ satisfying (4.15) and (4.18).

The existence theorem reads as follows

**Theorem 4.1** . Let $f^0 \in L^1(\Omega \times \Re^3)$ be such that

$$\int f^0(1+|\xi|^2)d\xi dx < \infty; \qquad \int f^0|\log f^0|d\xi dx < \infty. \tag{4.26}$$

Then there is a solution $f \in C(\Re_+, L^1(\Omega \times \Re^3))$ of the Boltzmann equation such that $f(.,0) = f^0$, which also satisfies relations (4.7) and (4.8).

For the proof I refer to the original paper (Arkeryd and Cercignani, 1993).

It is interesting to study the boundary condition satisfied by these solutions and prove

**Theorem 4.2.** There is a solution as in Theorem 4.1, which satisfies

$$\gamma^+(f) \geq K(\gamma^- f) \quad \text{a. e. on } E^+ \quad , \tag{4.27}$$

For the proof, I refer again to the original paper. Please remark that the fact that we obtain an inequality is a consequence of the fact that we can only expect convergence of the traces of the approximating sequence to measures, as discussed above. In fact Eq. (4.27) follows by taking the completely continuous part of Eq. (4.16) .

Theorems 4.1 and 4.2 contain Hamdache's result and extend it. The extension is of interest for the study of the solutions of the Boltzmann equation when the boundaries drive the gas out of equilibrium. In order to obtain a realistic result one needs to remove the cutoff, as shown in a recent paper (Arkeryd and Maslova, 1994) and discussed in the next section.

## 5. The results of Arkeryd and Maslova

In this section we study the results of a recent paper (Arkeryd and Maslova, 1994), where the authors introduce a class of boundary operators for which (1.2) (2.1-3) hold, by restricting the adjoint operator $K^*$, but are able to avoid the cutoff for large velocities.

A better control of mass, energy and entropy for the distributions emerging from the wall are provided by the following conditions:

$$\text{There exists } K_2 > 0 \text{ such that } K^*|\xi \cdot n(x)| \geq K_2 \quad \text{(spreading condition)} \tag{5.2}$$

$$\text{There exists } K_3 \leq \infty \text{ such that } K^*|\xi|^2 \leq K_3 \quad \text{(energy condition)} \tag{5.3}$$

$$\text{There exists } K_4 < \infty \text{ and } \alpha \in [0,1) \text{ such that, for every } f \in L^1(\Gamma^-) \text{ with} f \geq 0,$$

$$\text{it holds } < Kf, \log(Kf/(f,1)_-) >_+ -\alpha \hat{H}^- \leq K_4(q_2^- + q) \quad \text{(entropy condition)} \tag{5.4}$$

Here

$$\hat{H}^- =< f, \log(f/(f,1)_-) >_- \quad q_j^{\pm} =< f, |\xi|^j >_{\pm} \quad q =< f, |\xi \cdot n| >_+ + < f, |\xi \cdot n| >_- \tag{5.5}$$

These conditions are reasonable for a linear operator, except for (5.4), which appears a bit unusual, since it is nonlinear, albeit homogeneous of first degree, in $f$. In the next section we shall discuss how to dispense with that condition by using Lemma 0 in a suitable way.

The other conditions have the following role (Arkeryd and Maslova, 1994):

1) Eq. (5.2) (together with a proper use of momentum balance) gives a control on the mass flow hitting the boundary.

2) Eq. (5.3) (together with 1)) gives a control on $q_2^+$.

3) Using 1) and 2) together with energy balance one obtains an a priori bound upon the energy without using the entropy estimates.

At this point Arkeryd and Maslova use Eq. (5.4) to bound entropy and entropy source. They also obtain bounds on $< f, |\log(f/(f,1)_-)| >_\pm$ which are related to entropy flows. To bound the latters, however, one should remove the denominator $(f,1)_-$, which does not appear to be an easy matter.

The following lemma holds (Arkeryd and Maslova, 1994):

**Lemma 5.1.** Assume Eqs. (2.2-3) and (5.2-4), together with

$$(f_0, \log f_0) \in L^1(\Omega) \tag{5.6}$$

$$(Q, \log f) \leq 0, \quad (Q, \psi) = 0 \quad \text{for } \psi = 1, \xi, |\xi|^2. \tag{5.7}$$

Then $f$ satisfies the inequality

$$-< Q, \log f > +H(T) + < f, |\log(f/(f,1)_-)| >_\pm \leq C(T), \tag{5.8}$$

with $C(T) > 0$ depending only on $f_0$ and on $K_2, K_3, K_4$.

Having these a priori bounds the authors proceed more or less as in a previous paper (Arkeryd and Cercignani, 1993), the main change being that they prefer to avoid the semigroups that were used there, and finally arrive at

**Theorem 5.2** Assume that

$$(1 + |\xi|^2) f_0, \quad f_0 \log f_0 \in L^1(\Omega \times R^3), \qquad f_0 \geq 0.$$

and Eqs. (5.2-4). Then there exists an exponential solution of (1.1-3) satisfying

$$f \in C([0,T], L^1(\Omega \times R^3)), \quad f \geq 0, \quad < f, 1 >_t = < f_0, 1 >_0;$$

$$(1 + |\xi|^2)\gamma^\pm f \in L^{1\pm}\ ; \tag{5.9}$$

$$\sup_{t \leq T}[< f, \ln f >_t + < f, |\xi|^2 >_t] + < e(f), 1 > \leq C(T). \tag{5.10}$$

Here

$$e(f) = \frac{1}{4} \int_{R^3} \int_{B+} (f' f'_* - f f_*) \log \frac{f' f'_*}{f f_*} B(\xi - \xi_*, n) d\xi_* dn,$$

## 6. A new result and a generalization

In this section I want to prove a new result which relaxes one of the assumptions of Arkeryd and Maslova and a generalization to the case of moving boundaries.

The new result is that Lemma 5.1 and (as a consequence) Theorem 5.2 hold without assuming (5.4), but only the compatibility with a Maxwellian, Eq. (2.4). According to Lemma 0, the inequality of Darrozès and Guiraud then holds. I shall prove

**Lemma 6.1.** Assume Eqs. (2.2-4) and (5.2-3) , together with $\beta_w \leq C_o(T)$ and

$$(f_0, \log f_0) \in L^1(\Omega) \tag{6.1}$$

$$(Q, \log f) \leq 0, \quad (Q, \psi) = 0 \quad \text{for } \psi = 1, \xi, |\xi|^2. \tag{6.2}$$

Then $f$ satisfies the inequality

$$-<Q, \log f> + H(T) \leq C(T), \tag{6.3}$$

with $C(T) > 0$ depending only on $f_0$ and on $K_2, K_3$.

*Proof.* Using Green's formula (Ukai, 1986; Cercignani and Arkeryd, 1993; Arkeryd and Maslova, 1994) and approximation:

$$-<Q, \log f> + H(T) + <f, \log f>_- \leq H(0) + <f, \log f>_+ . \tag{6.4}$$

Because of Lemma 0 this becomes

$$-<Q, \log f> + H(T) + <\beta_w f, |\xi|^2>_+ \leq H(0) + <\beta_w f, |\xi|^2>. \tag{6.5}$$

Thanks to the fact that $\beta_w$ is bounded, energy balance now gives (Arkeryd and Maslova 1994) that the last term is bounded by some $C(T)$. Hence

$$-<Q, \log f> + H(T) + <\beta_w f, |\xi|^2>_+ \leq C(T)$$

which implies that the three quantities in the left hand side are separately bounded. In particular, Eq. (6.3) follows.

The only part of the thesis of Lemma 5.1, which does not follow from the new assumptions is the boundedness of the entropy flows $<f, |\log(f/(f,1)_-)|>_\pm$. This part of the lemma is never used in the proof of theorem (5.2) and thus one can prove

**Theorem 6.2**. Assume that

$$(1+|\xi|^2)f_0, \quad f_0 \log f_0 \in L^1(\Omega \times R^3), \qquad f_0 \geq 0.$$

and Eqs. (2.2-4) and (5.2-3). Then there exists an exponential solution of (1.1-3) satisfying

$$f \in C([0,T], L^1(\Omega \times R^3)), \quad f \geq 0, \quad <f,1>_t = <f_0, 1>_0;$$
$$(1+|\xi|^2)\gamma^\pm f \in L^{1\pm} ; \tag{6.6}$$

$$\sup_{t \leq T}[<f, \ln f>_t + <f, |\xi|^2>_t] + <e(f), 1> \leq C(T). \tag{6.7}$$

where $e(f)$ is given by Eq. (5.11).

**Remark**. The above result applies also to the inhomogeneous boundary condition

$$\gamma^+ f(x,\xi,t) = \alpha f_+ + (1-\alpha)K\gamma^- f \quad (0 \leq \alpha \leq 1) \tag{6.8}$$

where $f_+ \geq 0$ is assigned.

I pass now to problems with moving boundaries. These are of some interest and do not seem to have been considered so far. The main difference is that in Eq. (1.2) $E^+$ varies with time, because $\partial\Omega$ does. As remarked in Sect. 2, all the relations concerning the kernel $K$ hold in the reference frame of the wall. Then, when the Maxwellian $M_w$ has a drift velocity $u_w$, if one wants to adopt a reference frame, with respect to which the wall moves, then $\xi$ must be replaced by $\xi - u_w^2$ . In particular, the indices + and − refer now to $(\xi - u_w) \cdot n > 0$ and $(\xi - u_w) \cdot n < 0$, respectively.

When we integrate the Boltzmann equation to obtain a priori inequalities, we obtain a factor $(\xi - u_w)\cdot n$ in place of $\xi \cdot n$, so that most of the changes compensate. The main difference arises in the entropy inequality, where a factor $|\xi - u_w|^2$ appears in place of simply $\xi^2$. The extra terms can be easily controlled, however, by means of the momentum balance equation (after scalar multiplication by a smooth vector-valued function $u(x,t)$, which reduces to $u_w$ on the wall)

Then we have the following

**Corollary 6.3** . Theorem 6.2 holds in the presence of moving walls as well.

## 7. External problems

For applications to aerospace problems, the external case is of paramount importance. This aspect had been treated (Ukai, 1986; Ukai and Asano, 1980) only in the perturbation framework described in Sect. 1, till the relevant tools for dealing with external problems for large data in an $L^1$ framework were provided (Lions, 1994).

In this situation $\Omega = \overline{\mathcal{O}}^c$, where $\mathcal{O}$ is a smooth bounded set in $\Re^3$. One provides an initial condition (1.3), boundary condition (1.2) on the boundary $\partial\Omega \times \Re^3 \times (0,T)$ and a condition at infinity

$$f \to M \quad \text{as} \quad |x| \to \infty \tag{7.1}$$

where $M$ is any assigned Maxwellian.

The results proved by Lions are based on the following a priori estimates

$$\sup_{t\in[0,T]} \int_{\Omega\times\Re^3} dx d\xi (f \log \frac{f}{M} + M - f) \leq C \tag{7.2}$$

$$\sup_{t\in[0,T]} \sup_{x_0\in\Omega} \int_{\Omega\cap(x_0+B^1)} dx \int_{\Re^3} d\xi f(1 + |v|^2 + |\log f|) \leq C \tag{7.3}$$

whenever $f_0$ satisfies

$$\int_{\Omega\times\Re^3} dx d\xi (f_0 \log \frac{f_0}{M} + M - f_0) \leq C_0 \tag{7.4}$$

for some positive constant $C = C(C_0, T)$. In the above $B^1$ is the unit ball.

Lions gives a sketch of the proofs of the following results in the case when $\partial\Omega$ is specularly reflecting:

**Theorem 7.1**. Let $f_0 \geq 0$ satisfy (7.4). Then there exists a global renormalized solution satisfying Eq. (2.1).

The result can be extended to other boundary conditions with the usual difficulty of the trace control. In fact we can apply the results of previous sections to $\Omega^R \times \Re^3$, where $\Omega^R = B^R \cap \Omega$ with $B^R$'s radius $R$ large enough and the following boundary condition on the artificial boundary:

$$\gamma^+ f = M \quad \text{on} \quad (\partial B^R \cap \Omega) \times \Re^3 \times (0,T) \tag{7.5}$$

which is a particular case of (6.8) for $\alpha = 1$ and $f^+ = M$. Then using the compactness properties of sequences of approximated solutions (see Theorem IV.2 in Lions's paper, with obvious changes) we can pass to the limit as $R$ goes to $\infty$. This can be easily performed because the proof of the a priori bound (7.2) is only modified because of an additional boundary term of the form

$$\int_{\partial B^R \cap \Omega} d\sigma \int d\xi \xi \cdot n(f \log \frac{f}{M} + M - f) =$$

$$- \int_{\partial B^R \cap \Omega} d\sigma \int_{\xi \cdot n < 0} d\xi |\xi \cdot n| (f \log \frac{f}{M} + M - f) \tag{7.6}$$

which is obviously negative, contributes to the time derivative. Then the estimates remain valid with constants independent of $R$.

It is to be remarked that no restrictions on the number of space dimensions arises here, while the perturbation theory (Ukai, 1986; Ukai and Asano, 1980) does not hold in two dimensions, because of the analogue of the Stokes paradox in kinetic theory (Cercignani, 1967, 1988 and 1990a).

## 8. Concluding remarks

We have surveyed the initial-boundary value problem for the Boltzmann equation (including the recent results proved by P. L. Lions). We have stressed the role played in these problems by the inequality (2.5) introduced by Darrozès and Guiraud and proved two new results:

1) the existence theorem of Arkeryd and Maslova holds without assuming (5.4), but only the compatibility with a Maxwellian, Eq. (2.4), which together with mass conservation at the wall implies the inequality of Darrozès and Guiraud;

2) the same theorem can be extended to the case of moving walls.

It appears that the subject has reached a certain maturity. Only difficult problems (such as smoothness, uniqueness, asymptotic behavior for long times or small mean free paths, steady solutions), which seem to require significantly new ideas, appear to remain open.

Another problem, which might be easier to solve, is to find conditions under which equality holds in (4.18).

## References

Arkeryd, L. (1992): "On the strong $L^1$ trend to equilibrium for the Boltzmann equation", *Studies in Appl. Math.* **87**, 283-288.

Arkeryd, L., Cercignani C., Illner, R.(1991): "Measure solutions of the steady Boltzmann equation in a slab", *Commun. Math. Phys.* **142**, 285-296 .

Arkeryd, L., Cercignani, C. (1993): "A global existence theorem for the initial-boundary value problem for the Boltzmann equation when the boundaries are not isothermal", *Arch. Rat. Mech. Anal.* **125**, 271-288.

Arkeryd, L., Maslova, N. (1994): "On diffuse reflection at the boundary for the Boltzmann equation and related equations" , Preprint ISSN 03407-2809 (University of Göteborg, Göteborg).

Cercignani, C. (1967): "Stokes paradox in kinetic theory", *Phys Fluids* **11, 303-308.**

Cercignani, C. and Lampis, M. (1971): "Kinetic models for gas-surface interactions", *Transp. Th. Stat. Phys.* **1**, 101-114.

Cercignani, C. (1972): "Scattering kernels for gas-surface interactions", *Transp. Th. Stat. Phys.* **2**, 27-53.

Cercignani, C. (1988): *The Boltzmann Equation and its Applications* (Springer, New York).

Cercignani, C. (1990a): *Mathematical Methods in Kinetic Theory*, 2nd edition (Plenum Press, New York).

Cercignani, C. (1990b): "Equilibrium states and trend to equilibrium in a gas according to the Boltzmann equation", *Rend. Mat. Appl.* **10**, 77-95.

Darrozés, J.-S., Guiraud, J.-P. (1966): "Généralisation formelle du théorème $H$ en présence de parois. Applications", *C.R.A.S. Paris* **A262**, 1368-1371.

Desvillettes, L. (1990): "Convergence to equilibrium in large time for Boltzmann and BGK equations", *Arch. Rat. Mech. Analysis* **110**, 73-91.

DiPerna, R., Lions, P.-L. (1989): "On the Cauchy problem for Boltzmann equations", *Ann. of Math.* **130**, 321-366.

Guiraud, J.-P. (1975): "An $H$ theorem for a gas of rigid spheres in a bounded domain", in *Théories cinétiques classiques et rélativistes*, G. Pichon, Ed., 29-58, CNRS, Paris.

Hamdache, K. (1992): "Initial boundary value problems for Boltzmann equation. Global existence of weak solutions.", *Arch. Rat. Mech. Analysis* **119**, 309-353.,

Kato, T. (1966): *Perturbation Theory of Linear Operators* (Springer, New York).

Lions, P.-L. (1993): "Compactness in Boltzmann's equation via Fourier integral operators and applications. I." *Cahiers de Mathématiques de la décision* $n^0$ *9301* (CEREMADE, Paris).

Lions, P.-L. (1994): "Conditions at infinity for Boltzmann's equation" , Comm. Partial Diff. Equations **19**, 335-367 (1994).

Shizuta, Y., Asano, K. (1977): "Global solutions of the Boltzmann equation in a bounded convex domain", *Proc. Japan Acad.* **53A**, 3-5.

Truesdell, C., Muncaster, R. G. (1980): *Fundamentals of Maxwell's Kinetic Theory of a Simple Monatomic Gas* (Academic Press, New York).

Ukai, S. (1986): "Solutions of the Boltzmann equation" in *Pattern and Waves. Qualitative Analysis of Nonlinear Differential Equations,* 37-96 .

Ukai, S., Asano, K. (1980): "On the initial boundary value problem of the linearized Boltzmann equation in an exterior domain" *Proc. Japan Acad.* **56**, 12-17.

# Diffusion Approximation and Arnold's "Cat Map"

FRANÇOIS GOLSE
Université Paris VII
& Ecole Normale Supérieure, D.M.I.
45 rue d'Ulm, F75005 Paris

**Abstract**

In this article a diffusion equation is obtained as a limit of a reversible kinetic equation with a proper scaling. That the limiting dynamics be a diffusion comes mainly from two facts: a) the particles undergo many collisions per unit time with the boundary of the physical domain and b) the transformation of velocities in the collision process has strong mixing properties. The transformation used in the model presented here is based on Arnold's "Cat Map", whose ergodic properties can be investigated completely with elementary techniques of Fourier series.

## 1. Introduction

One of the main difficulties in the derivation of kinetic theory or hydrodynamics from hamiltonian dynamics (which is the corresponding "first principle" model) is that the latter can be considered a "reversible dynamics", whereas the former are not, for they satisfy Carnot's principle (roughly speaking, entropy is produced until the system reaches an equilibrium).

In the course of the present paper, we shall not abide by the classical definition of reversibility in Mechanics (that is, in a mechanical system evolved during a lap of time $t$, if the velocities of each point in the system are changed into their opposite while the positions of these points are left unchanged, and if the resulting system further evolves during the same lap of time $t$, its final configuration will be the initial configuration of the first system). In fact, for hamiltonian dynamics, the property of reversibility follows from the following mathematical structure: a) in most cases, the Hamilton equations generate a one-parameter group of transformations of the phase space which preserves the volume element according to Liouville principle (in fact, hamiltonian dynamics preserves the symplectic structure of the phase space, and the volume element is an external power of the symplectic form); b) in hamiltonian systems arising in Mechanics, the phase space is the cotangent bundle of the manifold of admissible positions and the Hamilton functions are even in the impulsion (fiber) variables. Hence, if $G_t$ denotes the one-parameter group generated by the Hamilton system, one has $G_{-t}(p,q) = G_t(-p,q)$.

Instead, we shall call "reversible" any system of differential equations generating a volume preserving one-parameter group (defined for all times). For example, it is clear that the heat equation is not reversible in this sense. Indeed, the solution of the Cauchy problem for the heat equation

$$\partial_t u(t,x) - \tfrac{1}{2}\Delta_x u(t,x) = 0\,, \qquad u(0,x) = f(x)\,,\ x \in \mathbf{R}^n$$

is given by

$$u(t,x) = \left(e^{t\Delta_x} f\right)(x) = \frac{1}{(2\pi t)^{n/2}} \int_{\mathbf{R}^n} e^{-\frac{\|x-y\|^2}{2t}} f(y)dy$$

and it is clear from this formula that for all $t > 0$, $e^{t\Delta_x}$ maps $L^2(\mathbf{R}^n)$ into $C^\infty(\mathbf{R}^n)$ (in fact the expression for $u(t,x)$ above can be extended to an entire holomorphic function of $x \in \mathbf{C}^n$) and hence cannot be extended as a strongly continuous one-parameter group on $L^2(\mathbf{R}^n)$. We mention in passing that the inversion problem of the heat semigroup $e^{t\Delta_x}$ has been considered by many authors in connection with microlocal analysis; the problem is to find an explicit formula for $f$ knowing the values of $e^{t\Delta_x} f$ for all $t > 0$ in a "tuboid" around $\mathbf{C}^n$ of the form $\{x + iy \mid \|y\| = \rho\}$ with some fixed value of $\rho$. It was solved by Lebeau (although it is known as "Lebeau's inversion formula for the Fourier-Bros-Iagolnitzer transform", see Lebeau 1985) in the case of $\mathbf{R}^n$, and by Golse-Leichtnam 1994 in the case of a compact, real-analytic riemannian manifold; see also Hörmander 1983, who does not refer to Lebeau for this formula. In any case, these formulas do not define "$e^{-t\Delta_x}$" in any way, for it is necessary to know the whole evolution of $f$ under the heat semigroup to recover $f$ itself. We shall therefore consider the heat equation as the paradigm of irreversible systems. The following property should also be noted: the function $t \mapsto \|e^{t\Delta_x} f\|_{L^2}$ is decreasing on $\mathbf{R}^+$; it cannot be constant on any interval of time unless $e^{t\Delta_x} f$ is constant for all $x \in \mathbf{R}^n$ and for any further period of time. Hence, $t \mapsto -\|e^{t\Delta_x} f\|_{L^2}^2$ should be considered as the relevant notion of entropy for the heat equation.

Now consider instead of the heat equation a Liouville equation of the form

$$\partial_t u(t,x) + \{H, u\}(t,x) = 0\,, \qquad u(0,x) = f(x)\,,\ x = (p,q) \in \mathbf{R}^{2n}$$

and suppose that the one-parameter group generated by the hamiltonian vector field $\{H, \cdot\}$ is globally defined for all times and all initial configuration $(p,q)$. Since the symplectic form $\omega = \sum_{k=1}^n dp_k \wedge dq_k$ is invariant under the one-parameter group generated by $\{H, \cdot\}$ (Liouville's theorem), the volume element $dpdq = (n!)^{-1}\omega \wedge \ldots \wedge \omega$ ($n$ times) also is, and therefore the function $t \mapsto \|u\|_{L^p}$ is constant for all times for all values of $1 \le p \le +\infty$ whenever $f \in L^p(\mathbf{R}^n)$.

We shall consider in this paper a model which looks very much like the latter, except that it is not a "smooth" hamiltonian system, but a "broken" hamiltonian system (for example in the sense of Hörmander 1983), which generates a volume-preserving one-parameter group globally defined for all times and on the whole phase space. It shall be proved that, under a convenient scaling of space and time, the solutions of this system converge to solutions of the heat equation.

In view of what has just been said on the preservation of volume by the hamiltonian flow and on the decay of the $L^2$ norm under the heat semigroup, such a result may seem paradoxical. However it is not, for the convergence holds only in some weak topology, ie. a topology for which only linear functional are continuous. In particular, the $L^2$ norm being not a linear functional, it follows that the $L^2$ norm of the limiting solution is not the limit of the (constant) $L^2$ norm of the solutions of the scaled Liouville equations. In other words, it is essential for our purpose to deal with these weak topologies: with our definition of reversibility, no reversible dynamics could admit the heat equation as limiting dynamics in a topology for which the $L^2$ norm is continuous, for the heat semigroup strictly contracts the $L^2$ norm of any nonconstant initial data. In particular, asymptotic expansions in power series of a small parameter cannot be used in this problem, for such expansions are, by their own structure, strongly convergent (that is, with respect to some $L^p$ norm) to their 0th order term.

To the best of our knowledge, there exist few examples where a reversible dynamics has been demonstrated to converge to a diffusion limit; probably the most interesting example is the Bunimovich-Sinai 1980 result, which shows that the process of positions for the periodic Lorentz gas converges to a brownian motion (see also Bunimovich-Sinai-Chernov 1991). Their analysis is based on an extremely complicated application of ergodic theory (in particular the coding of the system by a "symbolic" dynamics); however, the result is fascinating in itself, for the "ergodization" is produced only by the geometry of the system; in particular, there is no randomness introduced a priori (for example in the collision process as in Bardos-Dumas-Golse 1992). In the same spirit, one should consider Ratner's proof (see Ratner 1973) of the central limit theorem for geodesic flows on compact manifolds with negative curvature and Knauf's example (see Knauf 1987) of a quantum hamiltonian corresponding to a periodic array of singular potentials (which is similar to a Lorentz gas model, except that the interactions are not that of hard spheres). A noteworthy difference with Bunimovich-Sinai's result is that only the central limit theorem is established and that there is no claim to establish a limiting brownian dynamics in both Ratner's and Knauf's work.

The example below is also based on the ergodic properties of some transformation (Arnold's "Cat Map", see Arnold 1980), but these ergodic properties are easily established with Fourier analysis as in Katznelson's proof that hyperbolic automorphisms of tori are Bernoulli systems (Katznelson 1970), whereas the Bunimovich-Sinai system does not have any such obvious coding, and requires deep and complicated techniques.

To conclude this somewhat vague and lengthy introduction and before going into the description of the model that we shall consider, I wish to express my heartfelt thanks to J.-P. Guiraud for having stimulated my interest for Fluid Mechanics when I was a student and attended his lectures, that is, at an age when young mathematicians are usually attracted by abstraction "per se" and fail to consider "macroscopic" Physics a subject worthy of interest. It is my pleasure to offer him this modest contribution.

## 2. Description of the Model

The model considered here is that of a Knudsen gas (ie. a gas with no interparticle collisions) of like point particles with unit mass, confined between two infinite parallel plates. The vertical velocities of the particles are $\pm c$. Their horizontal velocities $ca(\omega)$ are parametrized by $\mathbf{T}^2 = \mathbf{R}^2/(2\pi\mathbf{Z})^2$. Whenever a particle hits the top or bottom plate, their vertical velocities are changed into their opposite while their horizontal velocities are modified by the right action of an hyperbolic automorphism of $\mathbf{T}^2$. More precisely the following notations are used : the space position of the particles is $(x,z) \in \mathbf{R}^d \times (0,h)$; the vertical component of the velocity of the particles is $\pm c$; the horizontal component of this velocity is given by $ca(\omega)$, $\omega \in \mathbf{T}^2$ with $a : \mathbf{T}^2 \to \mathbf{R}^d$ denoting a smooth enough mean zero vector field. The nonnegative functions $f_+(t,x,z,\omega)$ (resp. $f_-(t,x,z,\omega)$) represent the density of particles which at time $t$ and point $(x,v)$ have the velocity $(ca(\omega),+c)$ (resp. $(ca(\omega),-c)$).

The following hyperbolic automorphism $T$ of the torus (Arnold's "Cat Map") defined by

$$T\begin{pmatrix}\omega_1\\ \omega_2\end{pmatrix} = \begin{pmatrix}2&1\\ 1&1\end{pmatrix}\begin{pmatrix}\omega_1\\ \omega_2\end{pmatrix} \quad (\text{mod. } 2\pi). \tag{1}$$

will be the only case treated here; the method developped in the present paper would apply to any hyperbolic automorphism of $\mathbf{T}^n$. The map $T : \mathbf{T}^2 \to \mathbf{T}^2$ is one-to-one and $C^\infty$; it preserves the measure $d\omega_1 d\omega_2/4\pi^2$ and its inverse (which also is a $C^\infty$ map) is given by

$$T^{-1}\begin{pmatrix}\omega_1\\ \omega_2\end{pmatrix} = \begin{pmatrix}1&-1\\ -1&2\end{pmatrix}\begin{pmatrix}\omega_1\\ \omega_2\end{pmatrix} \quad (\text{mod. } 2\pi). \tag{2}$$

The densities $f^\pm$ satisfy the Liouville equations

$$\partial_t f^\pm + ca(\omega)\cdot\partial_x f^\pm \pm c\partial_z f^\pm = 0\,, \quad x \in \mathbf{R}^d\,,\ 0<z<h\,,\ \omega \in \mathbf{T}^2\,, \tag{3}$$

with the following boundary conditions on the plates:

$$f^+(t,x,0,\omega) = f^-(t,x,0,T\omega)\,, \quad x \in \mathbf{R}^d\,,\ \omega \in \mathbf{T}^2\,, \tag{4a}$$

$$f^-(t,x,h,\omega) = f^+(t,x,h,T\omega)\,, \quad x \in \mathbf{R}^d\,,\ \omega \in \mathbf{T}^2\,. \tag{4b}$$

and their value at $t=0$ is given by the following initial condition

$$f^\pm(0,x,z,\omega) = \phi(x)\,, \quad x \in \mathbf{R}^d\,,\ 0<z<h\,,\ \omega \in \mathbf{T}^2\,; \tag{5}$$

which is compatible with an approximation, as $h \to 0$, by an horizontal diffusion, and which avoids the appearance of an initial layer in the limiting process.

Since the densities $f^\pm$ satisfy equation (3), they are constant along the characteristic lines of the system; hence

$$f^\pm(t,x,z,\omega) = \phi(x - h\sum_{k=0}^{[ct/h]} a(T^k\omega)) + O(h)\,. \tag{6}$$

The asymptotic limit leading to an horizontal diffusion will be obtained by letting $h$ go to zero and observing the system for large positive times. A small parameter $\epsilon$ being introduced, $h$ is changed into $\epsilon h$ (thus letting the collision frequency going to $\infty$) and $t$ into $t/\epsilon$. The problem of interest (3)-(4a,b)-(5) becomes

$$\epsilon^2\partial_t f_\epsilon^\pm + c\epsilon a(\omega)\cdot\partial_x f_\epsilon^\pm \pm c\partial_z f_\epsilon^\pm = 0\,, \quad x \in \mathbf{R}^d\,,\ 0<z<h\,,\ \omega \in \mathbf{T}^2\,. \tag{7}$$

$$f_\epsilon^+(t,x,0,\omega) = f_\epsilon^-(t,x,0,T\omega)\,, \quad x \in \mathbf{R}^d\,,\ \omega \in \mathbf{T}^2\,, \tag{8a}$$

$$f_\epsilon^-(t,x,h,\omega) = f_\epsilon^+(t,x,h,T\omega)\,, \quad x \in \mathbf{R}^d\,,\ \omega \in \mathbf{T}^2\,, \tag{8b}$$

$$f_\epsilon^{\pm}(0,x,z,\omega) = \phi(x)\,, \quad x \in \mathbf{R}^d\,,\ 0 < z < h\,,\ \omega \in \mathbf{T}^2\,. \tag{9}$$

Its solution is given by (see (6))

$$f_\epsilon^{\pm}(t,x,z,\omega) = \phi(x - \epsilon h \sum_{k=0}^{[ct/\epsilon^2 h]} a(T^k\omega)) + O(\epsilon)\,; \tag{10}$$

therefore most of the analysis is reduced to studying the limit, as $\epsilon \to 0$, of the expression

$$\psi_\epsilon(t,x,\omega) = \phi(x - \epsilon h \sum_{k=0}^{[ct/\epsilon^2 h]} a(T^k\omega)). \tag{11}$$

## 3. Main Results

The following notation will be systematically used below:

$$\langle F \rangle = \frac{1}{4\pi^2}\int_{\mathbf{T}^2} F(\omega)d\omega\,.$$

The formula $U_T f = f \circ T$ with the mapping $T$ given by (1) defines an operator $U_T$ in the space $L^2(\mathbf{T}^2)$; this operator is unitary and therefore its adjoint is given by $U_T^* f = U_T^{-1} f = f \circ T^{-1}$.

**Definition 1.** A coboundary is an element of $\mathrm{Im}(I - U_T)$. Two functions $f$ and $g$ belonging to $L^2(\mathbf{T}^2)$ are said to be cohomologous if and only if $f - g$ is a coboundary and this equivalence relation will be denoted $f \sim g$.

The reader curious of the above terminology is referred to a classical source on the basics of group cohomology (J.-P. Serre's "Corps Locaux", for example — Serre 1962). We use this terminology just because it is the classical one in ergodic theory and we never have to appeal to any argument from homological algebra.

The next proposition describes the elementary properties of what will be, in the limit $\epsilon \to 0$, the diffusion coefficient. It is essentially based on the ergodic and mixing properties of the mapping $T$.

**Proposition 2.**
1) Any function $a \in L^2(\mathbf{T}^2)$ which satisfies the relation $a = a \circ T$ is constant and the subspace $\mathrm{Im}(I - U_T)$ is dense in the space of functions $a \in L^2(\mathbf{T}^2)$ such that $< a >= 0$ (notice that this space is invariant under $T$).

Let $s > 0$ and $a : \mathbf{T}^2 \to \mathbf{R}^d$ in the Sobolev class $H^s(\mathbf{T}^2)$, with mean value $\langle a \rangle = 0$
2) One has:

$$D(a) = \tfrac{1}{2}\langle a^2 \rangle + \sum_{k\geq 1}\langle a \circ T^k \otimes a \rangle = \tfrac{1}{2}\lim_{N\to\infty}\langle \left(\frac{1}{\sqrt{N}}\sum_{k=0}^{N-1} a \circ T^k\right)^{\otimes 2} \rangle \geq 0, \tag{12}$$

where the series

$$\sum_{k\geq 1}\langle a\circ T^k\otimes a\rangle$$

is normally convergent.
3) Let $\xi\in\mathbf{R}^d$ and $b\in H^s(\mathbf{T}^2;\mathbf{R}^d)$ such that $\langle b\rangle=0$. If $a\cdot\xi\sim b\cdot\xi$ then $\xi\cdot D(a)\xi=\xi\cdot D(b)\xi$.
4) For any $\xi\in\mathbf{R}^d$, the following properties are equivalent:
  i) $D(a)\xi=0$;
  ii) $\xi\cdot D(a)\xi=0$;
  iii) The sequence of functions $f_N\cdot\xi$ of $L^2(\mathbf{T}^2)$

$$f_N\cdot\xi=\sum_{k=1}^{N}(a\circ T^k)\cdot\xi$$

is (uniformly with respect to $N$) bounded in $L^2(\mathbf{T}^2)$;
  iv) The function $a\cdot\xi$ is a coboundary.

The main result of this paper is the following

**Theorem 3.** Let $s>d/2$, $a:\mathbf{T}^2\to\mathbf{R}^d$ be in the Sobolev class $H^s(\mathbf{T}^2)$ with mean value $\langle a\rangle=0$ and let $\phi\in C^2(\mathbf{R}^d)$ be an initial data. Denote by $u(t,x)$ the solution of

$$\partial_t u=hc\nabla_x\cdot(D(a)\nabla_x u)\,,\quad u(0,x)=\phi(x)\,. \tag{13}$$

Then the family of functions $f_\epsilon^\pm$ defined by (7), (8) and (9) converges to $u(x,t)$ as $\epsilon\to 0$ in the following sense: for any $\tau>0$ and any compact $K\subset\mathbf{R}_x^d$

$$\langle f_\epsilon^\pm(t,x,z,\omega)\rangle\to u(t,x)\,,\ C^0([0,\tau]\times K\times\mathbf{T}^2\times]0,h[))\,, \tag{14a}$$

$$f_\epsilon^\pm(t,x,z,\omega)\to u(t,x)\,,\ C^0([0,\tau],w^*-L^\infty(\mathbf{R}^d\times\mathbf{T}^2)) \tag{14b}$$

Furthermore with $\psi_\epsilon(x,t)$ defined by the formula (11) one has:

$$\|f_\epsilon^\pm-\psi_\epsilon\|_{L^\infty(\mathbf{R}\times\mathbf{R}^d\times(0,h)\times\mathbf{T}^2)}=O(\epsilon)\,. \tag{15}$$

An analogous result has been proven by Denker-Philipp 1984 for suspensions of finite type subshifts under hölderian maps. It would be theoretically possible to reduce the present analysis to this situation by coding the mapping $T$ with a Markov partition. However our goal is to produce a proof of a diffusion limit as explicit as possible and which in the present case uses only elementary techniques (Fourier series expansions instead of Markov partition as a coding of the system).

## 4. Method of Proof

Complete proofs of the statements above will be provided in Bardos-Golse-Colonna 1994. Here, I shall only insist on the most significant steps and outline the general method used in this reference.

*Proof of Proposition 2*

Proposition 2 is a compilation of rather classical results and does not contain anything really new, except for minor modifications.

As we said above, the ergodicity properties of $T$ can be studied with the help of Fourier series. Indeed the properties of the dynamical system induced by the map $T$ on the torus $\mathbf{T}^2$ are most conveniently translated into those of the dynamical system defined in $\mathbf{Z}^2$ (the dual group of $\mathbf{T}^2$) by the iteration of the matrix

$$M = \begin{pmatrix} 2 & 1 \\ 1 & 1 \end{pmatrix} .$$

The matrix $M$ is strictly hyperbolic, ie. has two (distinct) real eigenvalues given by

$$\lambda_+ = 1 + \theta , \quad \lambda_- = \lambda_+^{-1} , \quad \text{with } \theta = \frac{1+\sqrt{5}}{2} \tag{16}$$

and that the corresponding eigenvectors (generating the unstable and stable manifolds) are

$$e_+ = \mu \begin{pmatrix} \theta \\ 1 \end{pmatrix} , \; e_- = \mu \begin{pmatrix} 1 \\ -\theta \end{pmatrix} . \tag{17}$$

with $\mu = (1+\theta^2)^{-1/2}$. The vectors $(e_+, e_-)$ define an orthonormal basis. In this basis $M$ reduces the diagonal form

$$M \sim \begin{pmatrix} \lambda_+ & 0 \\ 0 & \lambda_- \end{pmatrix} .$$

In the next lemma we recall the following classical estimate (due to Kronecker).

**Lemma 4.** Let $\theta = \frac{1+\sqrt{5}}{2}$. Then:
(i) (best rational approximation for $\theta$)

$$\inf_{(p,q)=1} |\theta - \frac{p}{q}| \geq \frac{1}{(1+\sqrt{5})q^2} \tag{18}$$

(ii) (diophantine estimate for $\theta$)

$$\inf_{(p,q)\in\mathbf{Z}^2\setminus\{0\}} |q\theta - p| \geq \frac{1}{(1+\sqrt{5})\sup(|p|,|q|)} \tag{19}$$

**Proposition 5.** Let $0 \leq \chi(R)$ be a nonincreasing positive function such that

$$\lim_{R\to\infty} \chi(R) = 0. \tag{20}$$

Consider the class of functions:

$$H_\chi = \{f \in L^2(\mathbf{T}^2) \text{ st.} \sum_{|k_1|,|k_2|>R} |\hat{f}(k)|^2 \leq \chi(R)^2 \|f\|_2^2\} . \tag{21}$$

Then

i) **"Rate of Mixing":** For all $(f, g) \in H_\chi$ such that $\langle f \rangle = \langle g \rangle = 0$

$$|\langle f \circ T^n \cdot g \rangle| \leq \tfrac{1}{2\pi^2} \|f\|_2 \|g\|_2 \chi \left( C_0 \left( \tfrac{3+\sqrt{5}}{2} \right)^{n/2} \right), \text{ with } C_0 = \frac{1+\sqrt{5}}{\sqrt{2}}. \quad (22)$$

ii) **Property (H1): "Exponential mixing":** For all $s > 0$ and all $f \in H^s(\mathbf{T}^2)$ such that $\langle f \rangle = 0$, the self-correlation coefficient defined as

$$C_f(n) = \langle f \circ T^n \cdot f \rangle$$

satifies the decay estimate:

$$|C_f(n)| \leq C \|f\|_s^2 e^{-sn\alpha} \text{ with } \alpha = \log \left( \tfrac{3+\sqrt{5}}{2} \right). \quad (23)$$

The point 1) in proposition 2 follows from proposition 5 since mixing transformations are ergodic. Point 2 also follows from proposition 5 and the identity

$$\langle a^{\otimes 2} \rangle + 2 \sum_{m=0}^{N-1} \langle a \circ T^m \otimes a \rangle = \langle \left( \frac{1}{\sqrt{N}} \sum_{k=0}^{N-1} a \circ T^k \right)^{\otimes 2} \rangle + \frac{2}{N} \sum_{m=0}^{N-1} m \langle a \circ T^m \otimes a \rangle \quad (24)$$

in which one takes the limit as $N \to +\infty$. Hence

$$\langle a^{\otimes 2} \rangle + 2 \sum_{m=0}^{+\infty} \langle a \circ T^m \otimes a \rangle = \lim_{N \to +\infty} \langle \left( \frac{1}{\sqrt{N}} \sum_{k=0}^{N-1} a \circ T^k \right)^{\otimes 2} \rangle, \quad (25)$$

and this formula (classical in Statistical Physics under the name of "Green-Kubo formula") also demonstrates point 3).

As for point 4), it suffices to show that 4 ii) implies 4 iii) which in turn implies 4 iv) all other assertions being trivial. To prove that 4) ii) implies 4 iii) consider the function

$$r(z) = \tfrac{1}{2\pi} \sum_{k \in \mathbf{Z}} \langle \xi \cdot a \circ T^k \cdot \xi \rangle e^{ikz} \quad z \in \mathbf{T}^1, \quad \xi \in \mathbf{R}^d. \quad (26)$$

Observe that

$$r(0) = \tfrac{1}{2\pi} \sum_{k \in \mathbf{Z}} \langle \xi \cdot a \circ T^k \cdot \xi \rangle = \frac{1}{\pi} \xi \cdot D(a) \cdot \xi \quad (27)$$

and that, due to the decay estimate (23), $r(z)$ is holomorphic in the strip $B_{\alpha s} = \{z \in \mathbf{C} \text{ s.t. } |\Im z| < \alpha s\}$. Therefore one has:

$$\langle \left( \sum_{k=0}^{N-1} a \circ T^k \xi \right)^2 \rangle = \sum_{k=0}^{N-1} \sum_{l=0}^{N-1} \hat{r}(k-l) = \int_{-\pi}^{\pi} r(x) \left( \sum_{k=0}^{N-1} \sum_{l=0}^{N-1} e^{-i(k-l)x} \right) dx$$

$$= \int_{-\pi}^{\pi} \frac{r(x) + r(-x) - 2r(0)}{2 \sin^2 \frac{x}{2}} \sin^2(\tfrac{N}{2} x) dx + 2\pi N r(0). \quad (28)$$

The first integral term in the right hand side of (28) is bounded because $r \in C^2(\mathbf{T}^1)$. Therefore the relation

$$0 = \xi \cdot D(a) \cdot \xi = \pi r(0)$$

implies that the series

$$f_N \cdot \xi = \sum_{k=1}^{N} a \circ T^k \cdot \xi$$

is (uniformly with respect to $N$) bounded in $L^2(\mathbf{T}^2)$.

To prove that 4 iii) implies 4 iv), observe that the sequence $f_N \cdot \xi$ is, in any case, the "formal inverse" for the equation

$$a \cdot \xi = g^{\xi} - g^{\xi} \circ T\,. \tag{29}$$

Since the sequence

$$f_N \cdot \xi = \sum_{k=1}^{N} a \circ T^k \cdot \xi$$

is bounded in $L^2(\mathbf{T}^2)$, one can consider $f$, one of its weak $L^2(\mathbf{T}^2)$ limit points and, using lemma 4, one has, for any function $\phi \in C^{\infty}(\mathbf{T}^2)$

$$\langle (f_N - f_N \circ T)\phi\rangle - \langle \xi \cdot a\phi\rangle = -\langle \xi \cdot a \circ T^{N+1}\,\phi\rangle \to 0\,,\ \text{for } N \to \infty \tag{30}$$

which shows that $f - f \circ T = a \cdot \xi$.

*Proof of Theorem 3*

The proof of theorem 3 follows closely the proof the Ito formula for the Brownian motion, with however one major difference: the brownian motion has independent increments, which is not the case of the dynamics of interest here, for it is a purely deterministic dynamics. In any case the starting point is the Taylor formula at order three for the increment:

$$\begin{aligned}
&\langle \psi_\epsilon(t+\tau, x, \cdot, \cdot)\rangle - \langle \psi_\epsilon(t, x, \cdot, \cdot)\rangle \\
&= \langle \nabla\phi(x - \epsilon h \sum_{k=0}^{[ct/\epsilon^2 h]} a(T^k\omega)) \cdot \epsilon h \sum_{k=[ct/\epsilon^2 h]+1}^{[c(t+\tau)/\epsilon^2 h]} a(T^k\omega)\rangle \\
&+\tfrac{1}{2}\langle \nabla^2\phi(x - \epsilon h \sum_{k=0}^{[ct/\epsilon^2 h]} a(T^k\omega)) : \left(\epsilon h \sum_{k=[ct/\epsilon^2 h]+1}^{[c(t+\tau)/\epsilon^2 h]} a(T^k\omega)\right)^{\otimes 2} \rangle \\
&+O(\langle \left|(\epsilon h \sum_{k=[ct/\epsilon^2 h]+1}^{[c(t+\tau)/\epsilon^2 h]} a(T^k\omega))\right|^3 \rangle)\,.
\end{aligned} \tag{31}$$

The next task is to show that in the limit as $\epsilon \to 0$, one has the following decorrelation properties:

$$\langle \nabla\phi(x - \epsilon h \sum_{k=0}^{[ct/\epsilon^2 h]} a(T^k\omega)) \cdot \epsilon h \sum_{k=[ct/\epsilon^2 h]+1}^{[c(t+\tau)/\epsilon^2 h]} a(T^k\omega)\rangle$$

$$\simeq \langle \nabla\phi(x - \epsilon h \sum_{k=0}^{[ct/\epsilon^2 h]} a(T^k\omega))\rangle \cdot \langle \epsilon h \sum_{k=[ct/\epsilon^2 h]+1}^{[c(t+\tau)/\epsilon^2 h]} a(T^k\omega)\rangle = 0$$

(since $a$ is of mean 0), and

$$\tfrac{1}{2}\langle \nabla^2\phi(x - \epsilon h \sum_{k=0}^{[ct/\epsilon^2 h]} a(T^k\omega)) : \left(\epsilon h \sum_{k=[ct/\epsilon^2 h]+1}^{[c(t+\tau)/\epsilon^2 h]} a(T^k\omega)\right)^{\otimes 2} \rangle$$

$$\simeq \tfrac{1}{2}\langle \nabla^2\phi(x - \epsilon h \sum_{k=0}^{[ct/\epsilon^2 h]} a(T^k\omega))\rangle : \langle \left(\epsilon h \sum_{k=[ct/\epsilon^2 h]+1}^{[c(t+\tau)/\epsilon^2 h]} a(T^k\omega)\right)^{\otimes 2} \rangle$$

$$\simeq \tfrac{1}{2}\langle \nabla^2\phi(x - \epsilon h \sum_{k=0}^{[ct/\epsilon^2 h]} a(T^k\omega))\rangle : D(a)ch\tau\,.$$

In the case of the brownian motion, these decorrelation properties are obvious, for the brownian motion has independent increments. In the case dealt with here, independence is obtained only in the limit as $\epsilon \to 0$, and relies on the ergodic properties of the map $T$.

In order to demonstrate these decorrelation properties, we appeal to the following property of the map $T$, which is very similar to the Bernouilli property:

**Property (H2).** There exist two constants $\beta_0 > 0$ and $\beta_1$ such that for all $l, m \in \mathbf{N}$, $U \subset \{n, ..., n+l\}$, $V \subset \{n, ..., n+m\}$ and for all pair of trigonometric polynomials $P$, $Q$, of degree less than $R$, one has, for all $n \geq \beta_0 \log R + \beta_1$

$$\langle \prod_{k\in U} P \circ T^{-k} \prod_{k\in V} Q \circ T^k\rangle - \langle \prod_{k\in U} P \circ T^{-k}\rangle\langle \prod_{k\in V} Q \circ T^k\rangle = 0 \tag{32}$$

Our proof is based on the same method as that of proposition 5. The method is similar to the one used by Katznelson 1973, but here a more precise result is needed and proven.

In the course of the proof of the decorrelation properties above, and also in order to estimate the remainder in the Taylor expansion, one uses the following consequence of (H2).

**Property (H3)** For all $f \in W_\chi$ such that $\langle f\rangle = 0$

$$\frac{1}{\sqrt{N}} \sum_{k=0}^{N} f \circ T^k$$

is uniformly bounded (with respect to $N$) in $L^4(\mathbf{T}^2)$.

The proof follows the same line as the one indicated by Ratner 1973 (when the mapping $T$ is replaced by any $K$ system; here we derive it directly from property (H2)).

With properties (H1)-(H2)-(H3), it is relatively easy to give an elementary proof of Theorem 3, though at the expenses of somewhat lengthy and technical calculations.

## 5. Final Remarks

It would be extremely useful to have an explicit expression of $D(a)$. For $d = 1$ it can be easily obtained in term of the Fibonnacci sequence $F_n$

$$F_0 = 0,\ F_1 = 1,\ F_n = F_{n-1} + F_{n-2} \tag{33}$$

Observe that the minimal polynomial of the matrix $M$ is $X^2 - 3X + 1$ and introduce the matrix $P - I$ which has for minimal polynomial $X^2 - X - 1$, then on one hand one has

$$M = P + 1 = P^2 \tag{34}$$

and on the other hand

$$P^n = P^{n-1} + P^{n-2} \tag{35}$$

and this implies the formula:

$$M^n = \begin{pmatrix} F_{2n+1} & F_{2n} \\ F_{2n} & F_{2n-1} \end{pmatrix}$$

Therefore, with the diffusion matrix given by (12) one has, for any function $f : \mathbf{T}^2 \to \mathbf{R}$ in $H^s$ with $s > 0$:

$$D(f) - \tfrac{1}{2}\langle f^2\rangle = \sum_{n\geq 1}\langle f \circ T^n \cdot f\rangle = \tfrac{1}{4\pi^2}\sum_{n\geq 1}\ \sum_{k\in\mathbf{Z}^2-\{0\}} \hat{f}(M^n k)\hat{f}(-k)$$

$$= \tfrac{1}{4\pi^2} \sum_{k\in\mathbf{Z}^2-\{0\}}\ \sum_{n\geq 1} \hat{f}(F_{2n+2}k_1 + F_{2n+1}k_2, F_{2n+1}k_1 + F_{2n-1}k_2)\hat{f}(-k_1, -k_2) \tag{36}$$

In particular, since the Fibonnacci sequence is rapidly increasing this will provide and exact formula (involving a small number of non zero terms ) for $D(f)$ whenever $f$ is a trigonometric polynomial.

## References

Arnold V. (1980): "Chapitres supplémentaires des équations différentielles ordinaires" chap. 14, Mir.

Bardos C., Golse F., Colonna J.-F.(1994): preprint, submitted to Physica D.

Bardos C., Dumas L., Golse F. (1992): C.R. Acad. Sci. II 315, (1992), 1433-1437; and preprint submitted to Commun. in Math. Phys..

Bowen R. (1975): "Equilibrium states and the ergodic theory of Anosov diffeomorphisms" Lectures Notes in Math. 470, Springer Verlag.

Bunimovich L., Sinai Ya. (1980): Commun. in Math. Phys 78, (1980), 479–497.
Bunimovich L., Sinai Ya., Chernov N. (1991): Russian Math Surveys 46-4, (1991), 47-106.
Crawford J., Cary J. (1982): Physica D 6, (1982-83), 223-232.
Dautray R. (1989): "Méthodes probabilistes pour les équations de la physique", Masson.
Denker M., Philipp W. (1984): Ergodic Theory and Dyn. Syst. 4, (1984), 541–552.
Golse F., Leichtnam E. (1994): preprint, submitted to Ann. Scient. E.N.S..
Hörmander L. (1983): "The Analysis of Linear Partial Differential Operators I", sect. 9.6, Springer Verlag.
Katznelson Y. (1971): Israel J. of Math. 10, (1971), 186–195.
Knauf A. (1987): Commun. in Math. Phys. 110, (1987), 89.
Lebeau G. (1985): Ann. Inst. Fourier 35, (1985), 145–216.
Ratner M. (1973): Israel J. of Math. 16, (1973), 181–197.
Serre J.-P. (1962): "Corps Locaux", Hermann.

# Second Order Analysis of the Inner Shock Structure

J.M. Reese[1], L.C. Woods[1], F.J.P. Thivet[2] and S.M. Candel[3]

[1]Mathematical Institute, University of Oxford, 24-29 St Giles',
Oxford OX1 3LB, United Kingdom
[2]Centre d'Etudes et de Recherches de Toulouse, ONERA,
2, avenue Edouard Belin, BP 4025, 31055 Toulouse, France
[3]Laboratoire EM2C, CNRS, Ecole Centrale Paris,
92295 Chatenay-Malabry, France

**Abstract:** The internal structure of shock waves is of interest in hypersonics and also constitutes a basic test for modern kinetic theories. The shock profiles may be calculated from a second-order theory in the Knudsen number. The BGK expansion to the Boltzmann equation introduces additional tansport terms of momentum and energy. These fluxes, known as the Burnett equations, may be used to calculate shock profiles. It is found however, at least for some formulations that the solution breaks down at a critical Mach number. Also, certain terms in the Burnett equations allow unphysical effects in gas flow. A modified kinetic theory has been proposed by Woods which eliminates the frame dependence of the standard kinetic theory and corrects some of the second-order terms.

This article introduces a novel method for solving the time-independent conservation equations, including the second-order terms. The shock structure problem is calculated in one dimension with a finite difference global scheme (FDGS), in which a Newton procedure is applied to a discretized version of the governing equations and boundary conditions. The method is first tested on the Navier-Stokes shock equations. It is then used to integrate a modified version of the second-order fluxes derived by Woods for monatomic gases, up to a Mach number of 30. Results of calculations are in good agreement with experimental measurements for Argon gas flows characterized by Mach numbers up to 10.

# 1 Introduction

This paper is dedicated to Professor Jean-Pierre Guiraud, with our admiration for his scientific achievements.

The internal shock structure is a subject of standing interest. This topic is examined, for example, in the analysis of hypersonic flow in the continuum-transitional regime. The problem also constitutes a test for second-order theories based on the Boltzmann equation. A theory has been devised by Woods (1993) to replace the classical Burnett equations. This new formulation is employed in this article to examine shock wave profiles, determine the shock wave thickness as a function of the Mach number and compare this quantity with experimental measurements. The second-order shock wave equations are integrated using an original finite difference global scheme, based on a discretized version of the governing equations and boundary conditions, and a global Newton iteration procedure. The method is suggested by our experience in flame and reactive flow calculations (see e.g. Thivet *etal.* 1991). It uses a scheme initially proposed by Smooke (1982) but the application to the shock structure problem is not as straightforward as it may seem. To construct a successful integration scheme, it is important to understand some of the peculiar features of the problem. This may be achieved by examining the shock structure with the Navier-Stokes equations. While the Navier-Stokes equations yields incorrect shock profiles even at all Mach numbers, it provides a simple test problem. A phase plane analysis of this case defines guidelines for the finite difference global solution method, designated in what follows as the FDGS method. The NS shock structure is also used to check the solution procedure. This is achieved by first calculating the shock profiles with a more standard path integration and comparing these solutions with those obtained from the FDGS method.

The shock structure equations are presented in section 2. A phase plane analysis of the Navier-Stokes shock structure is carried out in section 3. The path integration scheme deduced from the phase plane analysis is then briefly described . The finite difference global solution method is introduced in section 4, and results obtained with the two integration schemes are compared in section 5. Shock profiles obtained from a modified version of the second-order formulation derived by Woods are discussed and the shock thickness deduced in this case are determined as a function of the upstream Mach number. A comparison with experimental measurements of shock thickness is also carried out.

# 2 Shock Structure Equations

## 2.1 The BGK Expansion

Balance equations for momentum and energy contain a viscous stress tensor $\mathbf{\Pi}$ and a heat flux vector $\mathbf{Q}$. The classical Chapman-Enskog expansion in Knudsen number of the Boltzmann equation may be used to determine these fluxes. One writes:

$$\mathbf{\Pi} = \mathbf{\Pi}^{(0)} + \mathbf{\Pi}^{(1)} + \mathbf{\Pi}^{(2)} + \dots, \tag{1a}$$

$$\mathbf{Q} = \mathbf{Q}^{(0)} + \mathbf{Q}^{(1)} + \mathbf{Q}^{(2)} + \dots. \tag{1b}$$

At zeroth-order, one obtains:

$$\mathbf{\Pi}^{(0)} = \mathbf{0}, \tag{2a}$$

$$\mathbf{Q}^{(0)} = \mathbf{0}. \tag{2b}$$

Substituting these results in the balance relations yields the Euler equations.
At the next order of approximation one finds:

$$\mathbf{\Pi}^{(1)} = -2\,\mu\,\mathring{\mathbf{e}}, \tag{3a}$$

$$\mathbf{Q}^{(1)} = -k\,\nabla T, \tag{3b}$$

where $\mathring{\mathbf{e}} \equiv \mathbf{e} - \frac{1}{3}\mathrm{tr}(\mathbf{e})\mathbf{1}$ and $\mathbf{e} = \frac{1}{2}\left(\nabla\mathbf{v} + {}^{\mathrm{t}}\nabla\mathbf{v}\right)$, and $\mu$ and $k$ are, respectively, the viscosity and the heat conductivity. For a monatomic gas modeled by point centers of force, kinetic theory indicates that the viscosity is proportional to $T^s$ and the Prandtl number $\mathrm{Pr} = \mu c_p/k$ is equal to $\frac{2}{3}$. The temperature exponent $s$ may be cast in the form:

$$s = \frac{1}{2} + \frac{2}{\nu - 1}, \tag{4}$$

In this expression $\nu$ is the power index of the inter-molecular force law. For Argon gas at NTP, $\nu = 7.5$ is given by Chapman and Cowling (1970). Lumpkin and Chapman (1991) suggest that $\nu = 9.0$ is a better approximation. This value is confirmed through systematic calculation of shock wave profiles. The conservation equations together with the fluxes (3) define the Navier-Stokes equations.

At second-order, the classical BGK expansion of the Boltzmann equation yields the accepted form of the Burnett equations. General expressions for the Burnett fluxes $\mathbf{\Pi}_{\mathrm{B}}^{(2)}$ and $\mathbf{Q}_{\mathrm{B}}^{(2)}$ and one-dimensional may be found in Reese *etal.* (1993).

The inner shock structure may be determined with these fluxes as shown by Lumpkin and Chapman (1991) who solve the time-dependent conservation equations subject to the Rankine-Hugoniot jump conditions satisfied at each end. The hyperbolic flow equations are iterated in time until steady profiles are reached . Zhong *etal.* (1991) shows that the Burnett equations applied in this manner are unstable to small oscillations in the solution. The problem is treated by augmenting the Burnett equations with special third-order terms. A stable solution can be shown to exist and the shock thickness values obtained for Argon are in reasonable agreement with experiment. The full third-order equations ("Super-Burnett") also feature instabilitie in the same way as the Burnett equations. Converged solutions are only assured by adding higher-order terms.The exact form of these extra terms is currently being investigated (see e.g. Welder *etal.* 1993).

Woods (1993) suggests that at least some of the difficulties derive from an inappropriate frame-dependent formulation of the BGK expansion. A close examination of the accepted form of the Burnett equations shows certain inconstistencies. One finds for example that some of these second-order terms are consequences of not properly distinguishing between convection and diffusion.

An alteranative approach to kinetic theory has been devised by Woods (1993). By incorporating fluid accelerations, it eliminates convective terms from the second-order transport equations. General expressions for viscous stress $\mathbf{\Pi}_{\mathrm{W}}^{(2)}$ and heat flux $\mathbf{Q}_{\mathrm{W}}^{(2)}$ are given in Woods (1993), Reese *etal.* (1993). One-dimensional versions of these terms are given below.

## 2.2 The One-Dimensional Problem Formulation

Consider the one-dimensional flow of a monatomic gas (Figure 1). A planar stationary shock wave is established between supersonic upstream conditions (subscript 1) and subsonic downstream conditions (subscript 2). The one-dimensional flow is governed by the following balance equations:

$$\rho u = m_0 , \tag{5a}$$

$$p + \rho u^2 + \pi = p_0 , \tag{5b}$$

$$\rho u \left( c_p T + \frac{u^2}{2} \right) + u\pi + q = m_0 h_0 , \tag{5c}$$

where $m_0$ is the mass flow rate, $p_0$ is the stagnation pressure, $h_0$ is the stagnation specific enthalpy and where $\pi$ and $q$ are the components of the stress tensor and of the heat flux in the flow direction $x$. Using Eq.(5a) and the perfect gas law ($p = \rho RT$), this set of equations becomes :

$$-\frac{u}{m_0}\pi = RT + u^2 - \frac{p_0}{m_0}u , \tag{6a}$$

$$-\frac{1}{m_0}q = c_v T - \frac{u^2}{2} + \frac{p_0}{m_0}u - h_0 . \tag{6b}$$

The Navier-Stokes transport terms are in one dimension:

$$\pi^{(1)} = -\frac{4}{3}\mu\frac{du}{dx}, \tag{7a}$$

$$q^{(1)} = -k\frac{dT}{dx}. \tag{7b}$$

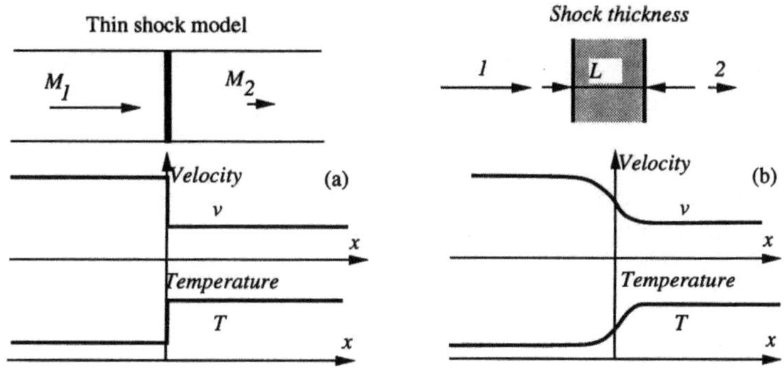


**Fig. 1** (a) Shock described as a flow discontinuity. (b) The inner shock wave structure.

The second order fluxes derived from Woods' formulation can be written:

$$\pi^{(2)}_{\mathrm{W}} = \frac{2}{3}\frac{\mu^2}{p}\left[\left(\tilde{\omega}_1 + \frac{1}{3}\tilde{\omega}_6\right)\left(\frac{\partial u}{\partial x}\right)^2 + \tilde{\omega}_3 R\frac{\partial^2 T}{\partial x^2} + \tilde{\omega}_5\frac{R}{T}\left(\frac{\partial T}{\partial x}\right)^2\right], \tag{8a}$$

$$q^{(2)}_{\mathrm{W}} = R\frac{\mu^2}{p}\left[\left(\theta_1 - \frac{5}{3}\theta_2 + 2\theta_5\right)\frac{\partial u}{\partial x}\frac{\partial T}{\partial x} - \frac{2}{3}\theta_2 T\frac{\partial^2 u}{\partial x^2}\right]. \tag{8b}$$

where the constants $\omega$ and $\theta$ have the following values :

$$\begin{array}{llll} \tilde{\omega}_1=\frac{4}{3}(4-s), & \tilde{\omega}_3=3, & \tilde{\omega}_5=3s, & \tilde{\omega}_6=6, \\ \theta_1=\frac{15}{4}\left(\frac{9}{2}-s\right), & \theta_2=\frac{45}{4}, & \theta_5=\frac{35}{4}+s. & \end{array} \tag{9}$$

Details on the derivation of these expressions may be found in Reese *etal.* (1993)
In what follows, the shock structure will be analyzed using expressions (8) in combination with the balance relations (6) and the first-order fluxes (7). It is worth noting that the energy equation takes a form which is reminiscent of the Korteweg-de Vries equation. It is known that the KdV equation exhibits shock profiles modulated by oscillations (see e.g. Karpman (1975). Under these circumstances one expects that the shock profiles obtained through solution of these equations will feature some mild oscillations.

## 2.3 The Non-Dimensional Set of Equations

It is convenient to replace the balance equations written in the previous subsection by dimensionless forms. The following minimal set of reference quantities may be used to this purpose: the speed of sound $c_1$, the pressure $p_1$, the heat at constant pressure $c_{p_1}$ and the viscosity $\mu_1$.

A complete set of reduced variables is then defined by:

$$\begin{aligned} u &= c_1\,\overline{u}\,, \quad p = p_1\,\overline{p}\,, \quad \rho = \tfrac{p_1}{c_1^2}\,\overline{\rho}\,, \quad T = \tfrac{c_1^2}{R}\,\overline{T}\,, \\ \mu &= \mu_1\,\overline{\mu}\,, \quad k = \tfrac{\mu_1 c_{p_1}}{\mathrm{Pr}}\,\overline{\mu}\,, \quad x = \lambda_1\,\overline{x}\,, \end{aligned} \tag{10}$$

where the mean free path on the upstream side $\lambda_1$ is used as a reference length-scale. A suitable expression for $\lambda_1$ is:

$$\lambda_1 = \frac{\mu_1}{p_1}\sqrt{2RT_1}\,, \tag{11}$$

This is close to the standard Maxwellian definition of mean free path.

Dimensionless fluxes may be defined by :

$$\pi = \sqrt{\frac{\gamma}{2}}\,p_1\,\overline{\pi}\,, \tag{12a}$$

$$q = \frac{\gamma}{\gamma-1}\,\frac{c_1 p_1}{\mathrm{Pr}}\sqrt{\frac{\gamma}{2}}\,\overline{q}\,, \tag{12b}$$

with:

$$\overline{\pi} = -\frac{4}{3}\overline{\mu}\frac{d\overline{u}}{d\overline{x}} + \frac{2}{3}\frac{\overline{\mu}^2}{M_1\sqrt{2\gamma}}\frac{\overline{u}}{\overline{T}}\left[\left(\tilde{\omega}_1 + \frac{1}{3}\tilde{\omega}_6\right)\left(\frac{d\overline{u}}{d\overline{x}}\right)^2 + \tilde{\omega}_3\frac{d^2\overline{T}}{d\overline{x}^2} + \frac{\tilde{\omega}_5}{\overline{T}}\left(\frac{d\overline{T}}{d\overline{x}}\right)^2\right], \tag{13a}$$

$$\overline{q} = -\overline{\mu}\frac{d\overline{T}}{d\overline{x}} + \frac{\gamma-1}{\gamma}\frac{\mathrm{Pr}\,\overline{\mu}^2}{M_1\sqrt{2\gamma}}\frac{\overline{u}}{\overline{T}}\left[\left(\theta_1 - \frac{5}{3}\theta_2 + 2\theta_5\right)\frac{d\overline{u}}{d\overline{x}}\frac{d\overline{T}}{d\overline{x}} - \frac{2}{3}\theta_2\overline{T}\frac{d^2\overline{u}}{d\overline{x}^2}\right]. \tag{13b}$$

With these definitions, the set of equations (6), becomes :

$$-\frac{1}{M_1\sqrt{2\gamma}}\,\overline{u}\,\overline{\pi} = \overline{T} + \overline{u}^2 - \tilde{p}_0\overline{u}\,, \tag{14a}$$

$$-\frac{1}{\mathrm{Pr}\,M_1}\sqrt{\frac{\gamma}{2}}\,\overline{q} = \overline{T} - \frac{\gamma-1}{2}\,\overline{u}^2 + (\gamma-1)\,\tilde{p}_0\overline{u} - \tilde{h}_0\,, \tag{14b}$$

where $M_1 = u_1/c_1$ is thethe upstream Mach number, and where:

$$\tilde{p}_0 = \frac{1}{\gamma M_1}\left(1 + \gamma M_1^2\right)\,, \tag{15a}$$

$$\tilde{h}_0 = 1 + \frac{\gamma-1}{2}M_1^2\,. \tag{15b}$$

# 3 Phase Plane Analysis of the Navier-Stokes Shock Structure

To understand the specific difficulties of the shock structure problem, it is worth examining the shock equations in the phase plane. Such an analysis was conducted in the early work of Ludford (1951) and the main results are synthesized by Hayes (1958) and Cabanne (1970). Our goal is to supply the conclusions which are useful in the numerical solution of the present problem. A complete analysis is contained in Reese *etal.* (1993) Retaining only the first-order (Navier-Stokes) fluxes (3), the system (14) becomes:

$$\frac{4}{3M_1\sqrt{2\gamma}}\overline{u}\mu\frac{d\overline{u}}{d\overline{x}} = \overline{T}+\overline{u}^2-\tilde{p}_0\overline{u}\,, \tag{16a}$$

$$\frac{1}{\Pr M_1}\sqrt{\frac{\gamma}{2}}\mu\frac{d\overline{T}}{d\overline{x}} = \overline{T}-\frac{\gamma-1}{2}\overline{u}^2+(\gamma-1)\,\tilde{p}_0\overline{u}-\tilde{h}_0\,. \tag{16b}$$

Thus, the monotonic solutions of system (16) are also solutions of the equation:

$$\frac{d\overline{T}}{d\overline{u}} = \frac{4\Pr}{3\gamma}\,\overline{u}\,\frac{g(\overline{u},\,\overline{T})}{f(\overline{u},\,\overline{T})}\,, \tag{17}$$

where $f(\overline{u},\,\overline{T})$ and $g(\overline{u},\,\overline{T})$ are the right hand sides of equations (16a) and (16b) respectively. In the phase plane $(\overline{u},\,\overline{T})$, the solutions of system (16) are connected by integral curves defined by the first-order differential equation (17). The spatial evolution may be retrieved by the integration of equation (16a):

$$\frac{d\overline{x}}{d\overline{u}} = \frac{4}{3M_1\sqrt{2\gamma}}\,\frac{\overline{\mu}\overline{u}}{f(\overline{u},\,\overline{T})}\,. \tag{18}$$

Now, the upstream and downstream states are stationary points of the system (16) because:

$$f(\overline{u}_i,\,\overline{T}_i)=0 \quad \text{and} \quad g(\overline{u}_i,\,\overline{T}_i)=0 \quad \forall i=1,2\,, \tag{19}$$

To start the integration of equations (17) or (18) it is necessary to expand around one of the states 1 or 2. The perturbation analysis is carried-out in Hayes (1958), Cabanne (1970), Reese *etal.* (1993). It is found that the upstream point is an unstable node while the downstream point is a saddle point (Figure 2 (a)). Hence any integration method must begin with the only stable solution of the problem. The initial direction at the downstream point is defined by the eigenvector $(\delta\overline{u},\,\delta\overline{T})^t$ corresponding to the negative eigenvalue of the system obtained by linearising equations (16a) .

General solutions corresponding to finite values of the Prandtl number are limited by two particular cases (Figure 2 (b)). One of them is the non-conductive case, for which the Prandtl number is infinite. Equation (16b) degenerates into an algebraic relation which gives the temperature directly as a function of the velocity. Equation (18) is still valid and describes the spatial evolution of the variables.

The nonviscous solution is the other limit. For a certain range of upstream Mach numbers, this solution features an isothermal jump as shown in Figure 2 (b)

The methods which may be used to compute the two limit cases only require some adaptations of a general solution scheme. In essence the path integration is started at the downstream point in the direction of the eigenvector associated with the negative eigenvalue (see Reese *etal.* 1993 for details). A typical Navier-Stokes shock structure is shown in Figure 3.

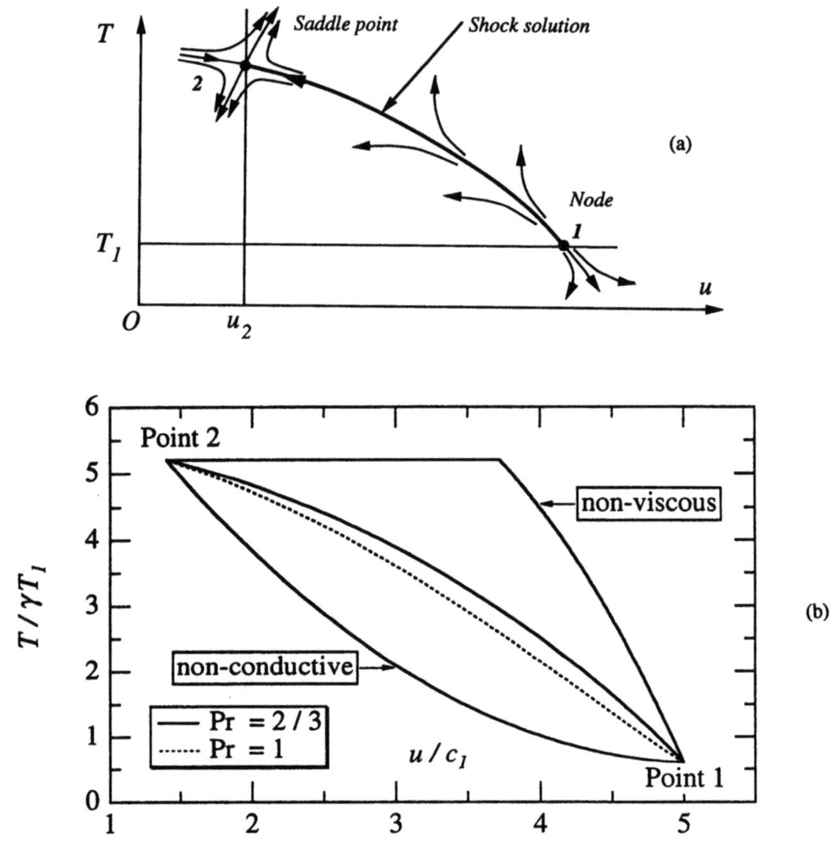


**Fig. 2.** (a) The phase plane geometry for shock structure analysis. (b) Representation of inner shock structures in the phase plane for different values of the Prandtl number ($M_1 = 5$).

# 4 The Finite Difference Global Solution Method

The previous method is not easily applied to the second-order equations, because the order of the problem precludes a phase plane analysis. While the perturbation analysis carried out near the downstream point may provide the starting direction of integration, the results obtained exhibit unwanted oscillations (see e.g. Sherman and Talbot 1960). It is more adequate to solve the nonlinear system of equations with a finite difference global solution scheme with well-posed boundary conditions.

The system at hand features two ordinary differential equations involving the first and second derivatives of velocity and temperature. Hence two boundary conditions are needed for the velocity, and two for the temperature. It can be noted that there is nothing in the downstream nor in the upstream conditions to fix the location of the shock. In other words, if $\left[\overline{u}(\overline{x}),\, \overline{T}(\overline{x})\right]$ is a solution of the system (14) which converges to the downstream conditions when $\overline{x} \to -\infty$ and to the upstream conditions when $\overline{x} \to +\infty$, then $\left[\overline{u}(\overline{x}_o + \overline{x}),\, \overline{T}(\overline{x}_o + \overline{x})\right]$ is another solution for all real values of $\overline{x}_o$. Thus, to ensure the uniqueness of the solution, another condition must be added and one of the boundary conditions has to be eliminated. Finally, the following conditions are associated with the system (14):

$$\lim_{\overline{x} \to -\infty} \overline{u}(\overline{x}) = \overline{u}_1 , \tag{20a}$$

$$\lim_{\overline{x} \to +\infty} \overline{u}(\overline{x}) = \overline{u}_2 , \tag{20b}$$

$$\lim_{\overline{x} \to +\infty} \overline{T}(\overline{x}) = \overline{T}_2 , \tag{20c}$$

$$\overline{T}(\overline{x}_o) = \frac{1}{2}\left(\overline{T}_1 + \overline{T}_2\right) , \tag{20d}$$

where $\overline{x}_o$ is an arbitrary real value. When considering the Navier-Stokes case, the conditions (20a) and (20c) overdetermine the system (14), since the second-order derivatives of the temperature and velocity are not included in the fluxes; hence these conditions are removed. This choice is suggested by the phase plane analysis of the Navier-Stokes shock structure and the remark on the uniqueness of the solution.

The spatial domain is taken to be sufficiently wide to contain the entire shock structure (typically 30 mean free paths) and it is discretized with a mesh ($\hat{x}_n$, $n = 1, \ldots N$) which is refined near the center of the domain $\hat{x}_m$, $m = N/2$. The arbitrary real value $\overline{x}_o$ is chosen equal to $\hat{x}_m$.

The set of unknowns may be then be treated as a vector $\mathbf{U} = (\mathbf{U}_n)^t$, $n = 1, \ldots N$, whose elements are 2-component vectors $\mathbf{U}_n = \left(\hat{u}_n,\, \hat{T}_n\right)^t$. The discrete problem consists of solving a non-linear system:

$$\mathbf{F}(\mathbf{U}) = 0, \tag{21}$$

where $\mathbf{F}$ is an $N-$vector $\mathbf{F} = (\mathbf{F}_n)^t$, $n = 1, \ldots N$, whose elements are 2-component vectors $\mathbf{F}_n(\mathbf{U}) = [F_{n1}(\mathbf{U}), F_{n2}(\mathbf{U})]^t$ representing discrete versions of the governing equations and of the boundary conditions. The discretization process is guided by the phase plane analysis (see Reese *etal.* (1993) for details).

The discrete system is solved in an iterative manner by the Newton method. The jacobian of the non-linear function $\mathbf{F}$ is evaluated numerically. Starting from an initial set of values $\mathbf{U}^o = \left(\hat{u}^o,\, \hat{T}^o\right)^t$, the following system has to be solved at each Newton step:

$$\frac{d\mathbf{F}}{d\mathbf{U}}\left(\mathbf{U}^i\right)\left(\mathbf{U}^{i+1} - \mathbf{U}^i\right) = -\mathbf{F}\left(\mathbf{U}^i\right) , \tag{22}$$

until the $L_2$ norm of $\mathbf{F}$ is small enough. Each block of the jacobian is inverted directly, and a recursive algorithm is used to solve the tridiagonal system (22).

The initial solution consists of hyperbolic tangent profiles for $\hat{u}^o$ and $\hat{T}^o$ connecting the upstream and downstream conditions. The mesh is adapted to obtain constant jumps in velocity between two successive points.

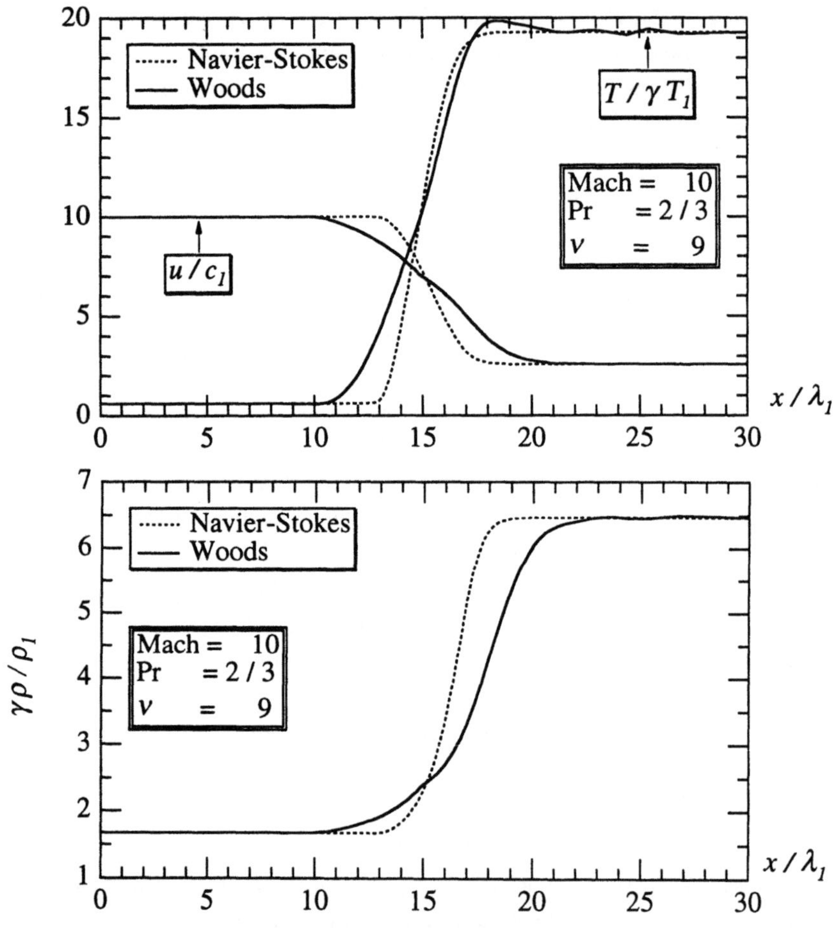


**Fig. 3.** Comparison of the first and second-order shock structures for an upstream Mach number equal to 10. The solutions are obtained with the FDGS method.

## 5 Results and Discussion

The FDGS scheme is first tested in the Navier-Stokes case. As the governing equations (16) are only first-order, two boundary conditions must be eliminated. It is found

that ten iterations suffice to decrease the $L_2$ norm of the residuals by ten orders of magnitude. The spatial convergence is reached with $N = 101$ mesh points.
Studies of shock structure are generally validated by comparing the reciprocal density thickness with experimental measurements. It is standard to define this thickness as:

$$L = (\rho_2 - \rho_1) \;/\; \left(\frac{d\rho}{dx}\right)_{\max} . \tag{23}$$

Figure 4 displays the results concerning the reciprocal density thickness. It compares the solutions of the path integration and FDGS methods and also shows experimental measurements from Alsmeyer (1976). The first conclusion is that both methods give very close results for the Navier-Stokes shock structure. The test confirms the validity of the FDGS scheme. It is also found that the Navier-Stokes results markedly differ from the measurements for Mach numbers exceeding 1.5. This confirms that the first-order Navier-Stokes fluxes do not adequately describe the inner shock wave structure. Second-order equations are needed, as is shown below.

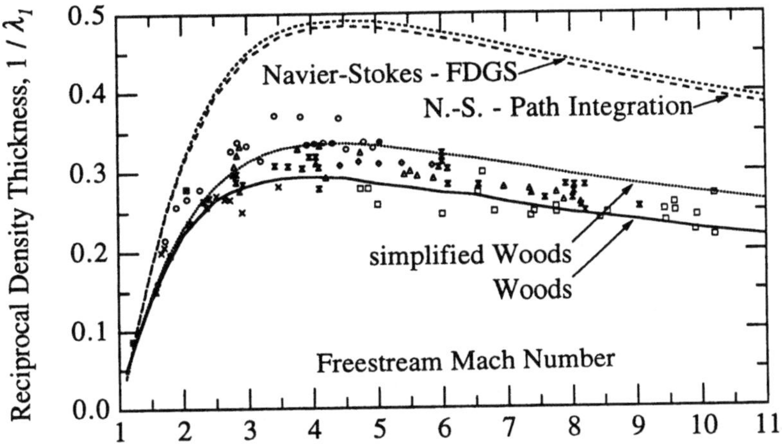


**Fig.** 4 Shock reciprocal density thickness versus the upstream Mach number. The experimental data are taken from Alsmeyer (1976).

The FDGS method is now applied to the second-order equations (14). As for the Navier-Stokes equations, convergence is reached within ten Newton iterations. For Mach numbers larger than 8, the computation must be initialized with an extrapolation of the solution obtained at smaller Mach numbers. The calculation remains quite easy even for large values of the Mach number (computations have been performed up to a Mach number of 30). The solution obtained for an upstream Mach numbers equal to 10 is plotted in Figures 3. It is found that the second-order theory predicts thicker shocks than those determined from the Navier-Stokes equations.

As expected, the shock profiles feature some oscillations at the downstream end, mainly in the temperature profile. These oscillations do not depend on the mesh size and discretization, and must be regarded as part of the solution of the second-order equations. They amount to some 3–4% of the the total difference between upstream and downstream temperature, but considerably less for the velocity profile. At the far downstream, the oscillations decrease in magnitude until the stationary state is reached. These oscillations are not reported in experiments, however the data are accurate to within 1–2% (see Alsmeyer 1976) and may not be tuned to identify downstream oscillations in the temperature.

As shown in Figure 4, the shock reciprocal density thickness is very close to the experimental measurements up to the Mach number where measurements are available (Mach 10). Above this limit our solution follows the experimental trend. The results obtained in this case demonstrate that the Woods formulation as defined in this article constitutes a useful second-order expansion.

It is worth inquiring if one may devise a model which would not exhibit slight oscillations. An eigenvalue analysis of the second order fluxes indicates that the sources of oscillations in the shock solution are the second-order derivatives.

For example, the second-order derivative of the temperature gives rise to a third-order term in the energy equation which takes a form analogous to the Korteweg-de Vries equation. As already mentioned, the KdV equation exhibits shock profiles modulated by oscillations (see e.g. Karpman 1975). This behavior is avoided by prescribing $\tilde{\omega}_3 = 0$ and $\theta_2 = 0$.

When the second-order fluxes are specified in this way oscillation-free solutions are obtained as exemplified in Figure 5. The reciprocal density thickness calculated with the simplified second-order fluxes is displayed in Figure 4 with the label "simplified Woods". These results are as good as those of the complete Woods formulation but the theoretical foundation of the simplified fluxes is less firmly based. This analysis indicates that: (1) the second-order derivatives induce the oscillations in the shock solution and (2) these terms only weakly influence the shock wave structure, and they may be neglected if one wishes to obtain oscillation-free shock profiles. For practical purposes (like for full 3D rarefied gas numerical computations), one would get nearly the same results without the difficult second-order derivative terms. It is expected that the simplified fluxes will pose fewer numerical problems. These forms may be recommended on this basis and also because they are easier to implement.

# 6 Conclusions

A second-order theory is used to describe inner shock structures. Balance equations obtained are solved with a finite difference global scheme which was not previously used in this context. A set of simplified second-order fluxes is also devised and recommended for three-dimensional numerical calculations of hypersonic flows in the continuum transitional regime. The results obtained in one dimension with the exact and simplified fluxes are close to the experimental measurements of the shock thickness in a monatomic gas for upstream Mach numbers below 10. Beyond this value no experimental data exists at this time, but a good theoretical trend is obtained. Extension to a diatomic gas is quite easy, but further theoretical effort are needed to deal with reacting mixtures.

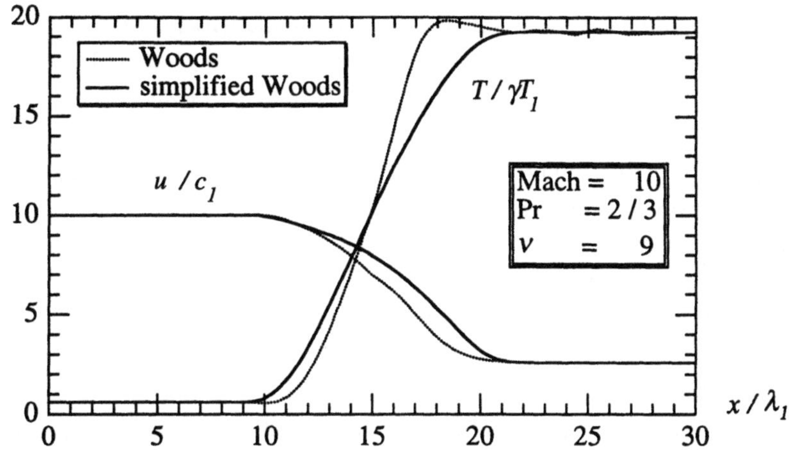


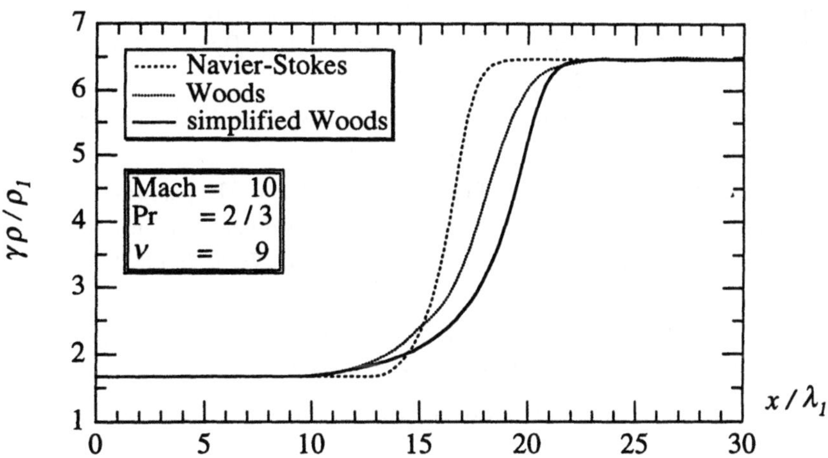


**Fig. 5.** Shock wave structure as described by the simplified Woods fluxes, compared to the results obtained with the complete Woods fluxes at Mach 10.

# References


Alsmeyer, H. (1976): Density Profiles in Argon and Nitrogen Shock Waves Measured by the Absorption of an Electron Beam, *J. Fluid Mech.* **74**, 497.

Cabanne, H. (1970): *Theoretical Magnetofluid Dynamics,* (Academic Press, New York)

Chapman, S. and Cowling, T. G. (1970): *The Mathematical Theory of Non-Uniform Gases* (Cambridge University Press, Cambridge 3rd ed.).

Hayes, W. D. (1958): "The Basic Theory of Gasdynamic Discontinuities", in *High Speed Aerodynamics and Jet Propulsion, Volume III: Fundamentals of Gas Dynamics*, edited by H. W. Emmons (Princeton University Press, New Jersey), p. 416.

Karpman, V. I. (1975): *Non-linear Waves in Dispersive Media* (Pergamon Press, Oxford), pp. 101–105.

Ludford, G. S. S. (1950): The Classification of One-Dimensional Flows and the General Shock Problem of a Compressible, Viscous, Heat-Conducting Fluid, *J. Aeronaut. Sc.* **18**, 830.

Lumpkin, F. E. and Chapman, D. R. (1991): Accuracy of the Burnett Equations for Hypersonic Real Gas Flows, *AIAA paper 91–0771.*

Reese, J. M. , Woods, L. C. , Thivet, F. J. and Candel, S. M. (1993): A second order description of shock structure, *submitted to J. Comp. Physics* .

Salomons, E. and Mareschal, M. (1992): Usefulness of the Burnett Description of Strong Shock Waves, *Phys. Rev. Letters* **69**, 269.

Sherman, F. S. and Talbot, L. (1960): "Experiment Versus Kinetic Theory for Rarefied Gases", in *Proceedings, 1st Intl. Symp. on Rarefied Gas Dynamics*, edited by F. M. Devienne (Pergamon Press, New York), pp. 161–191.

Smooke, M. D. (1982) Solution of Burner Stabilized Premixed Laminar Flames by Boundary Value Methods, *J. Comp. Phys.* **48**, 72–105.

Thivet, F. J. P. , Candel, S. M. and Perrin, M.-Y. (1991): "Hypersonic Viscous Shock Layer in Thermodynamic Nonequilibrium", in *Proceedings, 1st Intl. Symp. on the Aerothermodynamics of Space Vehicles, Noordwijk, The Netherlands*, ESA SP-318, pp. 325–341.

Welder, W. T. , Chapman, D. R. and MacCormack, R. W. (1993): Evaluation of Various Forms of the Burnett Equations, *AIAA paper 93–3094* .

Woods, L. C. (1993) *An Introduction to the Kinetic Theory of Gases and Magnetoplasmas* (Oxford University Press, Oxford).

Zhong, X. , MacCormack, R. W. and Chapman, D. R. (1991): Stabilization of the Burnett Equations and Applications for High-Altitude Hypersonic Flows, *AIAA paper 91–0770.*

# TRANSIENT BEHAVIOUR OF A GASEOUS CAVITY ATTACHED TO A PROJECTILE IN A TWO PHASE FLOW


A. DYMENT

ONERA-IMFL, 5bd Painlevé 59000 Lille, France
and
Laboratoire de Mécanique de Lille, USTL
59655 Villeneuve d'Ascq Cedex, France



The cavities we consider can be created by the entry of a body into water or during the vertical launch of a projectile by means of an expanding gas.

When the characteristic Froude number $F_0$ is large, the resulting two phase flow exhibits an elongated cavity. Within the framework of the slender bodies theory, it can be shown that the ratio of the cavity length to a transverse dimension of the body is of order $F_0^{2/3}$.

The significant parameter concerning the gas motion is $\delta_a F_0^{4/3}$, $\delta_a$ being the order of magnitude of the gas density. When this parameter is small, the gas inside the cavity may be considered as homogeneous. This event generally occurs when a projectile is driven out of an underwater silo by an expanding gas. A bubble then forms between its base and the launching tube. Two new parameters appear :one is related to the expansion of the gas and the other to the difference between the surrounding hydrostatic pressure and the gas pressure at the exit of the projectile base.

The interest of our asymptotic appraoch is obvious. At first, it yields a substantial economy in testing, because the parameters brought to light are less numerous than those resulting from a classical dimensional analysis. Secondly, the asymptotic equations can be solved by simple calculations in order to obtain usefull results, able to appraise the evolution of the phenomena versus the detected dominant parameters.

Experimental results obtained up to cavity sealing are in good agreement with our theoretical predictions.


## Introduction

Our paper is devoted to a family of two phase flows created by a body which, when moving through a liquid, is followed by a slender gaseous cavity, the development of which is the main unknown of the problem. In general, the fluids are air and water. The most common example of such a phenomenon is the penetration of a projectile into water.

The case of the submarine vertical launch of a projectile by means of a compressed gas will also be examined.

Apart from their unsteady character, the main difficulties of these transient flows lie, from the theoretical point of view, in the existence of a free boundary, while the experimentalist's main concern is their brevity. In the model we present here, a simplification is brought, in which the gaseous bubble is considered as slender. This, in turn, imposes some conditions to be satisfied by the initial values.

Through the asymptotic analysis that was used, the dominant similarity parameters have been put into evidence, from which the experimental work is strongly simplified. Furthermore, in some cases, a relatively simple calculation of the time evolution of the bubble can be performed up to sealing.

## 1. Impact on water

### 1.1 Projectile immersion phase

When a projectile penetrates water from the atmosphere, it leaves a trail behind it.

Strong and fast phenomena occur at impact. First a domain is created around the nose of the projectile where pressure is very high and the compressibility of water must be taken into account.

If the nose is pointed, rounded or streamlined, only a very minute domain is submitted to a very high overpressure: penetration remains of hydrodynamic nature and the drag is of order $\rho_e U_0^2 R_0^2$, with $R_0$ being a characteristic dimension of the body, $\rho_e$ the water density and $U_0$ the velocity at impact. Then, from the kinetic energy theorem, the velocity of the projectile remains of order $U_0$ and even is close to $U_0$ granted the projectile density is much larger than that of water, which usually is the case.

However, if the projectile is flat-nosed, the liquid, which cannot part fast enough, is compressed on an important fraction of the fore-body. This results in a large drag, so that the projectile very rapidly slows down and even stops before it entirely penetrates the water. This case will be left aside.

Let us come back to a streamlined nosed projectile around which water will easily flow at impact time. A thin film can be seen to separate from the body, forming some kind of a screen, which is not entirely impermeable and does not prevent air from being fed into the cavity in the wake. This secondary effect will be neglected. Since a deep penetration is possible only when the projectile velocity is high, the gaseous cavity is so much longer as the Froude number at impact $F_0 = U_0(gR_0)^{-1/2}$ is high. This case is of practical interest.

In the study of this phenomenon, we will neglect viscosity and surface tension. Thus the flow in the water is irrotational as it starts from rest. We consider the vertical impact of an axisymmetric projectile. The axis of symmetry is Ox, vertical and positive towards the bottom. Abscissae are counted from the undisturbed free surface. We suppose a thick projectile, $R_0$ being its radius amidship. Its volume $V_s$ is of order $R_0^3$. The atmospheric pressure is $p_a$, the respective relative densities of air and the projectile to water are $\delta_a$ and $\delta_s$. The flow is axisymmetric and the only space variables are the abscissa $x$ and the distance to the axis $r$. The time $t$ is counted from the moment when the projectile hits the free surface. The projectile velocity after impact is $U$, the pressure and velocity potential in the liquid are $P$ and $\phi$. The free surface is at abscissa $h$. The basic assumptions are (A. Dyment 1990 (1)) :

$$(1) \qquad \delta_s \sim 1\,, \qquad \delta_a \ll 1\,, \qquad F_0 \gg 1.$$

Let us go rapidly over the immersion phase, during which the scales for length, velocity and time are respectively $R_0$, $U_0$ and $R_0/U_0$. The scale for $\phi$ is $R_0U_0$ and that for $P - p_a$ is $\rho_e U_0^2$. In the equations which are written with variables normalized by their respective scales, the parameters $F_0$ and $\delta_s$ are seen to appear. This means that the solution of the asymptotic equations corresponding to $F_0 \gg 1$ only depends on $\delta_s$ and the shape of the projectile. Gravity has no influence. This is a classical result on which we will not spend any more time, as this problem has often been treated (See Korobkin et al, 1988, for an extensive bibliography).

As for the projectile shape, only its front part which is in direct contact with water, may have any influence. For a blunt based body such as a cone, the immersion phase can be considered as finished when the water reaches the level of the base. For a rounded, sphere-like projectile, experiment shows that the water bathed part rapidly stops increasing: as soon as this happens, the immersion phase may be considered as over.

### 1.2 The deep cavity flow

Let us consider now the following phase, when the front of the projectile is in contact with water and its back is in a gaseous cavity. Let $X$ be the abscissa, close to that of the body widest section, where water separates and $r = R$, $0 < x < X$ the equation of the meridian line of the cavity. $R_1$ will be the value of $R$ at abscissa $X$ and $S$ is the area $\pi R^2$.

Since the cavity is slender, the gas flow in it can be considered as quasi one-dimensional : the pressure $p$ of the gas and the x-component of its velocity $u$, the only one we have to consider, only depend on $x$ and $t$.

We introduce a small parameter $\varepsilon = R_0/L$, where $L$ is the order of magnitude of the cavity length. The projectile-cavity set is seen by the flow in the liquid as a single slender body of length $L$, since the projectile length is very small compared to that of the cavity. The potential for the liquid flow $\phi$ is of the order of $\varepsilon^2 U_0 L$. The pressure induced by this potential, i.e. by the projectile penetration, is of the order of $\rho_e \varepsilon^2 U_0^2$, while that due to gravity is of the order of $\rho_e g L$.

What precisely distinguishes this phase from the previous immersion phase is the fact that now gravity must play a role, under which the cavity will finally close. We must thus have $\rho_e \varepsilon^2 U_0^2 \sim \rho_e g L$, which in turn yields, since $L$ is only defined to a multiplicative constant:

$$(2) \qquad \varepsilon^3 F_0^2 = 1 \qquad L = R_0 F_0^{2/3}.$$

Thus a simple discussion about orders of magnitude gives $L$.

We then introduce the following starred dimensionless quantities:

$$(3)\qquad \begin{cases} x = Lx^*, & r = Lr^*, & U_0 t = Lt^*, & X = LX^* \\ R = \varepsilon LR^*, & S = \varepsilon^2 L^2 S^*, & h = \varepsilon^2 Lh^*, & U = U_0 U^* \\ u = U_0 u^*, & p - p_a = \rho_e g L p^*, & P - p_a = \rho_e g L P^*, & \phi = \varepsilon^2 U_0 L \phi^*. \end{cases}$$

This normalization is not valid in a small domain at the nose which size is at most $R_0$ and thus is very small compared to $L$. For the moment, we leave this domain aside.

Taking (2) into account and writing $f_\alpha$ for the partial derivative of $f$ with respect to $\alpha$, we have :

$$(4)\qquad r^*\phi^*_{x^*x^*} + (r^*\phi^*_{r^*})_{r^*} = 0, \qquad x^* > 0,$$

$$(5)\qquad \phi^*{}_{t^*} = F_0^{-4/3} h^*, \qquad h^*_{t^*} = \phi^*{}_{x^*}, \qquad x^* = 0,$$

$$(6)\qquad r^*\phi^*{}_{r^*} = R^* R^*_{t^*}, \qquad r^* \to 0 \qquad 0 < x^* < X^*,$$

$$(7)\qquad p^* = -\phi^*{}_{t^*} - \frac{1}{2} R^{*\,2}_{t^*} + x^*, \qquad r^* = F_0^{-2/3} R^*, \qquad 0 < x^* < X^*,$$

$$(8)\qquad S^*{}_{t^*} + (u^* S^*)_{x^*} = 0, \qquad 0 < x^* < X^*,$$

$$(9)\qquad \delta_a F_0^{4/3} (u^*{}_{t^*} + u^* u^*{}_{x^*}) + p^*{}_{x^*} = 0, \qquad 0 < x^* < X^*,$$

$$(10)\qquad R^* \simeq 1, \qquad u^* \simeq U^*, \qquad x^* = X^*,$$

$$(11)\qquad p^* + \delta_a F_0^{4/3} \frac{u^{*2}}{2} = 0, \qquad x^* = 0.$$

Equation (4) indicates that the potential $\phi$ is harmonic. The free surface conditions (5) have been linearized in a classical way, the choice for $h^*$ having been made in order to obtain the least degeneracy of these conditions. The kinematic condition (6) on the boundary of the cavity has been linearized according to the slender body theory. Relation (7) results from the dynamic condition $p = P(x, R, t)$ and Bernoulli theorem. Mass and momentum conservation lead to equations (8) and (9). When writing (10), we account for the fact that separation occurs close to the body's widest section. Equation (11) merely states that the atmosphere is a large reservoir for the gas flow inside the cavity.

A single parameter remains in the above equations which is (A. Dyment, 1990 (1)):

$$(12)\qquad D = \delta_a F_0^{4/3}$$

If this parameter is of order 1, the gas flow in the cavity must be taken into account. On the other hand, if $D \ll 1$, it may be neglected and the gas pressure is everywhere equal to $p_a$.

The asymptotic system is made of equations (4) to (11), in which the right hand side of the first condition (5) is made zero, but in which also care must be taken in writing (7) at $r^* \to 0$ because of the singularity of the potential on the axis.

This set of equations must be made complete by some examining of the flow in the vicinity of the projectile. Let $\Delta_e$ be the domain in the liquid close the nose where $P - p_a$ is of order $\rho_e U_0^2$ and let its size be $\lambda R_0$.

On the back, if the body is not blunt based, we must consider a small domain $\Delta_a$ as wide as the cavity but very short compared to $L$, where the gas flow cannot be quasi one-dimensionnal. The pressure in $\Delta_a$ is such that $p - p_a \equiv \delta_a \rho_e U_0^2$, just as in the remaining of the cavity. From that, we deduce that $\Delta_e$ and $\Delta_a$ cannot have a common boundary : they necessarily are separated by a liquid domain where the solution satisfies (4) to (7) and where $P - p_a$ is of order $\varepsilon^2 \rho_e U_0^2$. In the following $\Sigma_a$ and $\Sigma_e$ will indicate the boundaries between $\Delta_a$, $\Delta_e$ and the body.

The motion of the projectile is governed by

$$\rho_e \delta_s V_s \frac{dU}{dt} = \rho_e \delta_s V_s g + \int_{\Sigma_e} (P - p_a) n d\Sigma + \int_{\Sigma_a} (p - p_a) n d\Sigma$$

where n is the x-component of the interior normal to the body surface.

A simple calculation leads to

$$\rho_e \delta_s V_s \frac{dU}{dt} = \rho_e \delta_s V_s g - \rho_e V_e g - \rho_e \pi R_1^2 X g + \int_{\Sigma_e} (P - p_a - \rho_e g x) n d\Sigma + \int_{\Sigma_a} (p - p_a) n d\Sigma \tag{13}$$

where $V_e$ is the immersed part of $V_s$. If $\Delta U$ is the variation of $U$ on the length $L$, the order of magnitude of the different terms of (13) are respectively $\rho_e \delta_s R_0^3 \frac{U_0^2}{L} \frac{\Delta U}{U_0}$, $\rho_e \delta_s R_0^3 g$, $\rho_e R_0^3 g$, $\rho_e R_0^2 L g$, $\rho_e U_0^2 \lambda^2 R_0^2$ and $\delta_a \rho_e U_0^2 R_0^2$.

After dividing by $\rho_e \delta_s R_0^3 \frac{U_0^2}{L}$ and taking (2) and $D \sim 1$ into account, we get, keeping the terms in the same order as in (13) :

$$\frac{\Delta U}{U_0} = \mathcal{O}\left(F_0^{-4/3}\right) + \mathcal{O}\left(F_0^{-4/3}\right) + \mathcal{O}\left(F_0^{-2/3}\right) + \mathcal{O}\left(\lambda^2 F_0^{2/3}\right) + \mathcal{O}\left(F_0^{-2/3}\right).$$

Each quantity in the right hand side of this equation is very small except the penultimate as long as $\lambda \sim F_0^{-2/3}$. In this case, as a first approximation, equation (13) reduces to

$$\rho_e \delta_s V_s \frac{dU}{dt} = \int_{\Sigma_e} (P - p_a - \rho_e g x) n d\Sigma. \tag{14}$$

As the force acted upon the body by the gas is negligible, $\delta_a$ disappeared from the equation and $D$ thus remains the only parameter including together $\delta_a$ and $F_0$.

If, by any chance, the matching with the domain $\Delta_e$ showed $\lambda$ to be very small compared to $F_0^{-2/3}$, equation (13) would be replaced by $\frac{dU}{dt} = 0$. But $\delta_s$ would not however disappear, as the immersion phase must not be forgotten.

Finally, it can be seen that the only parameters for this problem are $D$ and $\delta_s$.

### 1.3 Validation of the penetration model

There is a particular case for which the various difficulties we mentionned above do not appear, that is to say there exists neither an immersion phase nor a particular domain around the nose nor again any coupling with the projectile movement.

This case, which then represents a "pure" sealing of a slender cavity, occurs when the blunt base of a semi-infinite partly immersed vertical cylinder penetrates water at a constant velocity (fig 2). The solution for $\phi^*$ is given by the distribution of sources of intensity ${S^*}_{t^*}$ on the axis of the cavity and its mirror image in the free surface :

$$-4\phi^* = \int_0^{t^*} \left\{ \left[(x^* - \eta)^2 + r^{*2}\right]^{-1/2} - \left[(x^* + \eta)^2 + r^{*2}\right]^{-1/2} \right\} {S^*}_{t^*}(\eta, t^*) d\eta.$$

On the base $x = U_0 t$, we have $R = R_0$ but also $R_x = 0$, because water particles leave the cylinder without deviation : thus $R_t = 0$ too.

The expression for $\phi^*$ yields the principal part of $p^*$ with the help of (7) and, when included in (9), this gives, together with (8) a system for $S^*$ and $u^*$ that can only be numerically solved.

If $D$ is very small, the air flow may be ignored. Then $p^* = 0$ , from which

$$\left[ {S^*}^2_{t^*} \log \left( F_0^{-4/3} S^* \right) \right]_{t^*} = 8x^* {S^*}_{t^*}.$$

A double integration, taking the condition in $x^* = t^*$ into account, gives $S^*$ and so the locus where $S^*$ becomes zero:

$$(8x^*)^{1/2}(t^* - x^*) = \int_0^1 \left[ (\alpha - 1)^{-1} \log(F_0^{-4/3}\alpha) \right]^{1/2} d\alpha.$$

The cavity seal $(x_c, t_c)$ is the first such point where $S^*$ reduces to zero. From that :

$$3x_c^* = t_c^*,$$

$$t_c^{*3/2} = \left( \frac{27}{32} \right)^{1/2} \int_0^1 \left[ (\alpha - 1)^{-1} \log(F_0^{-4/3}\alpha) \right]^{1/2} d\alpha.$$

These results are compared to experiments made at IMFL (figure 2)(J.P. Flodrops *et al* 1991). The agreement is quite satisfactory.

Another comparison can be done with May's experimental results (A. May, 1952), obtained with steel spheres for very large values of $F_0$, so that the air flow in the cavity cannot be neglected. According to our model,

$$\frac{S}{S_0} = \mathcal{H}\left( \delta_s, D, R_0^{-1} F_0^{-2/3} x, R_0^{-1} F_0^{-2/3} U_0 t \right).$$

The abscissa $x_c$ and time $t_c$ where cavity seal occurs are then tied by

$$R_0^{-1} F_0^{-2/3} x_c = \mathcal{K}\left( \delta_s, D, R_0^{-1} F_0^{-2/3} t_c \right).$$

For a given $\delta_s$, this is the equation of a curve, the projection of which in the plane $(R_0^{-1} F_0^{-2/3} x, D)$ is compared to May's results. Figure 3 shows $\frac{x_c}{R_0}$ versus $F_0$ for different values of $\delta_a$ corresponding to $0,005 < D < 0,75$. The same experimental results are shown with our model's coordinate system. A good gathering of the experimental points can be seen to occur around a single curve of approximate equation $x_c = 0.6\delta_a^{-3/8} F_0^{1/6} R_0$

## 2. Submarine launch

### 2.1 Projectile departure from the launch tube

We consider the submarine vertical launch of an axisymmetric slender blunt based projectile (A. Dyment 1991).

Initially, the projectile is in a water-tight tube. Outside, at the deck depth $H$, the hydrostatic pressure is $P_\infty = p_a + \rho_e g H$, with $p_a$ being the atmospheric pressure at the free surface. The projectile is expelled out of the tube by a compressed gas, initially at rest, either in the tube itself or in an auxiliary tank. In the first case, as soon as the projectile is freed, the gas expansion pushes it out of the tube. The second possibility is more interesting, since the launch can in this case be piloted: the initial pressure in the tube is $P_\infty'$ and connecting the reservoir to the tube sets the projectile into motion.

The end of the departure phase corresponds to the projectile base passing the deck. At this time, the projectile velocity is $U_0$. We now take the $x$ axis as positive upwards, with its origin in the deck (figure 4).

This first phase can be made computable owing to some simplifications. We will consider the more complex case when an auxiliary tank is used.

The gas used to expell the projectile is supposed to be a perfect gas undergoing an adiabatic evolution from the reservoir conditions which are supposed to be constant. The part of the tube directly affected by the injection is supposed to be small compared to the total gaseous volume. Furthermore, the projectile velocity is much smaller that the sound velocity. Thus, in almost the whole tube, the gas can be considered as homogeneous : its pressure $p$ is a function of time.

The gap between the cylindrical part of the projectile (of radius $R_0$) and the tube wall is extremely small, so that water cannot enter the tube during the emergence phase.

Finally, friction between tube and projectile is neglected.

The movements induced in the water by the exiting projectile are small and the corresponding velocity potential is of the order of $\tau^2 U_0 L_0$, with $\tau = R_0/L_0$, $L_0$ being the length of the projectile. According to Bernoulli theorem, the pressure induced by the projectile in the water is of the order of $\rho_e \tau^2 {U_0}^2$. This quantity is small compared to $p - P_\infty$, so that the drag of the projectile may be neglected in the force balance.

We then are brought to solving the problem of the filling of a tube with a moving ceiling, in this instance the projectile base. The movement of the ceiling is coupled to the evolution of the pressure inside the tube. This merely comes to integrating a set of non-linear ordinary differential equations for pressure and projectile base abscissa.

By correctly adjusting the initial values, it is possible to obtain, at projectile uncorking, a quasi constant velocity $U_0$ and a pressure in the tube $p_0$ differing little from $P_\infty$, so that the uncorking produces as little as possible perturbations in the water.

### 2.2 Base bubble : approximations

Once the projectile is out of the tube, it goes on freely in a vertical motion through water, with a gaseous bubble attached at its base. As it moves off, this bubble stretches out and finally breaks (figure 4). Subsequently to the bubble breaking re-entering jets may appear.

In the schematic model that we adopt, friction and surface tension are neglected. The projectile keeps the velocity $U_0$ it had at uncorking and this will be justified later. The projectile and the bubble are considered as slender bodies. The cylindrical part of the projectile is long enough for the perturbation induced by the nose to be negligible in the vicinity of the base. In this region, the flow of the water is only caused by the deformations of the gaseous bubble. The corresponding potential $\phi$ is of order $\varepsilon^2 U_0 L$, with $\varepsilon = R_0/L$, where $L$ is the order of magnitude of the bubble length. The corresponding induced pressure is of order $\rho_e \varepsilon^2 U_0^2$. Only if it is comparable to the imposed initial pressure $P_\infty - p_0$, which is supposed to be small, will this pressure variation have any influence. Furthermore, as the bubble stretches out, the variations of the hydrostatic pressure, which are of order $\rho_e g L$, will start interfering.

We thus have three pressure scales. In the most general case, these three effects are equivalent so that (A. Dyment 1990 (2)) :

$$\rho_e \varepsilon^2 U_0^2 \sim \rho_e g L \sim |P_\infty - p_0| \tag{15}$$

$L$ is given by the first equation in (15). We obtain exactly the same result as (2) :

$$L = R_0 F_0^{2/3}, \qquad \varepsilon = F_0^{-2/3}, \tag{16}$$

$F_0$ being the Froude number. As $\varepsilon$ has been supposed to be small compared to 1, $F_0$ must be large.

The second relation of (15) introduces the parameter $G$ which is defined by

$$G = \frac{P_\infty - p_0}{\rho_e g L} = \frac{P_\infty - p_0}{\rho_e g R_0 F_0^{2/3}}$$

or again

$$G = \frac{P_\infty - p_0}{\rho_e (g R_0 U_0)^{2/3}}. \tag{17}$$

It may be noted that the pressure difference at uncorking does not appear alone but as a combination with the Froude number.

To study the bubble deformations in as simple as possible conditions, we will, at first, give a look at a few problems concerning the flow of the gas and the motion of the projectile.

For the gas flow in the cavity, the same evaluations as previously made for the water impact problem have the parameter $\delta_a F_0^{4/3}$ appear, in which $\delta_a$ represents the order of magnitude of the relative density of the gas to water. In what follows, we will suppose that $F_0$ is large enough, so that $F_0 \gg 1$, but not too much in order to satisfy $\delta_a F_0^{4/3} \ll 1$. Then the movements of the gas in the cavity may be disregarded.

In order to describe the motion of the projectile, we have to evaluate the force $f$ to which it is submited . A simple calculation using both the Bernoulli theorem and the fact that $\phi_x$ and $\phi_r$ are zero at the base yields

$$f = -\rho_e \int_\theta \left( \phi'_t + \frac{1}{2}\phi'^2_x + \frac{1}{2}\phi'^2_r \right) n d\Sigma + \rho_e V_s g - \rho_e S_0 \phi_t(X, R_0, t)$$

where $\theta$ is the nose ogive, $V_s$ is the body volume, $X$ the abscissa of the base, $\phi'$ the potential due to the ogive and $S_0 = \pi R_0^2$.

The projectile velocity is $U$ and the equation for its motion is

$$\delta_s V_s \frac{dU}{dt} + (\delta_s - 1)V_s g + \int_\theta \left( \phi'_t + \frac{1}{2}\phi'^2_x + \frac{1}{2}\phi'^2_r \right) n d\Sigma + S_0 \phi_t(X, R_0, t) = 0. \tag{18}$$

When we evaluate the acceleration on a distance of order $L$ and if $\Delta U$ is the corresponding velocity variation, the different terms in (18) respectively are of the same order as $\frac{S_0 L_0 U_0}{L} \Delta U$, $(\delta_s - 1)S_0 L_0 g$, $\tau^2 U_0^2 S_0$, and $\varepsilon^2 U_0^2 S_0$. After dividing by $S_0 L_0 F_0^2$, (18) may be written, with the terms in the same order as previously :

$$\frac{\Delta U}{U_0} + \mathcal{O}\left( \frac{\delta_s - 1}{\varepsilon F_0^2} \right) + \mathcal{O}\left( \frac{\tau^3}{\varepsilon} \right) + \mathcal{O}(\tau\varepsilon) = 0.$$

The assumption made at the end of paragraph 2.1 that the projectile acceleration at uncorking is very small imposes a relation between $F_0$ and $\tau$. Indeed, at this time, with a null value of this acceleration, the equation for the projectile motion is approximately :

$$\rho_e(\delta_s - 1)V_s g + \rho_e \int_\theta \left( \phi'_t + \frac{1}{2}\phi'^2_x + \frac{1}{2}\phi'^2_r \right) n d\Sigma + (P_\infty - p_0)S_0 = 0$$

Taking (16) and (17) into account, we get, with the terms in the same order:

$$\mathcal{O}\left( \frac{\delta_s - 1}{\tau F_0^{2/3}} \right) + \mathcal{O}\left( \tau^2 F_0^{4/3} \right) + \mathcal{O}(G) = 0.$$

There follows $\tau \sim F_0^{-2/3}$ ( this means that $L$ is generally of the same order as $L_0$ !). This being so and coming back to the projectile motion in the second phase, we may verify that $\Delta U/U_0$ is of order $F_0^{-4/3}$ and is thus negligible, so that the projectile velocity remains roughly constant at its uncorking value $U_0$.

### 2.3 Bubble deformation

We now come to dealing with the problem of the gas expansion. The incoming flow of gas into the launch tube has been stopped before uncorking, so that

$$p\left( V_0 + \int_0^{U_0 t} S dx \right)^\gamma = p_0 V_0^\gamma \tag{19}$$

where $V_0$ is the launch container volume and $S$ the cross section of the bubble which radius is $R$. If the bubble volume is far smaller than that of the container, which we assume, (19) may be linearized:

$$p = p_0 - \frac{\rho_e g E}{S_0} \int_0^{U_0 t} S\, dx \tag{20}$$

with

$$E = \frac{\gamma p_0 S_0}{\rho_e g V_0}. \tag{21}$$

Having made the simplifications and knowing the various orders of magnitude, we now normalize the equations, which will put forward the main dimensionless parameters. We thus set

$$\begin{cases} x = Lx^*, & r = Lr^*, & U_0 t = Lt^*, & R = R_0 R^*, \\ S = S_0 S^*, & P_\infty - P = \rho_e g L P^*, & P_\infty - p = \rho_e g L p^*, & \phi = \varepsilon^2 U_0 L \phi^*. \end{cases} \tag{22}$$

We want to obtain a harmonic function, the gradient of which vanishes at infinity and which verifies ${\phi^*}_{x^*} = 0$ for $x^* = 0$ together with $r^* {\phi^*}_{r^*} \to R^* {R^*}_{t^*}$ for $r^* \to 0$, $0 < x^* < t^*$, which is the simplified condition at the bubble surface. The solution to this problem is:

$$-4\phi^* = \int_0^{t^*} \left\{ \left[ (x^* - \eta)^2 + {r^*}^2 \right]^{-1/2} + \left[ (x^* + \eta)^2 + {r^*}^2 \right]^{-1/2} \right\} {S^*}_{t^*}(\eta, t^*)\, d\eta. \tag{23}$$

On the deck $x^* = 0$, we may impose $S^* = 1$ but this is not mandatory : indeed, the gas may very well expand along the deck (but in this model water may not penetrate the tube).

At the base $x^* = t^*$, we impose the matching $S^* = 1$ but also the continuity for the tangent plane, that is ${S^*}_{x^*} = 0$. This means that the water flowing along the cylindrical part of the projectile leaves it without any deviation. Since the derivative following the motion of $S^*$ is zero, we also have ${S^*}_{t^*} = 0$ at $x^* = t^*$. When writing Bernoulli theorem on the bubble surface where $P^* = p^*$, we replace ${\phi^*}_{r^*}$ by ${R^*}_{t^*}$ and we neglect the term in ${\phi^*}_{x^{*2}}$ which is of order $\varepsilon^2$, yielding $-p^* + x^* + \frac{1}{2} {R^*}^2_{t^*} + {\phi^*}_{t^*}(x^*, \varepsilon R^*, t^*) = 0$. In normalized variables, (20) is written $p^* = G + E \int_0^{t^*} S^* dx^*$. Using an asymptotic expression for ${\phi^*}_{t^*}(x^*, \varepsilon R^*, t^*)$ we obtain after some calculations :

$$8 {S^*}_{t^*}(x^*, t^*) \left[ G - x^* + E \int_0^{t^*} S^*(\eta, t^*)\, d\eta \right] = \left\{ {S^*}^2_{t^*}(x^*, t^*) \log \left[ F_0^{-4/3} S^*(x^*, t^*) \right] \right\}_{t^*}, \quad 0 < x^* < t^*, \tag{24}$$

with

$$S^* = 1, \quad {S^*}_{x^*} = 0, \quad {S^*}_{t^*} = 0, \qquad x^* = t^*. \tag{25}$$

Among the three parameters $G$, $E$ and $F_0$that take place in (24) both $G$ and $E$ are dominant, whereas $F_0$ plays a minor role as it only appears through its logarithm.

### 2.4 Validation of the asymptotic model

The asymptotic formulation developped above has an important practical consequence for the experiments. Indeed, in order to realize small scale experiments the results of which may be transposed, at least as a good first approximation, to the full size case, it is only required to satisfy the invariance of the parameters $G$ and $E$. Owing to the fact that $g$ and $\rho_e$ do not change, this amounts to keeping constant the values of $\frac{(P_\infty - p_0)^3}{R_0^2 U_0^2}$ and $\frac{\gamma S_0 p_0}{V_0}$, which affords the experimentalist great liberty. It must be emphasized that a simple approach by mere dimensionnal analysis leads to the invariance of $\gamma$, $V_0/R_0^3$, $U_0^2/R_0$ and $p_0/P_\infty$, and so to four similarity conditions instead of the two we obtained above.

Another interest of the asymptotic approach is obviously its relative simplicity which allows a fairly inexpensive numerical calculation. Some results have been compared with data obtained at IMFL using the same blunt-based cylinder as in 1.3. But in this case, the cylinder is drawn upwards out of a container the volume and the initial pressure of which may be varied.

Figure 5 shows some shadow pictures taken around the bubble sealing time. Figure 6 and 7 present $p^*$ versus $t^*$ for $G = 1$, $E = 6$, $F_0 = 8$ and $t_c^*$ versus $G$ for $E = 6$ $F_0 = 8$ (Paquet et al 1993) We may consider that the agreement is satisfying, owing to the fact that the value of $\varepsilon$, which is 0.25, is not really much smaller than 1! (smaller values of $\varepsilon$ cannot be achieved in the experimental setup).

## Conclusion

In spite of the phenomenon complexity, which is essentially due to its unsteady character and the existence of a free boundary, we have managed to establish a realistic model, which is fairly representative as can be seen from the comparison to various experiments, old or new. The relative simplicity comes from the fact that, since the Froude number is supposed to be large, we benefit from the properties of the flows around slender bodies, allowing some degree of linearization of the equations.

The model we propose does not entirely solve the problem. For instance, in the impact problem, only the phase of the phenomenon when a deep cavity exists has been modelized: now, in order to give some initial values to a computation, it would be required to have the solution to the immersion phase. This difficulty may be removed when the projectile itself is slender.

Generally speaking, in either actual problem we looked upon, the model is only valid up to cavity sealing. What happens next, with the possible existence of "re-entering" jets cannot be treated in the frame of the slender bodies theory and forms itself another research field.

## References

A. Dyment (1990 (1)) : " Modèle asymptotique de l'écoulement produit par l'impact d'un projectile dans un liquide", in Comptes Rendus Ac. Sci. **II**, pp. 1589–1594

A. Dyment (1990 (2)) : unpublished IMFL report n° 90/23

J.P. Flodrops, J.B. Paquet, A. Gobert, J.C. Pegneaux (1991) : unpublished IMFL report n° 91/48

A.A. Korobkin, V.V. Pukhnachov (1988) : "Initial stage of water impact", in Ann. Rev. Fl. Mech. **20**, pp. 159–185

A. May (1952) : "Vertical entry of missiles into water", in J. Appl. Physics **23**, pp. 1362–1372

J.B. Paquet, J.P. Flodrops, A. Dyment, A. Gobert (1993) : unpublished IMFL report n° 93/46

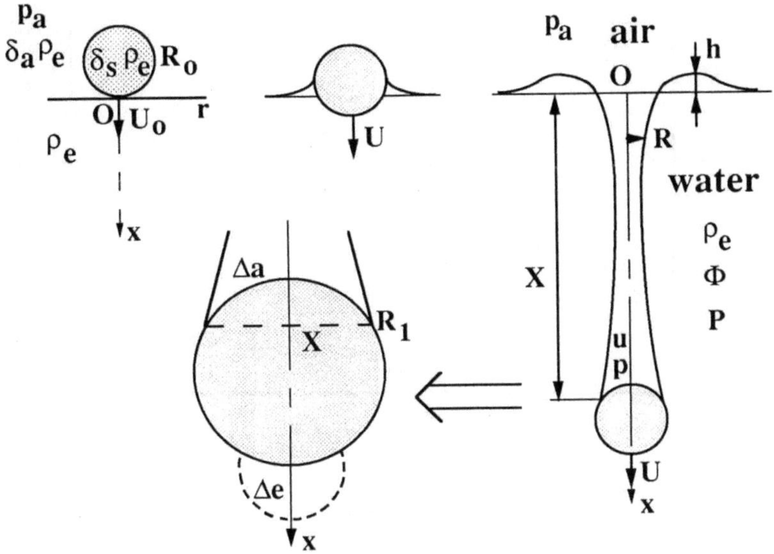


Figure 1

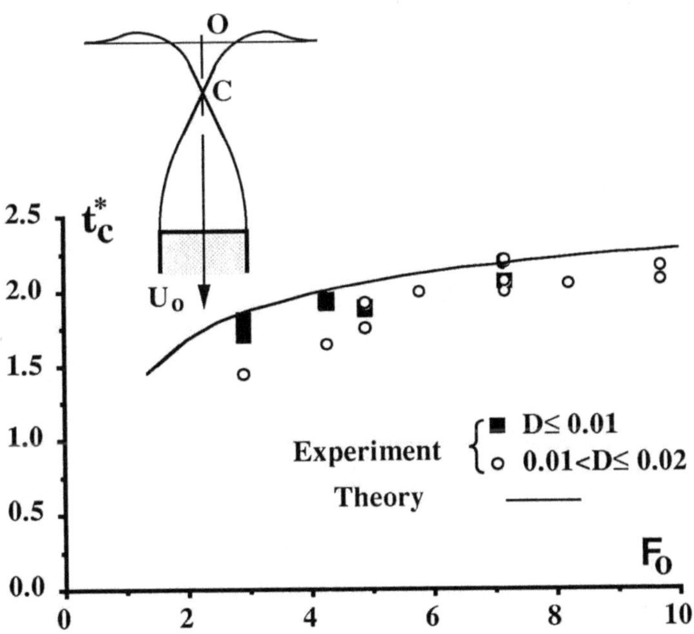


Figure 2

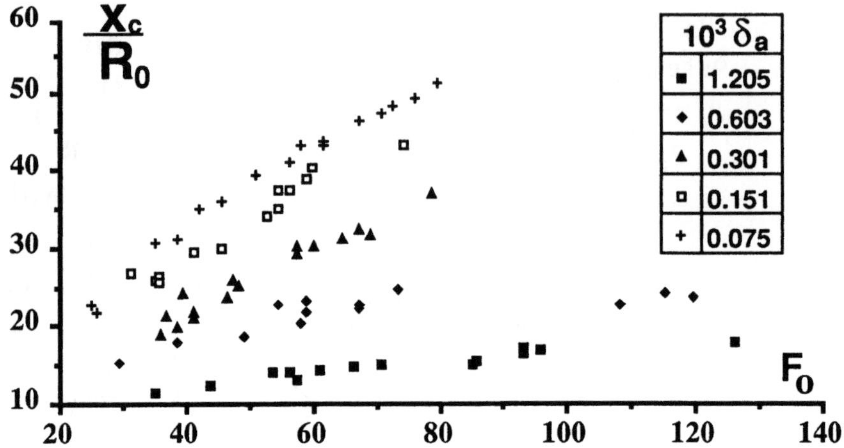

$\frac{x_c}{R_0}$
$10^3 \delta_a$
1.205
0.603
0.301
0.151
0.075
$F_0$
60
50
40
30
20
10
20
40
60
80
100
120
140


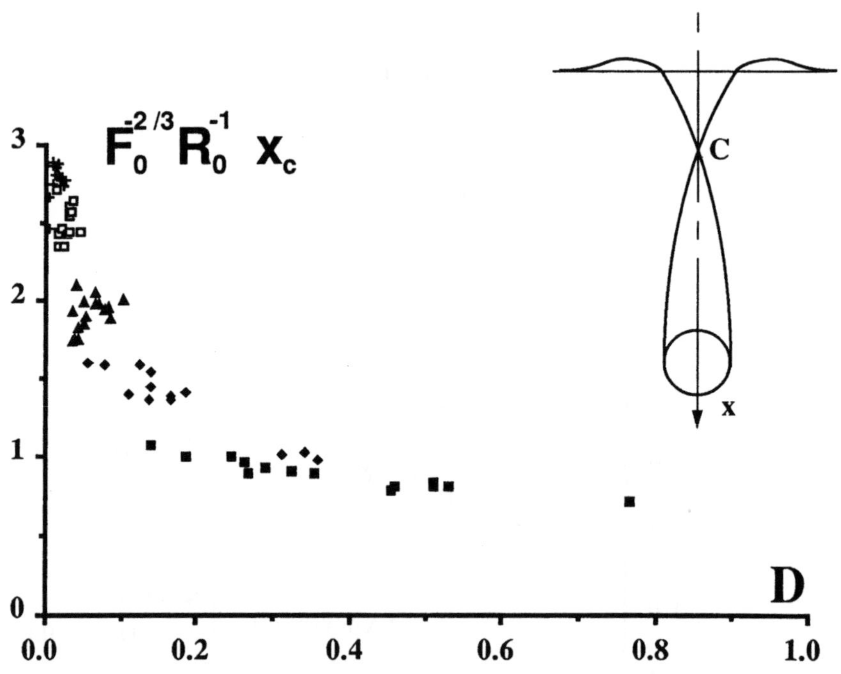

$F_0^{-2/3} R_0^{-1} x_c$
C
x
D
3
2
1
0
0.0
0.2
0.4
0.6
0.8
1.0


Figure 3

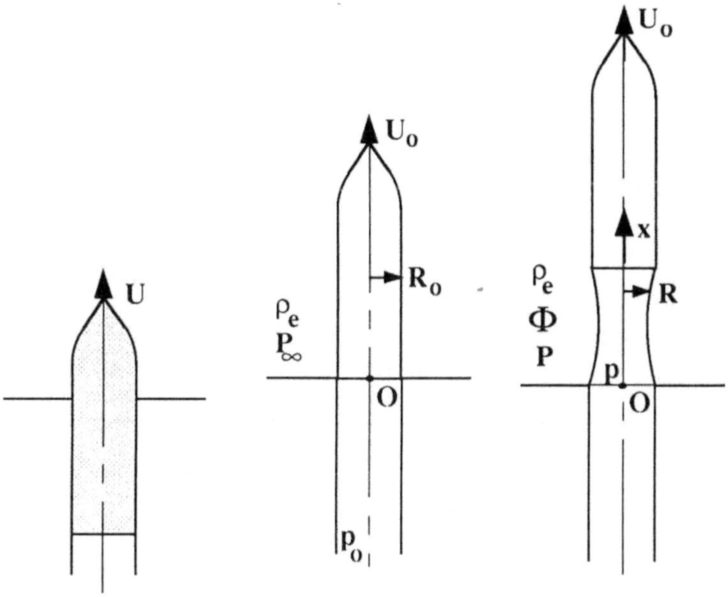


**Figure 4**

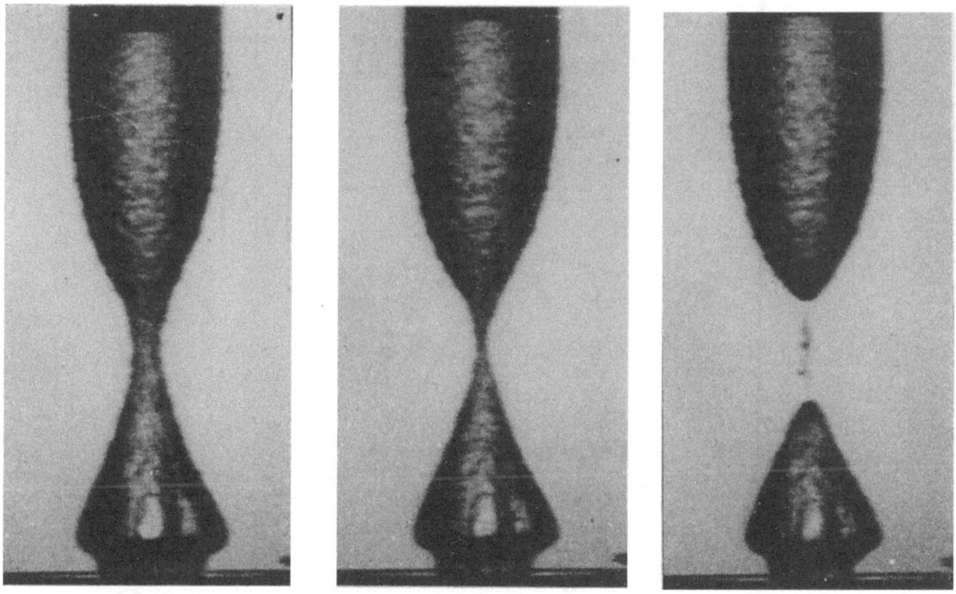

**Time interval between pictures : $10^{-3}$s**
**G = -5, E = 9, $F_0$ = 8**

**Figure 5**

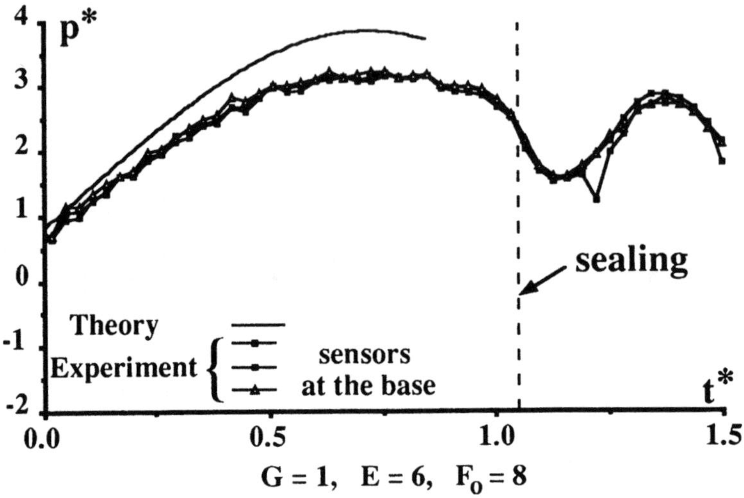


Figure 6

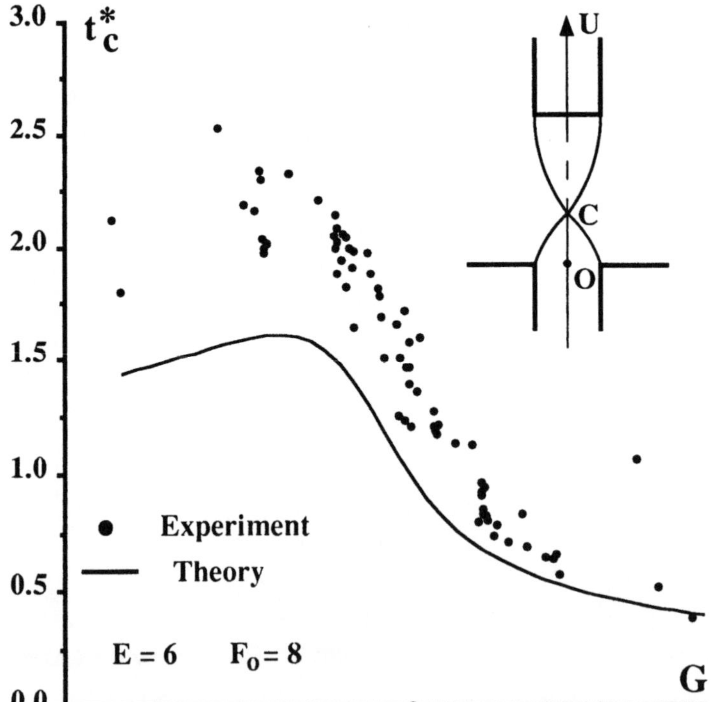


Figure 7

**From the analytical theory to hypersonic aircraft design**

A.Merlen

Laboratoire de Mécanique de Lille. Cité scientifique
59655 Villeneuve d'Ascq - France.

**Abstract:** Hypersonic flows around axisymmetrical power-law slender bodies are calculated for high, but finite, Mach numbers, and for low angles of attack. This is done by a small-perturbation expansion of self-similar solutions using the equivalence principle. The stream functions are found and a solidification principle is used in order to define hypersonic aircrafts.

# 1 Theory

## 1.1 Small disturbance equations

The gas in assumed to be inviscid and thermodynamically perfect, having the specific heat ratio $\gamma$. In a hypersonic configuration, the flow is contained in a thin layer between the shock and the body ; and if $(\gamma-1)$ is assumed to be of the order of unity, it is natural to assume that the physical quantities in this layer are of the order of their value on the shock. Except in the vicinity of the blunt nose, the angle $\sigma$ formed by the tangent to the shock in the plane $(\vec{V}_\infty, \vec{V}_c)$ (fig.1) and the direction of $\vec{V}_\infty$ verifies the following hypotheses:

$$\begin{aligned} \sin^2\sigma \ll 1 \\ M_\infty^2 \sin^2\sigma \gg 1 \end{aligned} \tag{1}$$

The equation of the shock $\Sigma$ is $r = R(x,\theta)$. Using the hypotheses (1) and assuming a low angle of attack defined by :

$$|\alpha| \ll 1, \tag{2}$$

the conditions on the shock can be written in a linearized form. So we can easily give the order of magnitude of the unknowns :

$$\rho \sim \rho_c \sim \rho_\infty \tag{3}$$

$$p \sim p_c \sim \rho_\infty V_\infty^2 \sigma^2 \tag{4}$$

$$\sqrt{u^2+v^2} \sim \sqrt{u_c^2+v_c^2} \sim V_\infty \sigma \tag{5}$$

$$w \sim w_c \sim V_\infty \sigma^2 \tag{6}$$

where we have let $\overrightarrow{V} = (V_\infty + w, u, v)$. With these orders of magnitude, we know that the system of equations can be put into degenerate form:

$$V_\infty \rho_x + \frac{1}{r}[r\rho u]_r + \frac{1}{r}[\rho v]_\theta = 0 \tag{7}$$

$$V_\infty u_x + u u_r + \frac{v}{r} u_\theta - \frac{v^2}{r} + \frac{1}{\rho} p_r = 0 \tag{8}$$

$$V_\infty v_x + u v_r + \frac{v}{r} v_\theta + \frac{uv}{r} + \frac{1}{\rho r} p_\theta = 0 \tag{9}$$

$$V_\infty \left(\frac{p}{\rho^\gamma}\right)_x + u\left(\frac{p}{\rho^\gamma}\right)_r + \frac{v}{r}\left(\frac{p}{\rho^\gamma}\right)_\theta = 0 \tag{10}$$

with one further decoupled equation for the longitudinal perturbation velocity:

$$w = -\frac{c^2}{(\gamma-1)V_\infty} - \frac{u^2+v^2}{2V_\infty} \tag{11}$$

It can be noticed that the variable change $t = \frac{x}{V_\infty}$ transforms the previous system into the unsteady axisymmetric Euler equations. That leads to the equivalence principle between explosions and hypersonic flows around slender bodies.

Figure 2 shows the structure of the flow when the nose of the body is blunted. It is clear that hypothesis (1) do not hold in area I and II which means that the small perturbation equations written above describe the outer development of the problem. The inner development for the bluntness effect was studied in [Guiraud et al 1965], particularly for power-law bodies $\Sigma'$, the equation of which is:

$$R' = \frac{\tau L}{2}\left(\frac{x}{L}\right)^n$$

The theoretical results can be summed up roughly by considering that the equivalence principle yields good results for pressure on the obstacle if the power-law exponent $n$ is given a value between 2/3 and 1. The experiments of Kubota [Kubota 1957] confirm these values. Hayes and Probstein [Hayes 1966] emphasized the poor description it provides of the blunting and of the consequences the blunting has on the density at the body surface. However, they do show that, if the stream function is known, the entropy really acquired at the passage through the shock can be reassigned to each streamline, to correct the density thusly.

## 1.2 Power law bodies

In order to obtain analytical solutions, we restrict the field of our research to power law bodies for which such solutions exist when the upstream Mach number $M_\infty$ is infinite and the incidence $\alpha$ is zero. Though these obstacle are particular, they present all the general features of axisymmetrical slightly blunted slender bodies. Consequently the results of our theory are qualitatively meaningfull for this whole type of obstacles.

The basic flow is assumed to be correctly given by the self-similar solutions provided by the equivalence principle, for which the equation of the shock is given by:

$$R = \frac{\tau L}{2\xi_0'}\left(\frac{x}{L}\right)^n \tag{12}$$

in which $\xi_0'$ is a positive constant, less than unity. Dimensional analysis and order of magnitude considerations show that $R$ can be written as:

$$R = R_0\left(1 + \frac{\alpha}{R_{0x}}\tilde{\xi}_1(\theta, n, \gamma) + \frac{1}{M_\infty^2 R_x^2}\tilde{\xi}_2(n,\gamma)\right)$$

As a matter of fact, the definition of $M_\infty$, $M_\infty^2 = \frac{\rho_\infty V_\infty^2}{\gamma p_\infty}$, shows that, under hypothesis (1), the conditions across the shock depend on $p_\infty$ only by way of a linear disturbance in $\frac{1}{M_\infty^2 R_x^2}$. It can also be observed that $\alpha$ appears in this equation only in the term $\frac{\alpha}{R_{0x}}$, $R_{0x}$ been the first approximation of the slope of the shock. This is a typical result of the hypersonic small disturbance theory.

The same result holds for all the unknowns of the problem, apart from the fact that they also depend on $r$, which introduces a dimensionless variable $\xi$ proportional to $\frac{r}{\tau L}\left(\frac{L}{x}\right)^n$.

The conditions on the shock can be used to determine explicitly the $\theta$-dependency of the function of subscript 1 [Merlen 1992].

Finally, by taking as variables :

$$\eta = \frac{\alpha x}{R_0} = n\frac{\alpha}{R_{0x}} = \frac{2\xi_0'\alpha}{\tau}\left(\frac{x}{L}\right)^{(1-n)}, \text{ and: } \zeta = \frac{n^2}{M_\infty^2 R_x^2}. \tag{13}$$

we write the solution in the form :

$$\begin{aligned}
R &= \frac{\tau L}{2\xi_0'}\left(\frac{x}{L}\right)^n (1 + \eta\xi_1\cos\theta + \zeta\xi_2)\\
R' &= \frac{\tau L}{2}\left(\frac{x}{L}\right)^n\\
\xi &= \frac{2r\xi_0'}{\tau L}\left(\frac{L}{x}\right)^n\\
u &= \frac{2nV_\infty}{\gamma+1}\frac{r}{x}(u_0 + \eta u_1\cos\theta + \zeta u_2)\\
v &= \frac{2nV_\infty}{\gamma+1}\frac{r}{x}(\eta v_1\sin\theta)\\
\rho &= \frac{\gamma+1}{\gamma-1}\rho_\infty(\rho_0 + \eta\rho_1\cos\theta + \zeta\rho_2)\\
p &= \frac{2n^2\rho_\infty V_\infty^2}{\gamma+1}\frac{r^2}{x^2}(p_0 + \eta p_1\cos\theta + \zeta p_2).
\end{aligned} \tag{14}$$

We see that $\xi$ varies between $1 + \eta\xi_1 \cos\theta + \zeta\xi_2$ on $\Sigma$ and $\xi'_0$ on $\Sigma'$. The terms with subscript 0, 1 and 2 are functions of $\xi$, except for $\xi'_0$, $\xi_1$ and $\xi_2$, which are constants, and $\xi_0$, which is introduced for the shock equation to be written $\xi = 1$ in the self-similar solution. Small disturbance hypotheses are expressed by :

$$\eta \ll 1 \text{ et } \zeta \ll 1 \tag{15}$$

Despite the low values of $\alpha$ considered here, the effect of the incidence is far from negligible.

# 2 Results

## 2.1 Solutions

The solution (14) must verify the general equations for any value of $\theta$, $\eta$ and $\zeta$. It is observed that the equations are identically verified for any $\theta$, which validates the form and reduces the problem to three systems of ordinary differential equations in $\xi$. The first of them gives the basic flow analogous to the self similar problem of the noninstantaneous explosion of an infinite rectilinear wire. The other systems (subscripts 1 and 2), too long to be quoted here, are given in [Merlen 1992] and give respectively the effect of the incidence and the conterpressure. The difficulty in solving them resides in the fact that the initial conditions of the integration are given on the shock, the position of which is one of the unknowns of the problem. To get round this difficulty, we use the variable change

$$\xi = \hat{\xi} + \eta\xi_1 \frac{\hat{\xi} - \xi'_0}{1 - \xi'_0} + \zeta\xi_2 \frac{\hat{\xi} - \xi'_0}{1 - \xi'_0} \tag{16}$$

In fact, $\hat{\xi}$ is the 0th-order approximation of $\xi$, equal to one on the shock and to $\xi'_0$ on the obstacle. The advantage is therefore that the conditions on the shock reduce to $\hat{\xi} = 1$ without changing the systems to be solved. The constant $\xi'_0$ is obtained by assigning the non penetration condition to the obstacle for the basic solution $u_0(\xi'_0) = \frac{\gamma+1}{2}$. The equivalence principle is used for calculating $\xi_0$. By expressing the identity between the released energy per unit length of the explosive wire and the obstacle drag for the self-similar solution, we get:

$$\xi_0^4 = \frac{(2n-1)(\gamma+1)}{2\pi n^3 \xi_0'^4 p'_0} \tag{17}$$

We get the missing equations for $\xi_1$ and $\xi_2$ by expressing the resultant of the forces along the $x$ axis on an element $d\Sigma'$ of the surface $\Sigma'$ bounded by two infinitely close planes $x = Const.$, and by a wedge of angle $d\theta$ the edge of which is parallele to $Ox$ (figure 3). This is calculated first by integrating the pressures over $d\Sigma'$ and then applying the momentum theorem to the volume element contiguous to $d\Sigma'$, bounded by the same wedge and planes as $d\Sigma'$ and by the shock (figure 3)

## 2.2 Aerodynamic coefficients

As an example, we limit this section to the presentation of the results on the lift coefficient. Starting with the definition, the lift is obtained by:

$$F_z = -\int_0^{2\pi}\int_0^L p'R'\cos\theta\, dx d\theta \tag{18}$$

where only the subscript -1 term is nonzero. With $S_{ref} = \frac{\pi\tau^2 L^2}{4}$, we get:

$$C_z = \frac{F_z}{\frac{1}{2}\rho_\infty V_\infty^2 S_{ref}} = -\frac{2np_1'\xi_0'}{\gamma+1}\alpha \tag{19}$$

which has the same sign as $\alpha$, since $p_1$ is negative. Experiments realized in ONERA R3Ch wind tunnel are in good agreement with these analytical results showing less than 5% discrepancy (fig 4).

## 2.3 Stream function

The method for finding the stream function for the self-similar flow can be found in reference [Sedov 1972]. This method, based on dimensional analysis, can be applied to find the complete solution, if the body is at zero incidence. The only difference stems from the fact that the dimensional considerations are no longer sufficient, since the flow is no longer self-similar. So the form of $\psi$ as a function of $\zeta$ must be anticipated. It is reasonable to look for $\psi$ in the forme $\psi = \psi_0 + \zeta\psi_2$. It is easy to show that:

$$\psi = \frac{\gamma+1}{\gamma-1}\rho_\infty r^2\left[\rho_0(u_0 - \frac{\gamma+1}{2}) + n\zeta\left(\rho_2(u_0 - \frac{\gamma+1}{2}) + \rho_0 u_2\right)\right]$$

is a stream function. As soon as the nose is placed at incidence, axisymmetry disappears and the streamlines are no longer two-dimensional except for the planes $\theta = 0$ and $\pi$, where $v$ is zero. For these special cases, the method described for $\alpha = 0$ again applies. We are led to :

$$\Psi_s = \left(\frac{x}{L}\right)^{2n}\left\{\rho_0(u_0 - \frac{\gamma+1}{2})\xi^2 + \frac{2n}{1+n}\eta\cos\theta\left[\xi^2\left(\rho_1(u_0 - \frac{\gamma+1}{2}) + \rho_0 u_1\right)\right]\right.$$
$$\left. + n\zeta\left[\xi^2\left(\rho_2(u_0 - \frac{\gamma+1}{2}) + \rho_0 u_2\right)\right]\right\} \tag{20}$$

in which $\Psi_s$ is the dimensionless form of the stream function in the plane of symmetry.

For the other values of $\theta$, the equation of the streamlines is obtained by integrating:

$$\frac{dx}{V_\infty} = \frac{dr}{u} = \frac{rd\theta}{v}$$

which, in dimensionless form, becomes:

$$\begin{aligned}\frac{d\xi}{\xi} &= \frac{2n}{\gamma+1}(u_0 - \frac{\gamma+1}{2} + \eta u_1\cos\theta + \zeta u_2)\frac{dx}{x}\\ d\theta &= \frac{2n}{\gamma+1}\eta v_1 \sin\theta\frac{dx}{x}.\end{aligned}$$

Any quantity $f$ that is conserved along a given streamline must therefore verify:

$$\frac{\partial f}{\partial x}+\frac{2n}{\gamma+1}(u_0-\frac{\gamma+1}{2}+\eta u_1\cos\theta+\zeta u_2)\frac{\xi}{x}\frac{\partial f}{\partial \xi}+\frac{2n}{\gamma+1}\frac{\eta}{x}v_1\sin\theta\frac{\partial f}{\partial \theta}=0 \quad (21)$$

We get the general function $\Psi$ by letting $\Psi = \Psi_s Q(\theta)$ and looking for the function $Q$ for which $\Psi$ verifies the previous equation. We can verify that $\Psi$ is a dimensionless lagrangian variable if $Q = (|\sin\theta|)^{\frac{2n}{1+n}}$. So for $\theta \neq 0$ or $\pi$:

$$\Psi = \left(\frac{x}{L}\right)^{2n}(|\sin\theta|)^{\frac{2n}{1+n}}\Psi_s \quad (22)$$

This function is constant along a streamline. At the zeroth order, the second stream function is obviously $\Phi = \theta$. We look for the form $\Phi = \theta + \eta\Phi_1(\xi)\sin\theta$. Writing (21) for this function $\Phi$, we get a first-order inhomogeneous differential equation for $\Phi_1$. Choosing $\Phi_1 = 0$ on the shock the solution is:

$$\Phi = \theta_i = \theta + \eta\Phi_1\sin\theta \quad (23)$$

with:

$$\Phi_1 = \frac{2}{1-\gamma}\left[\frac{2}{1-\gamma}\rho_0(u_0-\frac{\gamma+1}{2})\xi^2\right]^{\frac{1-n}{2n}}\int_{\xi}^{1}\frac{\rho_0 v_1\hat{\xi}}{\left[\frac{2}{1-\gamma}\rho_0(u_0-\frac{\gamma+1}{2})\hat{\xi}^2\right]^{\frac{1+n}{2n}}}d\hat{\xi}. \quad (24)$$

A more detailled analysis of this stream function is given in [Merlen 1992] including the Hayes-Probstein entropy correction.

# 3 Application to hypersonic aircraft design

Now, we are able to generate a large number of hypersonic aircraft by solidification of stream surfaces. Moreover, because pressure is known everywhere on these surfaces, we obtain analytically the aerodynamic coefficients. As an example, we show here a very simple aircraft based on the "waverider" concept. At first we choose a stream surface $\Phi = cste$ as lower surface. The upper surface is a cylinder starting from the shock intersection with the lower surface and generated by straight lines parallele to $\vec{V}_\infty$. Then, the pressure on the upper surface in supposed to be equal to $p_\infty$. Figure 6 shows one of the geometries obtain by this way.

# References

[1] Guiraud, J. P., Vallée D. and Zolver. R. (1965) "Bluntness effects in hypersonic small disturbance theory." in *Basic development in Fluid Dynamics* edited by Maurice Holt, (Academic Press. New York - London) , Vol. 1, pp.127-246

[2] Hayes, D., Probstein, R. F. (1966): "Application of similar solutions to steady flows", In *Hypersonic Flow Theory* (Academic press. New york - London) vol. 1, pp. 74-85

[3] Kubota. T. (1957): "Investigation of flow around simple bodies in hypersonic flow." (Guggenheim Aeronautical laboratory, California Institute of Technology) Memorandum No. 40.

[4] Merlen, A., Andriamanalina, D. (1992) "Analytical solutions for hypersonic flow past slender power-law bodies at small angle of attack" AIAA Journal Vol 30, number 11, pp. 2683-2693

[5] Sedov, L. (1972): "Algebraic Integrals for Self-Similar Motions" In *Similarity and Dimensional Methods in Mechanics* (Academic Press Inc., New York) pp.166-168


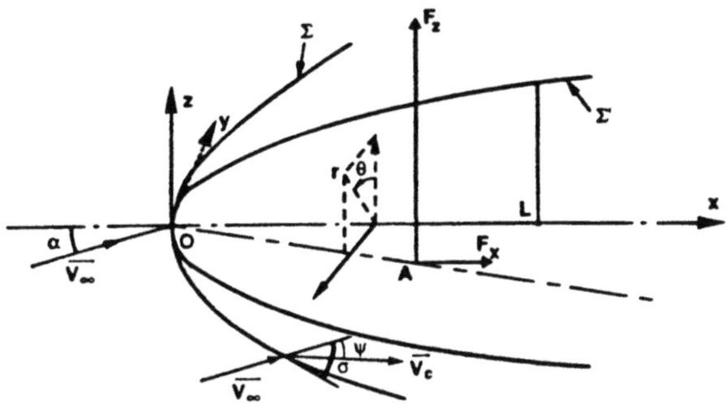

Fig 1 : Notations

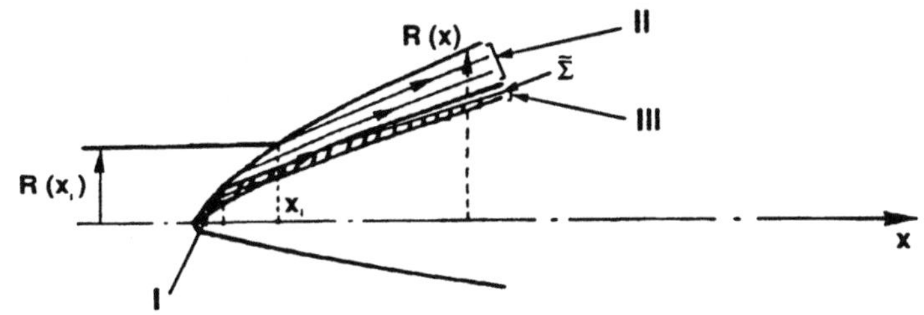

Fig 2 : Structure of the flow

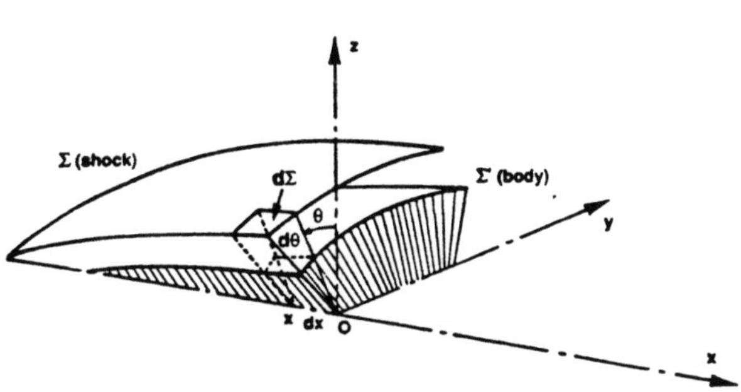

Fig 3 : Integration domain

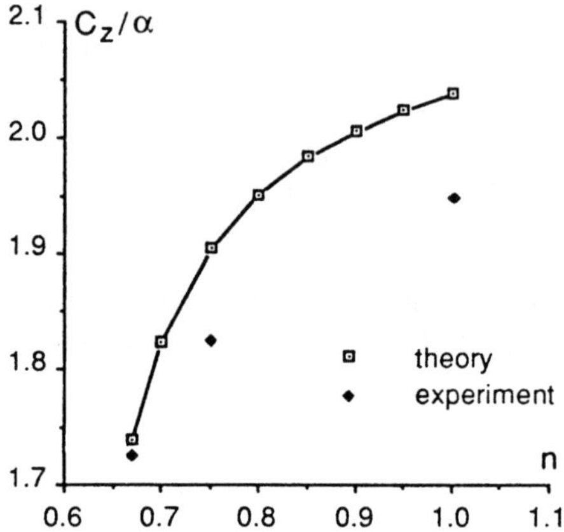


Fig 4 : Lift coefficient

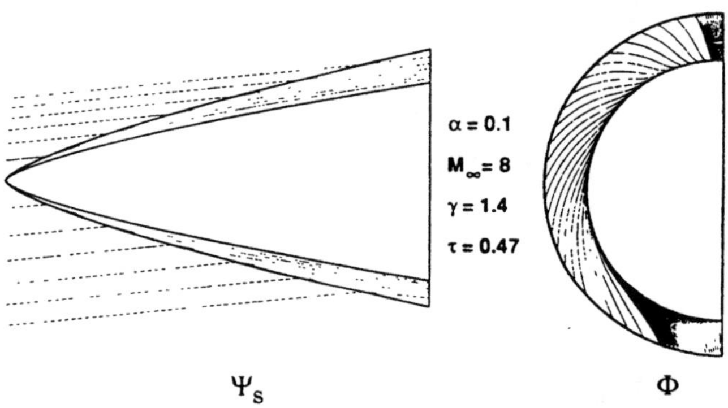


Fig 5 : Stream functions

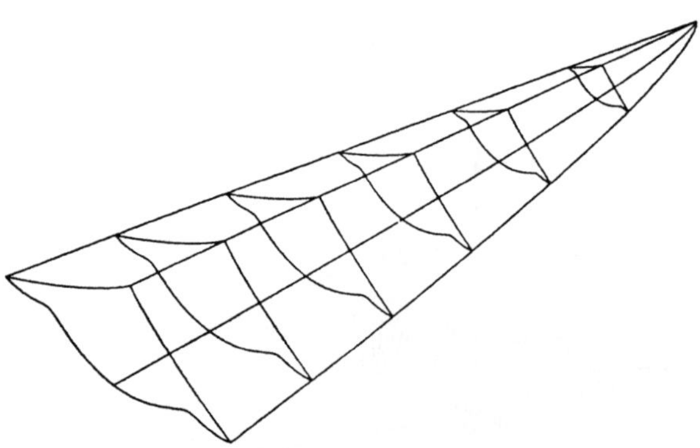

Fig 6 : Waverider

# Upstream influence in mixed convection at small Richardson number on triple, double and single deck scales

**Pierre-Yves Lagrée**

Laboratoire de Modélisation en Mécanique
URA CNRS 229 Université Paris 6, Boîte 162
4 Place Jussieu, 75252 PARIS CEDEX 5. FRANCE
e.mail: pyl@ccr.jussieu.fr

**Abstract:** We investigate the boundary layer flow with thermal effects on a flate plate perturbed by a bump or a thermal spot. We introduce the buoyancy effects in the framework of "triple deck" theory. This is a particular case of the mixed thermal convection problem which is known to lead to a breakdown when the plate is cooled, and when the governing parameter, the "Richardson number", is of order one. Here we try to study the onset of the phenomenon since we work at small Richardson number. For this simplified problem, the usual "triple deck" structure reduces to a "double deck" one when this number increases, and a new term in the pressure displacement relation, accounting for the hydrostatic pressure change across the main deck, appears from the analysis. The linearized and non linearized solutions of this new set of equation show upstream influence of a thermal spot or a bump on the boundary layer.

## 1. Introduction, state of the art

We investigate the flow régime which occurs when a laminar, uniform, stationary and two- dimensional flow of a Newtonian fluid meets a horizontal flat plate which is at a different temperature from the free stream. The plate is cooled (or heated) by natural convection by Archimedes force and by forced convection. This problem is known as "mixed convection" because both effect compete; the magnitude of this competition is the Richardson number defined here by $J = \frac{g\alpha\Delta T_0}{U_\infty^2}\frac{L}{\sqrt{Re}}$ (in fact our study deals with small $J$, and we will show that $\tilde{J} = JRe^{1/8}$ is more convenient). This number gauges the ratio of thermal coupling effect, modelled by Boussinesq approximation, versus dynamic effects. We suppose that viscosity does not change with temperature, see Méndez & Treviño (1992) for this influence, but without buoyancy.

The Reynolds number ($Re = \frac{U_\infty L}{\nu}$) is large, so, viscous and conductive effects in the fluid (Prandtl number of order one, but this is not restrictive) are localized in a small neighbourhood of the wall. This leads to the system (see next references) deduced from Prandtl boundary layer theory, with no slip condition, outer flow matching, and imposed temperature:

$$\frac{\partial u}{\partial x}+\frac{\partial v}{\partial y}=0,\quad (u\frac{\partial}{\partial x}+v\frac{\partial}{\partial y})u=-\frac{\partial}{\partial x}p+\frac{\partial^2}{\partial y^2}u, \qquad (1a,1b)$$

$$\frac{\partial}{\partial y}p=J\theta,\quad (u\frac{\partial}{\partial x}+v\frac{\partial}{\partial y})T=\frac{\partial^2}{\partial y^2}T. \qquad (1c,1d)$$

Relation (1c) defines $J$. Unfortunately, it has been shown by Schneider & Wasel (1985), Daniels & Gargaro (1993) and Daniels (1992) (for example) that boundary layer equations, when coupled with energy equation by buoyancy become abruptly singular at a (relatively small) fixed abscissa.

A very simple way to understand this, is to over simplify Daniels's work (1992). Suppose that there is an algebraic departure from the non buoyant solution for the displacement thickness $\delta \sim x^a$; then, the pressure ($\partial p/\partial y = J\theta$), after integration through boundary layer behaves like $p \sim \delta \sim x^a$; but the speed ($u\partial u/\partial x \sim -\partial p/\partial x$) is now $u \sim x^{(a/2)}$; next we obtain the boundary layer thickness ($u\partial u/\partial x \sim \partial^2 u/\partial y^2$) as $\delta \sim x^{((2-a)/4)}$. This gives $a = 2/5$, a more cumbersome calculation gives $a = 0.43...$ Another over simplified point of view will be explained in the last paragraph.

Thus, we propose to revisit this problem of "mixed convection" on the basis of "triple deck" (Neiland (1970) and Brown *et al.* (1975)), we will not smooth the singularity but we will show that this approach captivates some features like: strong self induced upstream influence (in case of cold wall), no upstream influence and Tollmien Schlichting (Smith (1979)) waves (in case of hot wall). Linearized and non linear computed (by a finite difference scheme) solutions will be presented in the case of bump or thermal spot.

Our problem is different from the problem studied by Sykes (1978), because he considers that the scale of temperature stratification is the scale of the upper deck, while we take it to be smaller and of order of the boundary layer thickness; it is different to from El Hafi (1994) who studies a vertical natural convection (previous papers referred to horizontal convection) boundary layer perturbed by a bump with similar asymptotic methods.

## 2. Basic flow

We suppose, for the sake of simplicity, that the basic flow, driven by the free stream uniform velocity, is a classical Blasius boundary layer. This is not restrictive, because all we need is a basic velocity profile $U_O(y)$. We study a localised disturbance at the distance $L$ of the leading edge. At this point, the boundary layer is of thickness $Re^{-1/2}L$. The incoming fluid is at temperature $T_\infty$, and the plate at $T_0$, (say: $\Delta T_0 = T_0 - T_\infty$). Pure thermal convection is relevant as long as the transverse gradient from equation (1c) is small. It means that $1 >> J >> Re^{-1}$. So, under those hypotheses, the forced thermal boundary layer is of the same thickness as the dynamic one.

Let us remark that we impose the scale $L$, but Schneider & Wasel (1985) or Daniels & Gargaro (1993) do not because, in the real mixed convection problem, the natural longitudinal scale is built with Richardson number. It is the length that gives unit Richardson number. Nevertheless, the calculations they did, show that the breakdown occurs for a rather small abscissa, that means that the Richardson number built with this abscissa is smaller than one. It is then interesting to investigate small $J$ behaviour.

## 3. Triple deck interaction: the $J \sim Re^{-1/8}$ régime

### 3.1.1 Main deck

The classical triple deck tool (Neiland (1970) and Brown *et al.* (1975)) is applied to cope with a localized disturbance gauged by $x_3 L$ longitudinally. Perturbation of speed in the main deck is $\varepsilon U_\infty$, where $\varepsilon$ is unknown, so we recover:

$$U = U_0(y) + \varepsilon A(x) U_0(y), v = -\frac{\varepsilon\delta}{x_3} A(x) U_0(y), \quad and \quad T = \theta_0(y) + \varepsilon A(x)\theta_0'(y), \qquad (2)$$

where $U_0(y)$ and $\theta_0(y)$ are of course the undisturbed basic speed and temperature profiles. For the temperature, as for the speed, there is a matching between the top of the main deck and the bottom of the upper deck, and it is the same for the lower deck. We see that the temperature behaves as the Stewartson $S$ function in hypersonic flows (Brown *et al.* (1975)). We recover in this deck the unknown displacement function of the boundary layer $-A(x)$.

### 3.1.2 Crossing the main deck

This perturbation of temperature gives rise to a transverse change of pressure through the "main deck", we develop it in power of $\varepsilon$ as (neglecting $0(\varepsilon^3)$ terms):

$$\frac{\partial p_0}{\partial y} + \varepsilon \frac{\partial p_1}{\partial y} + \varepsilon^2 \frac{\partial p}{\partial y} = \varepsilon \tilde{J}(\theta_0(y) + \varepsilon A(x)\theta_0'(y)), \qquad (3)$$

where we have anticipated the fact that the pressure in the lower deck is of order $\varepsilon^2$, and that $x_3$ is of order $\varepsilon^3$ and where, for convenience, we have defined a reduced Richardson number: $\tilde{J} = \frac{J}{\varepsilon}$.

Looking at each power of $\varepsilon$, we see that the first term is zero (as we supposed); the second one shows that there is a pressure stratification coming from basic temperature profile ($\tilde{J}\int_0^\infty \theta_0(y)dy$ ), it does not depend on $x$ at the short scale $x_3$, and it will then not be useful; the third one integrates (using $\theta_0(\infty) = 0; \theta_0(0) = 1$ by definition) as:

$$p(x, y \to \infty) - p(x, y \to 0) = \tilde{J}A(x)(\theta_0(\infty) - \theta_0(0)) = -\tilde{J}A(x), \qquad (4)$$

where $p(x, y \to \infty)$ splices with upper deck and $p(x, y \to 0)$ with lower deck.

### 3.1.3 The upper deck

The matching with the "upper deck" for the pressure ($p(x, y \to \infty)$ is written for short $p(x)$) gives Hilbert integral (we recover the classical linear perturbed perfect fluid in the subsonic case): $(\frac{\varepsilon\delta}{\varepsilon^3})\frac{1}{\pi}\int \frac{-A'}{x-\xi}d\xi$, the usual gauge $\varepsilon = \delta^{-1/4} = R^{-1/8}$ and the new pressure displacement relation (Lagrée (1994)) is:

$$\frac{1}{\pi}\int \frac{-A'}{x-\xi}d\xi = p - \tilde{J}A. \qquad (5)$$

The effect of the temperature is to add a new term proportional to the displacement function $A$, it may be interpreted as a hydrostatic variation of pressure.

### 3.2 Double deck interaction: the $Re^{-1/8} \ll J \ll 1$ régime

It is worth noticing that as the Richardson number increases, in order of magnitude, from $Re^{-1/8}$, the perturbations in the upper deck become smaller and smaller and the longitudinal gauge, while remaining much more smaller than one, grows; so we find that for moderate values of $J$, the good choice for $\varepsilon$ is rather $\varepsilon = |J|$ and the relation becomes: $\left(|J|^{-4}\, Re^{-1/2}\right) \frac{1}{\pi} \int \frac{-A'}{x-\xi} d\xi = (p \pm A)$ and ultimately the pressure displacement relation degenerates in the form (Lagrée (1994)):

$$p = -A, \tag{6a}$$

for a cold wall, and in the form:

$$p = A, \tag{6b}$$

for a hot one, with: $Re^{-1/8} \ll |J| \ll 1$. This shows that the upper deck is not necessary for the interaction to take place, the same phenomenon exists in hypersonic flows (Brown *et al.* (1975)) for cold wall. The $p = -A$ relation is found in the birth of hydraulic jumps in Bowles & Smith (1992), Gajjar & Smith (1983) and in subsonic pipe flows by Ruban & Timoshin (1986) and leads to upstream influence. Note that the interaction relation $p = \pm A$ was independently found by Bowles (1994).

### 3.3 Lower deck

There are two possibilities for the lower deck depending on the temperature of the spot either (i) it is of magnitude $\varepsilon \Delta T_0$ either (ii) bigger.

#### i) no spot, or weak one

If there is no spot (but a hump) or if it is weak, in the lower deck, classical gauges are used: $u$ is of order $\varepsilon$, $p$ of order $\varepsilon^2$, and $x_3$ is $\varepsilon^3$, temperature matching with main deck gives the order $\varepsilon$ for the temperature, transverse variable is $\varepsilon$ times the transverse variable in the main deck.

We recover the basic lower deck set:

$$\frac{\partial u}{\partial x} + \frac{\partial v}{\partial y} = 0, \quad (u\frac{\partial}{\partial x} + v\frac{\partial}{\partial y})u = -\frac{d}{dx}p + \frac{\partial^2}{\partial y^2}u, \quad (u\frac{\partial}{\partial x} + v\frac{\partial}{\partial y})T = \frac{\partial^2}{\partial y^2}T. \tag{7}$$

Boundary conditions are no slip at the wall, $A(-\infty) = 0$ and for $y \to \infty$, $u \to y + A$, and $T \to y + A$.

#### ii) spot

But, if the spot is warmer (or cooler), the temperature in the lower deck must be rescaled by the temperature of the spot $T_w$. Say that on a small distance scaled by $x_3$ the temperature of the plate changes from $T_0$ to $T_0 + T_w$; then it is better to use $((T_w/\Delta T_0)\Delta T_0) = T_w$, instead of $(\varepsilon(\Delta T_0))$ as a new gauge of temperature variation (and $T_w >> \varepsilon \Delta T_0$ ).

The convective/ diffusive temperature equation is still the same, but there is now a transverse pressure gradient in the lower deck ($\frac{\partial p}{\partial y} = j\theta$) if $j = \tilde{J}T_w/\Delta T_0$ is of order one (this is a second Richardson number):

$$-\frac{\partial p(x,y)}{\partial x} = -\frac{dp(x)}{dx} + j\int_y^\infty \frac{\partial \theta(x,y)}{\partial x}dy, \tag{8}$$

where $p(x)$ is the pressure at the top of the lower deck (at the bottom of the main deck). Temperature is gauged by: $T = T_\infty + (T_0 - T_\infty)(1 + \frac{T_w}{T_0 - T_\infty}\theta) = T_0 + T_w\theta$. The matching between the lower and main decks becomes: $1 + \frac{T_w}{T_0 - T_\infty}\theta(y \to \infty) = 1 + \varepsilon(y + A(x))\theta_0'$, and since the temperature is strong at the wall ($T_w, >> \varepsilon\Delta T_0$.), the temperature $\theta$ is then zero at infinity. Hence, we obtain the lower deck problem: (which is very similar to Zeytounian (1991)):

$$\frac{\partial u}{\partial x} + \frac{\partial v}{\partial y} = 0; \quad (u\frac{\partial}{\partial x} + v\frac{\partial}{\partial y})u = -\frac{d}{dx}p + j\int_y^\infty \frac{\partial \theta}{\partial x}dy + \frac{\partial^2}{\partial y^2}u, \quad (u\frac{\partial}{\partial x} + v\frac{\partial}{\partial y})\theta = \frac{\partial^2}{\partial y^2}\theta. \tag{9}$$

Boundary conditions are no slip at the wall, $A(-\infty) = 0$ and for $y \to \infty$, $u \to y + A$, but $\theta \to 0$.

### 3.4 Various fundamental problems

Depending on the values of the Richardson number we may build various "fundamental problems of triple deck" by mixing the above lower deck problem and the pressure displacement relations we found.

### 3.5 Linearized solution, eigenvalue

The system admits the Blasius solution $u = y$ as a basic one. Perturbing it by a small amount, and taking the Fourier transform of the result, gives:

$$i\alpha\tilde{u} + \frac{\partial}{\partial y}\tilde{v} = 0, \quad i\alpha y\tilde{u} + \tilde{v} = -i\alpha\tilde{p} + j\int_y^\infty i\alpha\tilde{T}dy + \frac{\partial^2}{\partial y^2}\tilde{u}, \quad i\alpha y\tilde{T} = \frac{\partial^2}{\partial y^2}\tilde{T}. \tag{10}$$

The temperature ( $\tilde{T} = Ai(y(i\alpha)^{1/3})$) is an Airy function. Perturbation of skin friction $\tilde{\tau} = \frac{\partial \tilde{u}}{\partial y}$ verifies an Airy differential equation forced by an Airy function, this is solved as:

$$\frac{\partial}{\partial y}\tilde{u}(0) = 1 + \frac{(i\alpha)^{2/3}\tilde{p}}{Ai'(0)} - j(i\alpha)^{1/3}(\frac{Ai(0)}{3Ai'(0)} - Ai'(0)), \tag{11}$$

then we find the displacement function ($\tilde{F}$ is the Fourier transform of the bump):

$$\tilde{A} + \tilde{F} = \frac{(i\alpha)^{1/3}\tilde{p}}{3Ai'(0)} - j(Ai(0) + \frac{1}{9Ai'(0)}). \tag{12}$$

In this expression (written for short $\tilde{A} + \tilde{F} = \beta^*\tilde{p} + j\beta_j$, where $\beta^*$ is the standard triple deck coefficient) with temperature effect, we see that, for a negative $j$, there is a decrease of skin friction and a positive displacement $-\tilde{A}$, so an increase of the boundary layer. The Fourier transform of the new pressure displacement relation reads:

$$\tilde{p} = (|\alpha| + \tilde{J})\tilde{A}. \tag{13}$$

Then we deduce, for the pressure at the top of the lower deck, the following expression:

$$\tilde{p} = \frac{(|\alpha| + \tilde{J})}{1 - (|\alpha| + \tilde{J})\beta*}(j\beta_j - \tilde{F}). \tag{14}$$

This expression will be used to compute the linearized response of the stratified boundary layer to a bump or/and a thermal spot as we will see in section 3.7.

On the other hand, the linear approach gives the behaviour for $x \to -\infty$. In the case of cold wall, ($p = -A$) we recover the same behaviour as in hypersonic flows (Brown *et al.* (1975) and Gajjar & Smith (1983)), thus, the Lighthill eigenvalue (from the growing exponential) may be found, which shows that there is upstream influence: $p = ae^{kx}$, with $k = (-3Ai'(0))^3$. This upstream influence may describe the phenomenon of "blocking" which is observed in stratified flows.

Note that in Daniels & Gargaro (1993) breakdown, in the vicinity of the singular point, the perturbed quantities behave with a $p = -A$ law.

### 3.6 Stability

Classical triple deck technique is applied, and we find that the hot wall case ($p = A$) is linearly unstable and marginal stability curve is given by $\omega = 2.29\alpha^{2/3}$ and $\tilde{J} = 1.001\ \alpha^{-1/3} - |\alpha|$ (by analogy with Tollmien- Schlichting waves with triple deck scales deduced by Smith (1979)), so there is a non viscous mode for $\tilde{J}$ negative and large, and a viscous mode of large wave length for $\tilde{J}$ positive and large (which corresponds to $p = A$).

### 3.7 Numerical results

The behaviour, computed with those relations will be shown in the case of a bump or a spot of shape $(1 - x^2)^2$, with various values of $\tilde{J}$. It shows upstream influence (as predicted), that may be interpreted, in stratified flows (Tritton (1988)), as "blocking", and downstream we observe small oscillations, interpreted as "lee waves", positive values of $\tilde{J}$ show indeed lowering and cancelling of upstream influence (see figure 1 and 2). Non linear results are presented on figure 3 and 4.

## 4. Single deck interaction?: the $J \sim 1$ régime

Numerical calculations have clearly shown that there is a singularity in the self interaction of the boundary layer at $J = O(1)$. We believe that this singularity is similar to the "branching solutions" obtained in supersonic inviscid- viscous interacting flows (Le Balleur (82)). An over simplification of the problem with integral methods, may be done as follow. First, let us suppose that the pressure is averaged through the boundary layer, $p \approx 1 - JRe^{1/2}\delta_1\alpha$ (introducing an integral parameter $\alpha \approx 1.7$ for $J = 0$). Second, the classical Kármán momentum equation with standard notations reads:

$$U_0^2(\frac{d(\delta_1/H)}{dx}) + \delta_1(1 + \frac{2}{H})U_0\frac{dU_0}{dx} = \frac{f_2 H}{Re\delta_1}U_0. \tag{15}$$

Combining those two equations, and without forgetting that $H$ may be written as a function of $\Lambda_1 = \delta_1^2 Re \frac{dU_0}{dx}$, the Pohlhausen description (near the Blasius solution $(\delta_0 Re^{-1/2}, \delta_0 \approx 1.7)$) gives $H \approx H_0 - H_p \frac{dU_0}{dx}$, with $H_0 \approx 2.58$, & $H_p \approx 0.533(1.7)^2$. We obtain:

$$J\alpha\delta_0 \frac{H_p}{H_0^2}\frac{d^2\delta_1}{dx^2} + \frac{d\delta_1}{dx}\left(\frac{1}{H_0} + J\alpha Re^{1/2}\delta_1\left(1+\frac{2}{H_0}\right)\right) = \frac{f_{20}H_0}{Re\delta_1}. \tag{16}$$

This development is exactly Schneider & Wasel (1985) ones, however we have added the second derivative term which is the most dominant one coming from the derivative $\partial(1/H)/\partial x$. When this term is omitted is easy to see that the forward integration leads to a singularity at a fixed abscissa whose value $x_s = (6H_0^4(\alpha J)^2(1+2/H_0)^2 f_{20})^{-1}$ is not so bad an estimation. But the forgotten second derivative term means, in our opinion, that another boundary condition must be imposed. This equation may be solved without difficulty with $\delta_1$ prescribed at each boundary of the domain. This over simplified heuristical approach may be a guideline for the numerical resolution of the single deck equations: they must be solved with a global technique with the two end values prescribed (work in this direction is in progress, Bowles (1994)), not by a marching technique from downstream to upstream.

## 5. Conclusion

We have presented the response of a boundary layer with small buoyant effects to a thermal spot or a bump. This is an extension of the triple deck approach, and we have seen that if the Richardson number is large enough, this is a double deck interaction. Stability and the understanding of the singularity could be investigated. We did not clear the singularity up, but we have shown that even at small values of $J$ there is upstream influence (from downstream) and self induced solutions which may be interpreted as birth of some thing being downstream which may be simply seen on the over simplified partial equation.

## Note

Recently, H. Steinrück ("Mixed convection over a cooled horizontal plate: non-uniqueness and numerical instabilities of the boundary layer equations", *J. Fluid Mech.*, vol 278, pp. 251-265) has shown that the mixed convection problem presents branching solutions (when numericaly solved in marching from up to downstream) associated to an unbounded sequence of eigenvalues that he found by asymptotic expansion (the one he obtains for small $x$, is exactly the Lightill eigenvalue coming from $p = -A$).

## References


R.I. Bowles & F.T. Smith (1992):"the standing hydraulic jump: theory, computations and comparisons with experiments", *J. Fluid Mech.*, vol 242, pp. 145-168

R. I. Bowles (1994): private communication

S.N. Brown, K. Stewartson & P.G. Williams (1975): "hypersonic self induced separation", *The Phys. of Fluids*, vol 18, No 6, June

P.G. Daniels (1992):" Asingularity in thermal boundary- layer flow on a horizontal surface", *J. Fluid Mech.*, vol 242, pp. 419- 440

P.G. Daniels & R.J. Gargaro (1993): " Buoyancy effects in stably stratified horizontal boundary- layer flow", *J. Fluid Mech.*, Vol. 250, pp. 233- 251

M. El Hafi (1994): "analyse asymptotique et raccordements, étude d'une couche limite de convection naturelle". thèse de l'Université de Toulouse

J. Gajjar & F.T. Smith (1983): "On hypersonic self induced separation, hydraulic jumps and boundary layer with algebraic growth", *Mathematika*, 30, pp. 77-93

P.-Y. Lagrée (1994): "Convection thermique mixte à faible nombre de Richardson dans le cadre de la triple couche", *C. R. Acad. Sci. Paris*, t. 318, Série II, pp. 1167- 1173, 1994

J.-C. Le Balleur (1982) "Viscid- inviscid coupling calculations for two and three- dimensional flows, Von Kármán Institute for Fluid Dynamics, Computational Fluid Dynamics, Lecture Series 1982-04

V. Ya Neiland (1969): "Propagation of perturbation upstream with interaction between a hypersonic flow and a boundary layer", *Mekh. Zhid. Gaz.,* Vol. 4, pp. 53-57

A. I. Ruban & S. N. Timoshin (1986): "Propagation of perturbations in the boundary layer on the walls of a flat chanel", *Fluid Dynamics*, No 2, pp. 74-79

W. Schneider & M.G. Wasel (1985): "Breakdown of the boundary layer approximation for mixed convection above an horizontal plate", *Int. J. Heat Mass Transfert* , Vol. 28, No 12, pp. 2307-2313

F.T. Smith (1979): "on the non parallel flow stability of the blasius boundary layer" *Proc. Roy. Soc. Lond.*, A366, pp. 91- 109

R.I. Sykes (1978): "Stratification effects in boundary layer flows over hills", *Proc. Roy. Soc. Lond.* A 361, pp. 225- 243

D.J. Tritton (1988): "Physical fluid dynamics", Oxford Science Publications

R. Zeytounian (1991):"Meteorogical fluid dynamics", lecture notes in physics m5, Springer Verlag

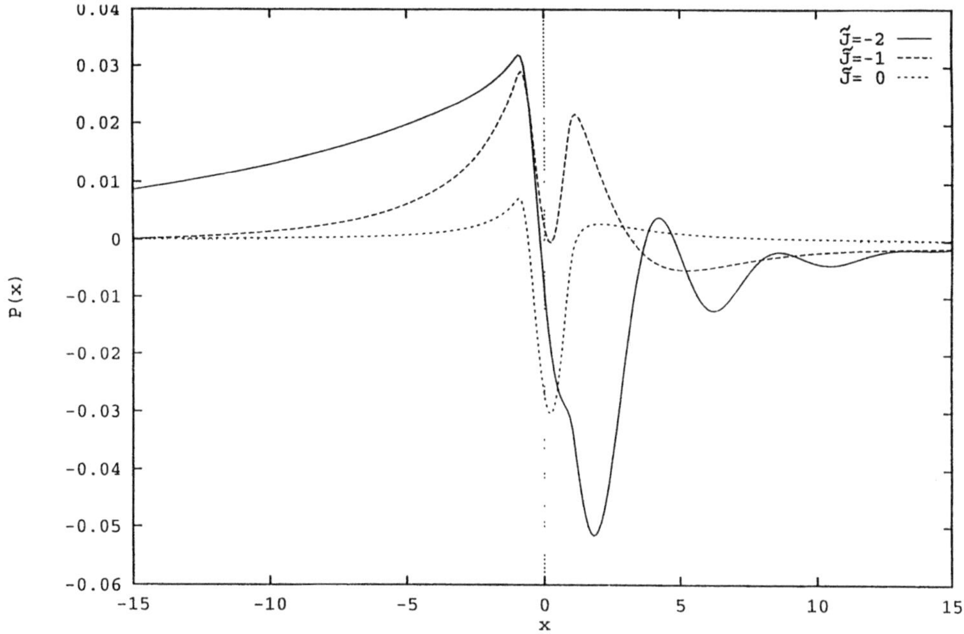


**Fig 1.** Cold spot $-1 < x < 1$, Pressure at the top of the main deck, linear results.

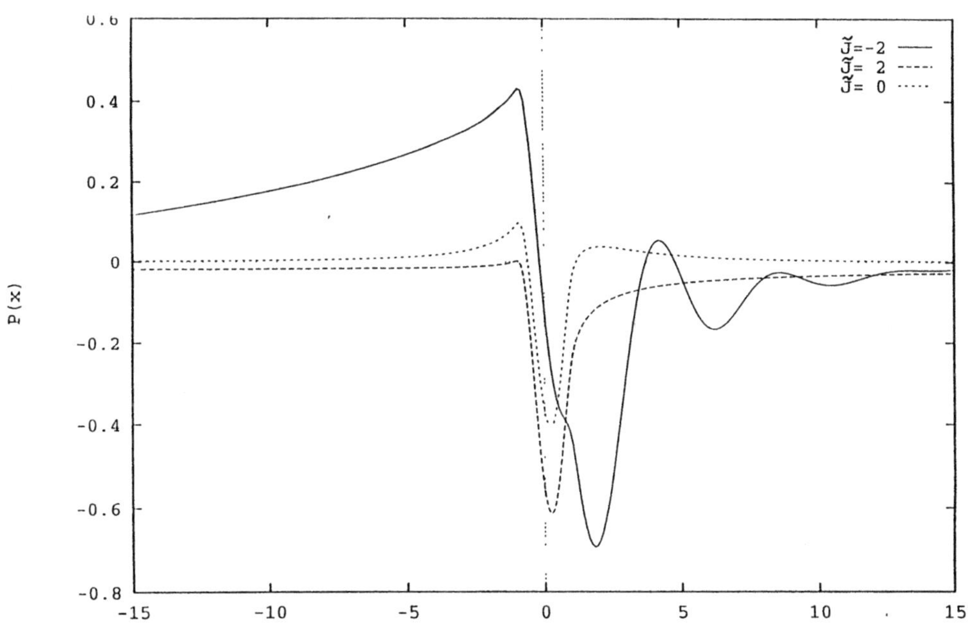


**Fig 2.** Bump $-1 < x < 1$, Pressure at the top of the main deck, liner results.

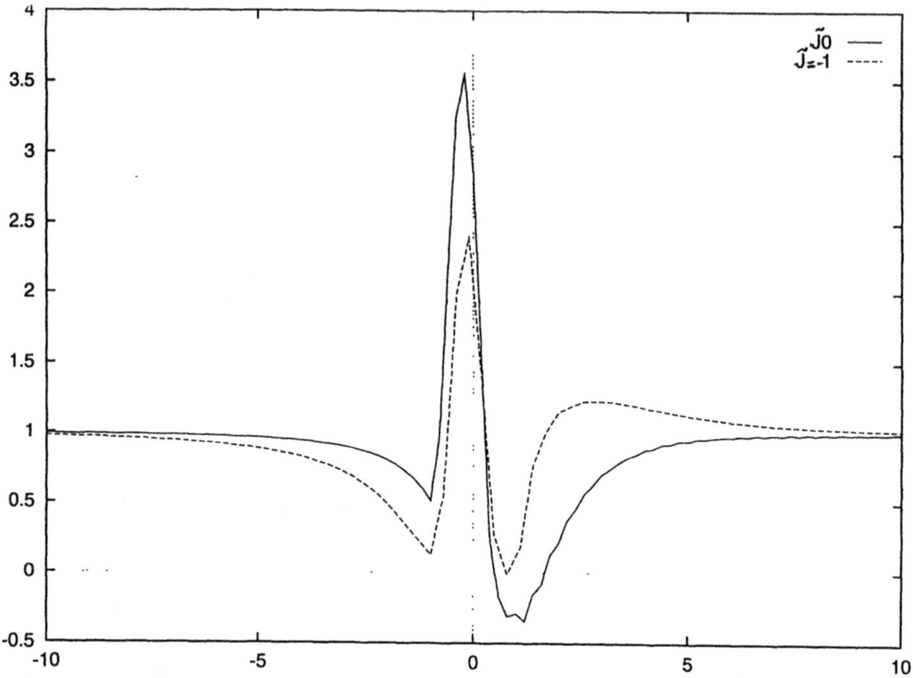


**Fig 3.** Bump $-1 < x < 1$, skin friction, non linear resolution $\tilde{J} = 0$, $\tilde{J} = -1$.

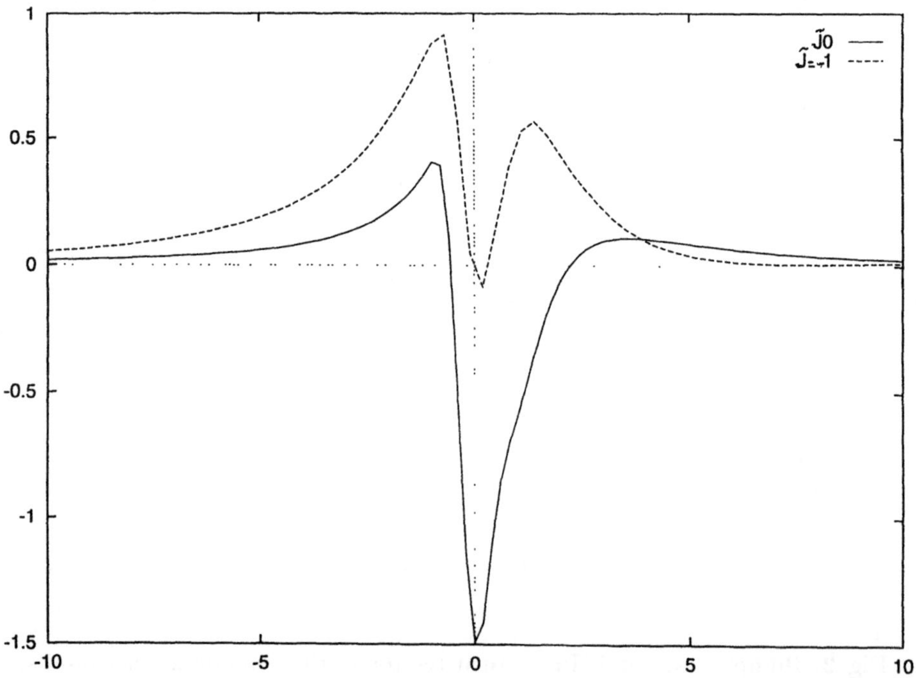


**Fig 4.** Bump $-1 < x < 1$, Pressure at the top of the main deck, non linear resolution $\tilde{J} = 0$, $\tilde{J} = -1$.

# ASYMPTOTIC MODELLING FOR SEPARATING BOUNDARY LAYERS

**Jacques MAUSS**

Laboratoire de Modélisation en Mécanique des Fluides, Université Paul Sabatier, 118, route de Narbonne, 31062 Toulouse, France

**Abstract** : The method of matched asymptotic expansions is used to explain how the triple deck structure in a boundary layer can be formed. In the context of a laminar steady flow of an incompressible fluid over a flat plate, a theory is developed to explain the separation over significant wall disturbances. In particular, we show that the triple deck structure is the first perturbation that can both displace the classical boundary layer and cause separation of the flow. Above this exist a serie of perturbations, smaller but "stronger", that cause a separation of the boundary layer without displacing it. This serie is limited by the smallest perturbation compatible with the hypothesis of the theory, thus leading to a theory in double deck.

## I. Introduction

Following the work of Lighthill (1953) on supersonic flows, three papers initiated the triple deck theory for boundary layers. The first is the well known work of Stewartson and Williams (1969) on self-induced separation. This was followed, rather simutaneously, by the work of Neiland (1969) and Messiter (1970). Since these pioneering developments, a lot of work has been done on this problem and a good review of the different contributions can be found in Smith (1982) and Lagree (1992).

However, it is remarkable how little has been done in the development of a proper description of the nature of the triple deck structure. Following the work of Nayfeh (1991), a systematic analysis of the Navier-Stokes equations has been performed to prove the triple deck theory (Mauss et al. 1992). It was also found that a second characteristic structure is present in the theory. This structure is smaller since it does not affect the external flow. Furthermore, it does not require a triple deck description even though its perturbation is able to cause boundary layer separation. This new description of the structure of the boundary layer was called a "Double Deck".

## II. Formulation

### II.1 Definition of the problem

As we are studying steady flows at high Reynolds number $R_e$, a small parameter $\varepsilon$ can be defined such that $R_e = \varepsilon^{-m}$ , where m is an arbitrary positive number that we will fix for clarity of presentation (for instance, in triple deck theory, we put m = 8).

The dimensionless equations of the problem can be written:

$$\bar{V} \cdot \nabla \bar{V} = -\nabla \pi + \varepsilon^m \nabla^2 \bar{V} \quad \text{and} \quad \nabla \cdot \bar{V} = 0, \tag{1}$$

where $\bar{V}$ is the velocity vector of components $(v_x , v_y)$ and $\pi$ is the pressure.

For simplicity we consider only a plane steady laminar flow of an incompressible fluid over a flat plate in a uniform parallel stream.

In cartesian coordinates (x , y), where x is parallel to the plate, the boundary conditions are:

$\bar{V} = 0$ on the flat plate $x > -1$, where x = -1 is the leading edge of the plate,
$\bar{V} \cdot \bar{y} = 0$ for $x < -1$, and
$\bar{V} = \bar{x}$ and $\pi = 0$ , far upstream.

When we are looking for significant perturbations which cause boundary layer separation, we assume that $\varepsilon^\alpha$ and $\varepsilon^\beta$ are respectively the scale perturbations in the longitudinal and the transversal directions that will cause boundary layer separation. It was found [Nayfeh 1991, Mauss 1992] that two significant structures appear:

1. The triple deck structure with $\alpha = \dfrac{3m}{8}$ and $\beta = \dfrac{5m}{8}$
2. The double deck structure with $\alpha = \dfrac{m}{2}$ and $\beta = \dfrac{2m}{3}$.

Moreover, we know that the basic theory is defined in the frame $0 < \alpha \le \dfrac{m}{2}$. One question arises: what happens when $\alpha$ is neither $\dfrac{3m}{8}$ nor $\dfrac{m}{2}$?

The answer to this question becomes straightforward when systematic application of the method of matched asymptotic expansions is used. For clarity, we will consider a small hump defined by $y = \varepsilon^\beta F\left(\dfrac{x}{\varepsilon^\alpha}\right)$, where f is a sufficiently smooth function.

## II.2 Middle layer or main stream

The basic theory lays on the following idea: we are looking for a couple $(\alpha,\beta)$ wih $\alpha > 0$ for which the classical boundary layer solution is not strongly modified. We therefore write:

$$v_X = u_o(Y) + \varepsilon^a\, u(X,Y,\varepsilon) \tag{2}$$

where $a > 0$ issues from the perturbation in the vicinity of $x = 0$. Also, we have introduced the variables:

$$X = \frac{x}{\varepsilon^\alpha},\ Y = \frac{y}{\varepsilon^{m/2}}$$

where X is the local natural variable and Y is the classical boundary layer variable. It is well known that with $u_o(Y) = f'(Y)$, we have the Prandtl-Blasius problem,

$$2f''' + f\,f'' = 0$$

with the boundary conditions :

$$f(0) = f'(0) = 0,\ f'(\infty) = 1.$$

For our purpose, the essential results of the numerical solution are contained in the expansions for small and large Y:

$$f(Y) = \lambda_o \frac{Y^2}{2} + O(Y^5), \qquad \lambda_o = 0.33206,$$
$$f(Y) \cong Y - \beta_o + TST, \qquad \beta_o = 1.7208.$$

In order for the continuity equation to be satisfied, we write:

$$v_y = \varepsilon^{a-\alpha+\frac{m}{2}}\, v(X,Y,\varepsilon)$$

and, without restricting the general case,

$$\pi = \varepsilon^b\, p(X,Y,\varepsilon),$$

where a, b and $\alpha$ have not yet been defined.

The Navier-Stokes equations (1) are now written in the following form:

$$u_o \frac{\partial u}{\partial X} + v\, u'_o = -\varepsilon^{b-a} \frac{\partial p}{\partial X} + O\left[\varepsilon^a, \varepsilon^{m-\alpha}, \varepsilon^{\alpha}\right],$$

$$u_o \frac{\partial v}{\partial X} = -\varepsilon^{b+2\alpha-a-m} \frac{\partial p}{\partial Y} + O\left[\varepsilon^a, \varepsilon^{m-\alpha}, \varepsilon^{\alpha}\right].$$

These equations induce a singular perturbation when

$$0 < \alpha < m$$

since they reduce to first order equations inside the middle layer or the main stream. In order to satisfy the boundary conditions, we introduce a boundary layer in the vicinity of the plate.

### II.3 Inner layer or viscous deck.

The boundary layer variable is clearly:

$$\tilde{y} = \frac{y}{\varepsilon^{\beta}} = \frac{Y}{\varepsilon^{\beta - m/2}}$$

Moreover, to ensure:

- a matching with the first term $u_o(y)$ of the middle layer,
- the conservation of mass (continuity equation), and
- a comparison, in the longitudinal equations, between pressure, viscous and inertial terms,

we obtain the fundamental relation:

$$\beta = \frac{\alpha}{3} + \frac{m}{2} \tag{3}$$

and we write:

$$v_x = \varepsilon^{\frac{\alpha}{3}}\, \tilde{u}, \quad v_y = \varepsilon^{\frac{m}{2} - \frac{\alpha}{3}}\, \tilde{v}, \quad \pi = \varepsilon^{\frac{2\alpha}{3}}\, \tilde{p} \tag{4}$$

The Navier-Stokes equations (1) are now written:

$$\tilde{u} \frac{\partial \tilde{u}}{\partial X} + \tilde{v} \frac{\partial \tilde{u}}{\partial \tilde{y}} = -\frac{\partial \tilde{p}}{\partial X} + \frac{\partial^2 \tilde{u}}{\partial \tilde{y}^2} + O\left[\varepsilon^{m - \frac{4\alpha}{3}}\right]$$

$$\tilde{u} \frac{\partial \tilde{v}}{\partial X} + \tilde{v} \frac{\partial \tilde{v}}{\partial \tilde{y}} = -\varepsilon^{\frac{4\alpha}{3} - m} \frac{\partial \tilde{p}}{\partial \tilde{y}} + \frac{\partial^2 \tilde{v}}{\partial \tilde{y}^2} + O\left[\varepsilon^{m - \frac{4\alpha}{3}}\right]$$

In the case where:

$$0<\alpha<\frac{3m}{4},$$

we can write the inner asymptotic expansions:

$$v_x=\varepsilon^{\frac{\alpha}{3}}\,\tilde{u}_1(X,\tilde{y})+\ldots\ ,\quad v_y=\varepsilon^{\frac{m}{2}-\frac{\alpha}{3}}\,\tilde{v}_1(X,\tilde{y})+\ldots$$

and
$$\pi=\varepsilon^{\frac{2\alpha}{3}}\,\tilde{p}_1(X,\tilde{y})+\ldots$$

The boundary layer problem can now be written:

$$\begin{aligned}&\tilde{u}_1\frac{\partial\tilde{u}_1}{\partial X}+\tilde{v}_1\frac{\partial\tilde{u}_1}{\partial\tilde{y}}=-\frac{\partial\tilde{p}_1}{\partial X}+\frac{\partial^2\tilde{u}_1}{\partial\tilde{y}^2}\\&\frac{\partial\tilde{u}_1}{\partial X}+\frac{\partial\tilde{v}_1}{\partial\tilde{y}}=0,\ \frac{\partial\tilde{p}_1}{\partial\tilde{y}}=0\end{aligned}\qquad(5)$$

and the boundary conditions are:

$$\tilde{u}_1=\tilde{v}_1=0\quad\text{for}\quad\tilde{y}=F(X),$$

The matching condition is:

$$\tilde{u}_1\equiv\lambda_o\,\tilde{y}+\ldots\ \text{when}\quad\tilde{y}\to+\infty\qquad(6)$$

while the initial condition is simply

$$\lim_{X\to\infty}\tilde{u}_1=\lambda_o\tilde{y}$$

We know that this problem is mathematically poorly posed [Audounet et al.1991]. In fact, in the matching condition (6), we must have more information from the middle layer.

### II.4 Return to the middle layer

In the middle layer, the asymptotic expansions are now:

$$\begin{cases}v_x=u_o(Y)+\varepsilon^a\,u_1(X,Y)+\ldots\\v_y=\varepsilon^{a-\alpha+\frac{m}{2}}\,v_1(X,Y)\ +\ldots\\\pi=\varepsilon^{\frac{2\alpha}{3}}\,p_1(X,Y)+\ldots\end{cases}$$

In fact, from (5), we can write:

$$b = \frac{2\alpha}{3} \quad \text{and} \quad \lim_{Y \to 0} p_1(X,Y) = \tilde{p}_1(X).$$

Moreover, the middle equations are:

$$\begin{aligned} & u_o \frac{\partial u_1}{\partial X} + v_1 u'_o = -\varepsilon^{\frac{2\alpha}{3} - a} \frac{\partial p_1}{\partial X} + \ldots \\ & u_o \frac{\partial v_1}{\partial X} = -\varepsilon^{\frac{8\alpha}{3} - a - m} \frac{\partial p_1}{\partial Y} + \ldots, \quad \frac{\partial u_1}{\partial X} + \frac{\partial v_1}{\partial Y} = 0 \end{aligned} \tag{7}$$

In the following, it will be shown that the guiding principle of least possible degeneracy, roughly applied, leads to wrong solutions.

Instead, we must find all possible solutions using a key analysis: i.e. study the scale terms in (7) using the following diagram:

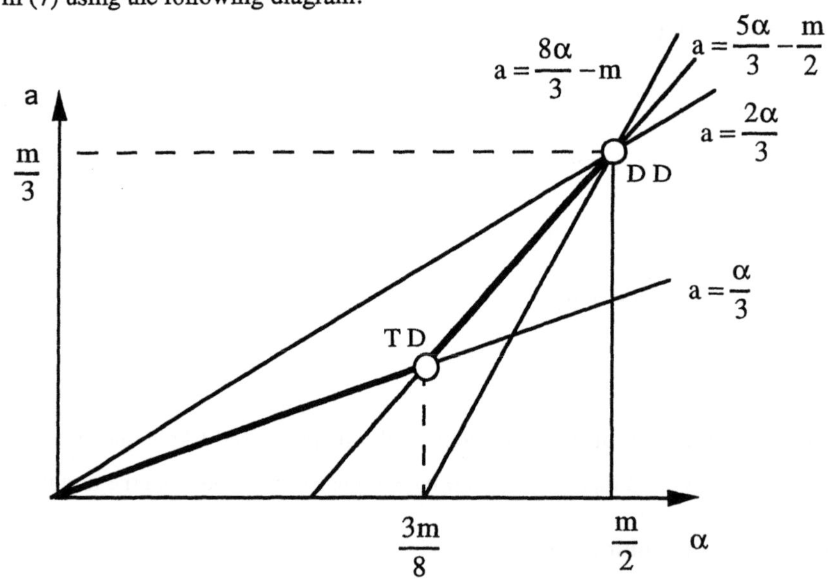


**Fig. 1.** Order of magnitude of the ratio of the perturbed velocity to the length of the hump.

The study of this diagram shows that solutions exists only in the triangle:

$$0 < a \leq \frac{2\alpha}{3}, \quad \frac{8\alpha}{3} - m \leq a$$

As "a" indicates the order of the induced perturbation, we first look at the case:

$$a < \frac{2\alpha}{3} \tag{8}$$

The velocity is easily obtained, and takes the following form:

$$u_1 = A(X)\, u'_o(Y), \quad v_1 = -A'(X) u_o(Y) \;, \tag{9}$$

where A(X) is an arbitrary function called the displacement, such that:

$$\lim_{X \to -\infty} A(X) = 0.$$

For the pressure, two cases appear:

i. $\dfrac{\partial p_1}{\partial y} = 0$ for $a > \dfrac{8\alpha}{3} - m$

ii. $\dfrac{\partial p_1}{\partial y} = -u_o \dfrac{\partial v_1}{\partial X}$ for $a = \dfrac{8\alpha}{3} - m$

The nature of the condition necessary to resolve the inner problem is found by performing a matching between $v_x$ in the inner (I) layer and the middle (M) layer.

## II.5 Middle layer and inner layer matching (MI matching)

Following Eckhaus (1979), we define the expansion operator M to the order $\varepsilon^a$ such that:

$$M\, v_x = u_o(Y) + \varepsilon^a\, A(X) u'_o(Y)$$

The behaviour of $M\, v_x$ for $Y \to 0$ in the inner layer is given by:

$$I * M\, v_x = \lambda\, \varepsilon^{\frac{\alpha}{3}}\, \tilde{y} + \lambda\, \varepsilon^a\, A(X) + \ldots \tag{10}$$

This is to be compared with the behaviour when $\tilde{y} \to \infty$ of the expansion operator in the inner layer to the order $\varepsilon^{\frac{\alpha}{3}}$ :

$$I\, v_x = \varepsilon^{\frac{\alpha}{3}}\, \tilde{v}_1 (X, \tilde{y}) \tag{11}$$

The method to obtain matching relations is based, after Van Dyke (1964), on what is called an asymptotic matching principle. Following once again Eckhaus (1979), matching is closely related to the problem of constructing a composite expansion. The key is to use Van Dyke's principle only with expansion operators defined <u>at the same order</u>, (Mauss, 1994).

Here, we must write, at the same order, $\varepsilon^{\frac{\alpha}{3}}$ or $\varepsilon^{a}$,

$$M\,I\,v_x \equiv I\,M\,v_x$$

Now, it is possible to construct a uniform approximation of $v_x$ such that:

$$A_p\,v_x = M\,v_x + I\,v_x - M\,I\,v_x$$

which is supposed to be at the considered order.

As for as we know, all counter examples to Van Dyke's rule can be removed with this special matching principle.

In our case, comparing first (10) and (11), we see that there is no possible matching for $a < \frac{\alpha}{3}$.

Then two cases appear:

i. $\quad I\,M\,v_x = \lambda_o\,\varepsilon^{\frac{\alpha}{3}}\,(\tilde{y} + A(X)) \qquad$ for $\qquad a = \frac{\alpha}{3}$

leading to the condition for the inner problem:

$$\lim_{\tilde{y}\to\infty} \left[\tilde{u}_1(X,\tilde{y}) - \lambda_o \tilde{y}\right] = \lambda_o\,A(X) \tag{12}$$

ii. $\quad I\,M\,v_x = \lambda_o\,\varepsilon^{\frac{\alpha}{3}}\,\tilde{y} \qquad$ for $\qquad a > \frac{\alpha}{3}$

with the condition

$$\lim_{\tilde{y}\to\infty} \left[\tilde{u}_1(X,\tilde{y}) - \lambda_o \tilde{y}\right] = 0 \tag{13}$$

Now, the boundary layer problem is well-formulated (Audounet et al. 1991). In case i, when $a = \frac{\alpha}{3}$, we need a coupling relation between $A(X)$ and $\tilde{p}_1(X)$; in case ii, when $a > \frac{\alpha}{3}$, $\tilde{p}_1(X)$ is known from numerical analysis.

## II.6 Exterior layer or upper deck

When we are looking to solution (9), it is easy to see that in the limit of $y \to \infty$, $v_1$ is not zero. We must therefore predict an exterior layer (E) with the exterior variable:

$$\bar{y} = \frac{y}{\varepsilon^c} = \frac{Y}{\varepsilon^{c-\frac{m}{2}}} \quad \text{with} \quad c < \frac{m}{2}.$$

We have to write, with $d > 0$, $v_x = 1 + \varepsilon^d\, \bar{u}(X,\bar{y})$ and $v_y = \varepsilon^{d+c-\alpha}\, \bar{v}(X,\bar{y})$ such that the continuity equation is satisfied.

Moreover, the pressure is not known. Thus, we write:

$$\pi = \varepsilon^e\, \bar{p}(X,\bar{y})$$

The Navier-Stokes equations are now written as:

$$\frac{\partial \bar{u}}{\partial X} = -\varepsilon^{e-d} \frac{\partial \bar{p}}{\partial X} + O\left[\varepsilon^d, \varepsilon^{m-\alpha}, \varepsilon^{m+\alpha-2c}\right]$$
$$\frac{\partial \bar{v}}{\partial X} = -\varepsilon^{e+2\alpha-d-2c} \frac{\partial \bar{p}}{\partial \bar{y}} + O\left[\varepsilon^d, \varepsilon^{m-\alpha}, \varepsilon^{m+\alpha-2c}\right]$$

There is no solution to this problem unless:

$$e = d \quad \text{and} \quad c = \alpha$$

The asymptotic exterior expansions are therefore:

$$\begin{cases} v_x = 1 + \varepsilon^e\, \bar{u}_1(X,\bar{y}) + \ldots \\ v_y = \varepsilon^e\, \bar{v}_1(X,\bar{y}) + \ldots \\ \pi = \varepsilon^e\, \bar{p}_1(X,\bar{y}) + \ldots \end{cases}$$

with the exterior equations:

$$\frac{\partial \bar{u}_1}{\partial X} + \frac{\partial \bar{v}_1}{\partial \bar{y}} = 0, \quad \frac{\partial \bar{u}_1}{\partial X} = -\frac{\partial \bar{p}_1}{\partial X}, \quad \frac{\partial \bar{v}_1}{\partial X} = -\frac{\partial \bar{p}_1}{\partial y}$$

and the exterior variable:

$$\bar{y} = \frac{y}{\varepsilon^\alpha} = \frac{Y}{\varepsilon^{\alpha - \frac{m}{2}}}$$

We observe that $\bar{v}_1$ and $\bar{p}_1$ are harmonic conjugated functions.

We now apply the preceeding matching principle successively for the pressure $\pi$ and the transverse velocity $v_y$.

## II.7 Exterior layer and middle layer matching (EM Matching) for $\pi$

Using the middle equations (9), two cases are obtained:

i. $\dfrac{\partial p_1}{\partial Y} = 0 \qquad \text{for } a > \dfrac{8\alpha}{3} - m$

ii. $\dfrac{\partial p_1}{\partial Y} = A''(X)u_o^2(Y) \quad \text{for } a = \dfrac{8\alpha}{3} - m$

In case ii, we easily obtain:

$$p_1(X,Y) = A''(X)\left[f(Y)f'(Y) + 2f''(Y) - 2\lambda_o\right] + \tilde{p}_1(X)$$

which leads, when Y is large, to:

$$p_1(X,Y) \cong A''(X)\left[Y - \beta_o - 2\lambda_o\right] + \tilde{p}_1(X) + TST$$

Thus, writing $M\pi = \varepsilon^{\frac{2\alpha}{3}} p_1(X,Y)$ to the order $\varepsilon^{\frac{2\alpha}{3}}$, we obtain:

$$E*M\pi = \varepsilon^{\frac{5\alpha}{3} - \frac{m}{2}} A''(X)\bar{y} + \varepsilon^{\frac{2\alpha}{3}}\left[\tilde{p}_1(X) - (\beta_o + 2\lambda_o)A''(X)\right] + \ldots$$

Comparison with $E\pi = \varepsilon^e \, \overline{p}_1(x,\bar{y})$ leads to:

$$e = \frac{5\alpha}{3} - \frac{m}{2} \text{ and } \overline{p}_1(x,0) = 0.$$

As $\overline{p}_1$ is harmonic in the half-plane $\bar{y} > 0$,

$$\overline{p}_1(X,\bar{y}) = 0$$

which is not a solution for our problem.

In case i, since $p_1(X,Y) = \tilde{p}_1(X)$, to the order $\varepsilon^{\frac{2\alpha}{3}}$, we can write

$$EM\pi = \varepsilon^{\frac{2\alpha}{3}} \tilde{p}_1(X)$$

leading to $e = \frac{2\alpha}{3}$ and to the required matching condition:

$$\overline{p}_1(X,0) = \tilde{p}_1(X) \tag{14}$$

The nature of the perturbation given by "a", together with the last matching condition, are found by matching the transverse velocity of the external layer and the middle layer.

## II.8 EM matching for $v_y$

The middle expansion operator for $v_y$, to the order $\varepsilon^{a-\alpha+\frac{m}{2}}$, is given by:

$$M\,v_y = -\varepsilon^{a-\alpha+\frac{m}{2}} A'(X)u_o(Y)$$

This expression written in outer variable and expanded gives:

$$E*M\,v_y = -\varepsilon^{a-\alpha+\frac{m}{2}} A'(X) + TST$$

This is to be compared with the exterior expansion operator to the order $\varepsilon^{\frac{2\alpha}{3}}$ :

$$E\,v_y = \varepsilon^{\frac{2\alpha}{3}}\, \overline{v}_1(X,\overline{y})$$

The following possibilities arise:

i. if $a - \alpha + \frac{m}{2} < \frac{2\alpha}{3}$, there is no possible matching,

ii. if $a - \alpha + \frac{m}{2} > \frac{2\alpha}{3}$, the matching condition is given by $\overline{v}_1(X,0) = 0$.

The second possibility leads to $\overline{v}_1(X,\overline{y}) = 0$ since $\overline{v}_1$ is harmonic in the half plane $\overline{y} > 0$. The only possible choice is then:

$$a = \frac{5\alpha}{3} - \frac{m}{2} \tag{15}$$

with the matching condition:

$$\overline{v}_1(X,0) = -A'(X) \tag{16}$$

## II.9 Missing cases

As we know, for $a \geq \frac{\alpha}{3}$, the preceeding result (15) shows the frame of the present theory, i.e.:

$$\frac{m}{8} \leq a \leq \frac{m}{3}$$

with

$$\frac{3m}{8} \leq \alpha \leq \frac{m}{2}$$

Therefore, from (3):

$$\frac{5m}{8} \leq \beta \leq \frac{2m}{3}$$

The case $a = \frac{2\alpha}{3}$ which has been forgotten in hypothesis (8) is easily treated. The preceeding analysis in regards to matching is the same, thus leading to the only possible case $a = \frac{m}{3}$ or $\alpha = \frac{m}{2}$ which is included in our theory.

For the case of perturbations such as:

$$0 < \alpha < \frac{3m}{8}$$

we have to relax the hypothesis (4) on the pressure gradient in the boundary layer. Therefore,

$$\pi = \varepsilon^b \, \tilde{p}(X, \tilde{y}) \text{ with } b > \frac{2\alpha}{3} \, .$$

The boundary layer equations can now be written as:

$$\tilde{u}_1 \frac{\partial \tilde{u}_1}{\partial X} + \tilde{v}_1 \frac{\partial \tilde{u}_1}{\partial \tilde{y}} = \frac{\partial^2 \tilde{u}_1}{\partial \tilde{y}^2}$$

$$\frac{\partial \tilde{u}_1}{\partial X} + \frac{\partial \tilde{v}_1}{\partial \tilde{y}} = 0 \, , \quad \frac{\partial \tilde{p}_1}{\partial \tilde{y}} = 0 \tag{17}$$

EM matching for pressure and transverse velocity leads to:

$$a = \frac{\alpha}{3} \text{ and } b = \frac{m}{2} - \frac{2\alpha}{3} \, .$$

## III. Conclusion

In this section, we give a brief description of the main results presented in this paper.

### The exterior flow

For $\bar{y} = \frac{y}{\varepsilon^{\alpha}}$, we obtained:

$$\begin{cases} v_x = 1 + \varepsilon^e \, \bar{u}_1(X,\bar{y}) + \ldots \\ v_y = \varepsilon^e \, \bar{v}_1(X,\bar{y}) + \ldots \\ \pi_z = \varepsilon^e \, \bar{p}_1(X,\bar{y}) + \ldots \end{cases}$$

with the following possibilities:

i. if $0 \le \alpha < \frac{3m}{8}$ then $e = \frac{m}{2} - \frac{2\alpha}{3}$

ii. if $\frac{3m}{8} \le \alpha \le \frac{m}{2}$ then $e = \frac{2\alpha}{3}$

From the matching conditions for the harmonic function $\bar{v}_1$ and $\bar{p}_1$, we obtain the coupling relations:

$$A'(X) = -\frac{1}{\pi}\int_{-\infty}^{\infty} \frac{\tilde{p}_1(\xi)}{X-\xi}\, d\xi \quad \text{or} \quad \tilde{p}_1(X) = \frac{1}{\pi}\int_{-\infty}^{\infty} \frac{A'(\xi)}{X-\xi}\, d\xi \qquad (18)$$

### The middle flow

For $Y = \frac{y}{\varepsilon^{\frac{m}{2}}}$, we obtained:

$$\begin{cases} v_x = u_o(Y) + \varepsilon^a \, u_1(X,Y) + \ldots \\ v_y = \varepsilon^{a-\alpha+\frac{m}{2}} \, v_1(X,Y) + \ldots \\ \pi = \varepsilon^b \, p_1(X,Y) + \ldots \end{cases}$$

with the following possibilities:

i. if $0 \le \alpha < \frac{3m}{8}$ then $a = \frac{\alpha}{3}$ and $b = \frac{m}{2} - \frac{2\alpha}{3}$

ii. if $\frac{3m}{8} \le \alpha \le \frac{m}{2}$ then $a = \frac{5\alpha}{3} - \frac{m}{2}$ and $b = \frac{2\alpha}{3}$

For $\alpha$ strictly smaller than m / 2 , the solution is known:

$$u_1 = A(X)u'_o(Y), \quad v_1 = -A'(X)u_o Y), \quad p_1 = \tilde{p}_1(X)$$

However, for the limiting case $\alpha = m/2$, where the equations can be written:

$$u_o \frac{\partial u_1}{\partial X} + v_1 u'_o = -\frac{\partial p_1}{\partial X} \; ; \qquad u_o \frac{\partial v_1}{\partial X} = -\frac{\partial p_1}{\partial Y}$$

and where the upper deck disappears, the solution is not known analytically.

This special limiting case is called the "Double Deck Theory". The corresponding perturbation is the smallest possible which involves separation of the flow.

**The interior flow (boundary layer)**

For $\tilde{y} = \dfrac{y}{\varepsilon^{\frac{\alpha}{3}+\frac{m}{2}}}$ , we obtained:

$$\begin{cases} v_x = \varepsilon^{\frac{\alpha}{3}} \tilde{u}_1(X,\tilde{y}) + \dots \\ v_y = \varepsilon^{\frac{m}{2}+\frac{\alpha}{3}} (X,\tilde{y}) + \dots \\ \pi = \varepsilon^b p_1(X,\tilde{y}) + \dots \end{cases}$$

The boundary layer equations are submitted to the matching conditions :

i. for $0 \le \alpha \le \dfrac{3m}{8}$, $\quad \lim\limits_{\tilde{y}\to\infty} \left[\tilde{u}_1 - \lambda\tilde{y}\right] = \lambda\, A(X)$

ii. for $\dfrac{3m}{8} < \alpha \le \dfrac{m}{2}$, $\quad \lim\limits_{\tilde{y}\to\infty} \left[\tilde{u}_1 - \lambda\tilde{y}\right] = 0$

When $0 \le \alpha < \dfrac{3m}{8}$, we have a classical boundary layer without separation and with $A(X) = -F(X)$. The solution for (17) is given by $\tilde{u}_1 = \lambda_o\,(\tilde{y} - F)$ and $\tilde{v}_1 = \lambda_o\,(\tilde{y}F' - FF')$ and the coupling relation gives:

$$\tilde{p}_1(X) = -\frac{1}{\pi}\int_{-\infty}^{\infty} \frac{F'(\xi)}{X-\xi}\, d\xi$$

When $\alpha = \frac{3m}{8}$, it is the "Triple Deck Theory", i.e. the first perturbation for which there is both displacement and separation of the flow.

The triple deck theory plays a central role for it is the only case where, in solving boundary layer problems, we have to use a coupling relation, equation (18).

When $\frac{3m}{8} < \alpha \leq \frac{m}{2}$, there is no displacement but separation of the flow is possible. The boundary layer solution gives $\tilde{p}_1(X)$ and the coupling relation A(X).

If we study the pressure gradient, putting:

$$\frac{\partial \pi}{\partial x} = O(\varepsilon^{\nu}),$$

figure 2 shows the growth of the adverse pressure gradient. When $\alpha$ increases, the adverse pressure gradient increases. It can be compared to the driving pressure gradient for

$$\alpha = \frac{3m}{10}.$$

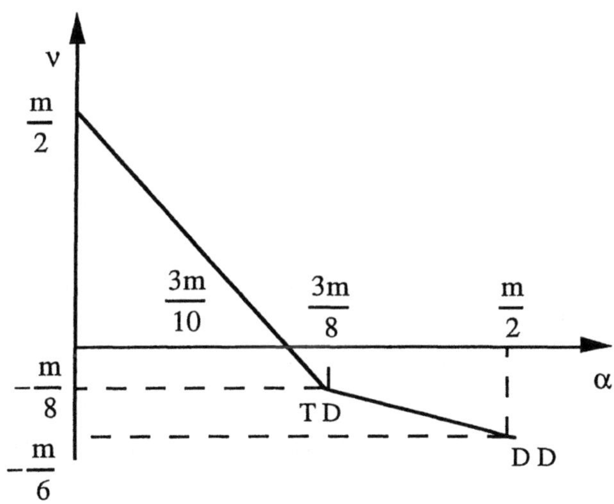


**Fig. 2.** Order of magnitude of the ratio of the pressure gradient to the length of the hump.

Then, becoming more important, for

$$\nu = -\frac{m}{8},$$

we have the Triple Deck (TD) and for

$$\nu = -\frac{m}{6}$$

we have the Double Deck (DD), which is the last case of the theory.

## References

Audounet, J., Mauss, J., Saintlos, S. (1991): "Sur la théorie des écoulements laminaires à grands nombres de Reynolds", MATAPLI, SMAI (Bulletin de liaison) n° 28, pp. 17-25.

Eckhaus, W. (1979): "Asymptotic analysis of singular perturbations", in Studies in Mathematics and its Applications, ed. by North-Holland.

Lagree, P.Y. (1992): "Structures interactives fluide parfait/couche limite en hypersonique. Variations sur le thème de la triple couche", Doctoral Thesis, Laboratoire de Modélisation en Mécanique, Université PARIS VI, France.

Lighthill, M.J. (1953): "On boundary layers and upstream influence", *Proc. Roy. Soc.* **1217**, pp. 478-507.

Mauss, J., Achiq, A., Saintlos, S. (1992): "Sur l'analyse conduisant à la théorie de la triple couche", *C. R. Acad. Sci. Paris* **T. 315**, série II, pp. 1611-1614.

Mauss, J. (1994): "Analyse asymptotique singulière", in Lecture Notes in Advanced Fluid Mechanics, Université de Toulouse III, France.

Messiter, A.H. (1970): "Boundary layer flow near the trailing edge of a flat plate", *SIAM J. Appl. Math.* **18**, pp. 241-257.

Nayfeh, A.H. (1991): "Triple deck structure", *Computers and Fluids* **20**, n°3, pp. 269-292.

Neiland, V.Y.A. (1969): "Towards a theory of separation of the laminar boundary layer on a supersonic stream" in Izv. Akad. Nauk, ed. by S.S.S.R., Meckh, Zhid, Gaza, n°4.

Smith, F.T. (1982): "On the high Reynolds number theory of laminar flows", *IMA J. of Appl. Math.* **28**, pp. 207-281.

Stewartson, K., Williams, P.G. (1969): "Self-induced separation", *Proc. Roy. Soc.* A **312**, pp. 181-206.

Van Dyke, M.D. (1964): *Perturbation methods in fluid mechanics* (Applied Mathematics and Mechanics, Academic Press).

# Going on with Asymptotics

# GOING ON WITH ASYMPTOTICS

Jean-Pierre Guiraud

Laboratoire de Modélisation en Mécanique, Université Pierre et Marie Curie & C.N.R.S., Case 162, 4, place Jussieu, F75252 Paris Cedex 05, France

This is not a review paper. It is, rather, a collection of thoughts on the role of asymptotics in a variety of fields, most of which are related to mechanics but are not restricted to fluids. The emphasis is more on the achievements of asymptotics, than on the description of the methods used to achieve these. The topics have been chosen so as to illustrate the versatility of asymptotic methods concerning the type of problems solved and the degree of sophistication required by each particular solution. No internal logic other than that illustrating this efficiency and versatility should be expected in this paper.

## 1. Why asymptotics ?

### 1.1 General comments

The purpose of a book like this one is to collect contributions from recognized scientists, on some more or less well defined scientific issues, and encourage readers to further reading, in order to learn more about the subject. I hope that this goal has been achieved here to a large extent. Comments on the contents of what precedes would be useless. My purpose is, rather, to skip all over that written matter and go ahead with the general comments made by Van Dyke in his introductory lecture. I would like to try to communicate enthusiasm to those who read the book. As with any human activity, enthusiasm is spurred by evidence of achievements. The lectures allow a rather wide spectrum of specific achievements to be grasped, while providing some details and some indications for further reading. My purpose at present, is to open the spectrum of achievements widely by skipping details, but give specific references.

Why asymptotics ? The answer that I shall try to emphasize is that a very wide number of significant achievements were realized through asymptotics. I hope the reader will be convinced that much remains to be discovered and that investing in this kind of mathematical activity will yield returns for scientists who intend to devote part of their efforts to solving problems of very wide interest that can be analyzed by mathematical models. I use the word "asymptotics", rather than writing "asymptotic methods", to stress two points : first, that the ultimate goal is to solve problems rather than simply improve known methods and discover new ones ; second, that asymptotics is often only part of the solution. My first point in no way means that research on methods should be prohibited. I hope the reader will be convinced by some of my arguments that such research has to be sustained and encouraged. I simply mean that methods should be used, extensively, even if one feels that they might be improved. Of course, this raises the problem of communication between those who conceive the methods and those who use them ; each has to make a step towards learning some of the language of the other. I hope to convince readers inexperienced with asymptotics that this is not only necessary but very rewarding for advancing in research.

My opinion is that asymptotics is more or less present, sometimes at an elementary level, in all scientific activities which may lead to a mathematical model. A discussion of this aspect may be found in Landahl 1981. Of course, as shown through numerous examples, some given by Landahl himself, a number of difficult and very important issues (turbulence, two-phase flows, granular materials, etc...) require purely ad hoc modeling. However, even under such circumstances, asymptotics may be used in building other models, considered simpler or more tractable, either analytically or numerically. As an example of such use of asymptotics, the reader may refer to Harris, Crighton 1994.

### 1.2 Comments on some deficiencies encountered with asymptotics

Why go any further with asymptotics when so many deficiencies have been emphasized, sometimes quite heavily ? Some concern a lack of mathematical rigor ; others argue that numerics is a very efficient substitute, much more powerful, and giving higher precision in the results. Another kind of drawback is the high degree of sophistication in many applications of asymptotics, associated with an apparent lack of deductive reasoning, the results being obtained sometimes in a way which appears obscure to the reader. A short excursion in the realm of some mathematical aspects of asymptotics will be made in Section 6, so I just need to make a comment on the lack of rigor. Mathematicians are perfectly right when addressing this remark to fluid dynamicists. The latter, when using asymptotics, mainly intend to obtain an "ansatz" on the form that the solution sought is expected to take, while the former cannot be satisfied before a proof of existence and an estimate of the error involved in the asymptotic approximation is got. Unless a lot of rigorous mathematics is completed, beyond the first step, that I like to call "naive asymptotics", there is no guarantee that the approximation obtained is within a controlled error of a true solution. An example may be found with the theory of Guiraud, Zeytounian 1977 on rolled vortex sheets, which gives an asymptotic representation of a so-called "solution" which probably does not exist in a mathematical sense. Needless to say that when rigorous mathematics can be completed, beyond "naive asymptotics", it is highly desirable that such rigorous mathematics be worked out. When this is not possible, we must admit that a plausible approximation is better than nothing at all. Many scientists feel that obtaining an approximation via asymptotics is an argument in favor of plausibility. Accuracy is another thing and, making the leap from plausibility to certainty is again a quite different issue. Often, but not always, the mathematics underlying accuracy and certainty are quite intimately related, but this is not my main concern here.

Concerning numerics as a substitute for asymptotics, my feeling is that one should not oppose each other but rather consider them as different tools for attacking either different problems or different aspects of the same problem. Often, the complete solution of quite a complex problem begins with asymptotics and ends with a numerical solution to some mathematical model issuing from the, previously worked out, asymptotic analysis. This occurs in many of the examples given below to illustrate our various topics. Here, we simply observe that in Zeytounian, Guiraud 1987 under the heading "The differential analyzer of numerical schemes" we suggested how asymptotics could provide assistance to scientists studying the basic properties of numerical schemes by heuristic techniques. It would be appropriate at this point to give a view on the asymptotic nature of error estimates which are the ultimate goal of rigorous mathematical achievements concerning numerical methods, but it is difficult to do this in a concise yet precise way and I lack place.

I now come to the last criticism I mentioned, namely the excessive obscureness of some asymptotic literature. This criticism should be considered seriously and dealt with accordingly. The first remark is that asymptotics occurs on many levels and that a number of problems may be treated, asymptotically speaking, in a way which looks natural, with the help of a slight amount of technicality. Then, why not simply try it after reading a little? Many books are useful for the beginner : Cole 1968, Van Dyke 1975, and for French readers Zeytounian 1994, especially chapters II to V inclusive. But this is not actually an answer to the question : why such sophisticated asymptotics ? It is true that some degree of ingenuity is needed to use of asymptotics when difficulties are involved. Ingenuity cannot always be explained in a logical and fully deductive way, especially in research papers. Scientists who feel frustrated, often with good reason, should consider that other frustrations occur with work on numerics (I use this example as it is often opposed to asymptotics), when the author cannot record all the unsuccessful attempts that have been tried before capturing the right technique which makes the numerical scheme work. It is safe to say that there is a big difference : asymptotics does not lead directly to the answer required, it is only a step towards it, while, generally, numerics does. My opinion is that this should not discourage scientists, inexperienced with the field, from becoming involved with asymptotics. Let me add that ingenuity can be learned to some degree through reading papers and books. It seems to me that Meyer, Wilson 1973 is especially illuminating. Last but not least, ingenuity is helped by knowledge of specific techniques (Matched Asymptotic Expansions, Multiple Scale Techniques, Strained Coordinates, WKB or Liouville-Green methods, Asymptotic Expansion of Integrals, etc...) which may be learned in a number of books. Some of them are listed below. Matched Asymptotic Expansions may be found in Van Dyke 1975, chapterV, in Lagerstrom, Casten 1972 and more extensively in Lagerstrom 1988. For Multiple Scale Techniques, see Cole 1968, chapter III, expanded in Kevorkian, Cole 1981, chapter III. Strained Coordinates are explained in Van Dyke 1975, chapter VI, and Crocco 1972. For WKB, one may use Bender, Orszag 1978, chapter X, or Dingle 1973, chapter XIII. Finally, Asymptotic Expansion of Integrals is treated extensively in Bleistein, Handelsman 1975, but also by Bender, Orszag 1978, chapter VI.

### 1.3 Asymptotics and industry related problems

It seems to me that a strong incentive for getting involved with asymptotics is the high demand of industry for mathematical modeling, ending, of course, with results given by numbers. I express the view that asymptotics will grow in importance in the near future for at least three reasons. First, it will allow numerics to be applied to simplified problems, even making untractable ones tractable. Secondly, it will allow a much larger range of parameters to be covered. Thirdly, with the help of some numerics, a number of problems will evolve from pure empiricism to scientific understanding, provided some asymptotics is involved in the process. Many of the examples I advocate in the following sections can serve as illustrations. Here, I want to emphasize some specific ones.

The first one is discussed in Favret, Cornil, Deneuve, Guiraud 1990. It concerns an attempt at understanding the instability observed in the operation of gas regulators included in network distributions. A crude model leads to a system of ten first-order ordinary differential equations. The problem could be dealt with by brute force computations, but there were so many parameters involved in the model that efficiency ruled out this approach. Some ingenuity led to discovering a small parameter which allows a drastic reduction in the order of the system, through an elementary application of multiple scaling analysis. This was not the end of the story because the new model obtained in this way was not amenable to linearization and could not be used for a straightforward stability analysis. Purely ad hoc engineering linearization was devised as a substitute and was used to make predictions of instability range, in rough agreement with general trends issued from accumulated observations. Then a Liapunov analysis on the nonlinear system by Adou 1993 confirmed the prediction of the linear ad hoc model. This is an example of use of very crude asymptotics, but efficient with respect to industrial requirements. Needless to say that brought to this point, the problem was far from being solved for a scientist, but it was raised from pure empiricism to an embryo of scientific understanding. The share of asymptotics in this achievement was very slight indeed, but my opinion is that it was nevertheless essential.

Two other examples of some asymptotics and industry are only opened perspective which have not led to work available in the literature and, accordingly, will be sketched briefly below. The first is concerned with sustaining and cooling a jet of molten metal by an air flow. There is a lot of asymptotics involved, dealing with thin jet modeling, cooling and solidification, through use of slender body decoupling, and, finally, calculating root mean square deviation of the molten metal trajectory at the end of sustaining. The asymptotics of the latter suggests a problem which is of interest by itself but might be ignored by scientists due to a lack of incentive. Consider a long circular cylinder subjected to an unsteady transversal flow with moderate Reynolds number, and a high frequency, probably random (due either to incomplete knowledge or to turbulence). Asymptotic calculation of the root mean square deviation requires obtaining space-time correlations for the aerodynamic forces on the cylinder with length separation and time delay, tuned in a suitable way with respect to each other. Those correlations may be obtained experimentally, if full turbulence is concerned, or from direct numerical simulations of the Navier-Stokes equations, if large coherent structures are involved. The second of my examples is concerned with obtaining an annulus of molten metal by very fast centrifuging. The process must be sufficiently rapid to obtain a film of uniform thickness before a significant start of cooling and solidification. How could this purely fluid mechanical process be decoupled from the thermal one ? How does solidification proceed in presence of a strong centrifugal field ? These are just two questions out of many others whose answer might prove useful for designing manufacturing processes. They seem manageable through asymptotic modeling followed by moderate numerical simulation.

I now refer to Lange, Weinitschke 1991 as an example accessible in the literature of using asymptotics for a problem issued from industry. A reacting material is confined within an infinitely long circular cylinder. The mathematical model arises from competition between thermal diffusion and chemical kinetics, with very high activation energy, as a source of heat. The process is controlled by the intensity of the exothermal character of the chemical reaction, which is included in the mathematical model, for the temperature $T$, namely :

$$\Delta T+\lambda \text{ Exp}\{T/(1+\mu T)\}=0, \tag{1}$$

by the parameter $\lambda$ , $\mu$ being related to the activation energy. The model is completed by a Robin (or Biot) boundary condition on the external cylinder whose radius is taken as unity. We remind the reader that a Robin condition states that the heat flux on the boundary is proportional to the difference between the temperature of that boundary and a prescribed temperature. The limit of infinite activation energy corresponds to $\mu=0$. Of course, $\Delta$ stands for the Laplacian. When $\mu=0$ and the condition at the boundary is a Dirichlet one, an explicit axisymmetric solution is known. Accordingly there exist two solutions when $0<\lambda<2$ : a cold one and a hot one. The value $\lambda=\lambda_\ell=2$ is a so-called limit point at which runaway (i.e. an explosion) may occur.

The asymptotics goes as follows. One puts inside the reacting material a small cylindrical concentric tube with radius $\varepsilon$ , maintained at a constant temperature for cooling, the same as that of the external boundary, for example. The limit $\lambda_\ell=2$ is changed by the cooling effect of the inner tube and asymptotics provides (for a Dirichlet external condition) the formula

$$\lambda_\ell=2+\delta(\varepsilon).2\text{ Log}4+\delta^2(\varepsilon).2(1+\text{Log}^2 2)+O(\delta^3)\ ,\ \delta(\varepsilon)=\text{Log}(1/\varepsilon). \tag{2}$$

The limit value $\lambda_\ell$ may be obtained through a numerical computation, which becomes very stiff when $\varepsilon$ is too small. For $\varepsilon=0.1$, formula (2) (neglecting $O(\delta^3)$ ) gives $\lambda_\ell=3.7625$ while the numerical code delivers $\lambda_\ell=3.8915$. At $\varepsilon=0.04$ one finds $\lambda_\ell=3.1472$ from (2) compared with 3.2003 from the code. For the same value $\varepsilon=0.04$, a two-term composite expansion for the temperature gives an error of 11 % at the maximum temperature, and this occurs at a radius of 0.4 (we remind the reader that the external radius is unity). The accuracy of the temperature distribution is relatively poor but the composite formula is so short ! On the other hand, the accuracy of $\lambda_\ell$, which is the most important value for industrial requirements, is 1.7 % which is not too bad.

A second way of using asymptotics arises in investigating how the limit point is affected by a high, but finite (that means $\mu$ close to zero) value of the activation energy. As a matter of fact, without cooling by the inner tube, there exist two limit points : an upper one $\lambda_{\ell u}$, beyond which the solution does not exist locally but may snap through towards an upper branch. This upper branch starts at a second, lower, limit point $\lambda_{\ell\ell}$ and then, this branch persists for $\lambda$ increasing to infinity. One may compute a critical value $\mu_c$ of $\mu$ such that at $\mu=\mu_c$, both limit points coalesce $(\lambda_{\ell u}(\mu_c)=\lambda_{\ell\ell}(\mu_c)=\lambda_c)$. For $\mu>\mu_c$, there is no longer any limit point, and the solution exists whatever $\lambda$ . The presence of the inner tube of radius $\varepsilon$ , at the same temperature as the external cylinder (Dirichlet boundary condition), changes the values $\lambda_c$, $\mu_c$ . A three-term expansion as with (2), gives $\lambda_c=5.5912$, $\mu_c=0.2468$, for $\varepsilon=0.1$ compared with the numerically computed values of $\lambda_c=5.7982$, $\mu_c=0.2451$ , which is not too bad. The authors report serious difficulties with the numerical code for $\varepsilon$ smaller than

0.01. At such values the asymptotic formula gives $\lambda_c$ with a relative error better than $0.78\ 10^{-2}$, while it is better than $1.2\ 10^{-3}$ for $\mu_c$.

### 1.4 Organization of the paper

I had already the opportunity to explain that my main purpose is to convince the nonspecialist that significant progress may be got by investing in asymptotics. As a consequence I am more interested in results than in the way one obtains them and, to a large extent, in rather recent results. The examples I have chosen are numerous and offer a wide variety, but there is no internal logic in the choice I offer to the reader. I shall mainly proceed by slight touches that I have tried to render self-consistent, provided one is satisfied simply in understanding the nature of the achievement. Nevertheless I decided to select two subjects in order to give a bit more about each. They occupy Sections 2 and 3. The first has been chosen because it is related to a breakthrough that occured in the very early 1970, concerning laminar separation of high Reynolds number flows, and which is expanding considerably presently and expected to expand in the future. Here I have tried to explain the ideas more than giving an account of the most recent results. The reason for this is that the topic dealt with in Section 2 is blurred by a combination of ingenuity and sophistication. My main purpose when writing this account was to make my own contribution to the struggle against an unfortunate circumstance which might discourage scientists to learn this subject which is so much rewarding when one has succeeded in moving up the first step. Section 3 is completely disconnected from the preceding one. I have chosen it because it concerns a wide subjet in solid mechanics, namely brittle fracture. The more specific topic that I consider is one in which asymptotics has contributed, recently, to make some significant progress and I hope that the account I am to give of it is self-consistent, not only at the level of the results but also concerning the way by which they have been got. So far I have tried to explain and with Section 4, I change by choosing my examples in order to convince. In order to meet this goal I insist on versatility giving up with explanation. Section 5 is the continuation of Section 4, by putting more emphasis on examples which show how asymptotics may help in understanding phenomena which look spurious. Finally, Section 6 has been written in order to show to the nonspecialist that rigorous mathematics may be worked out with topics mainly concerned with asymptotics. My main purpose is to explain why the reader who is not a mathematician may take benefit in understanding some results got by professional mathematicians. A short conclusion occupies Section 7. It is mainly devoted to trying to communicate enthusiasm for further work, and also with education, on asymptotics.

## 2. Asymptotics of the triple deck

### 2.1 The early roots of the theory

The triple-deck theory deserves a special mention among the numerous applications of high Reynolds numbers $(R_e)$ laminar flows. It is somewhat misleading to associate high Reynolds numbers with laminar régime, although such an association is not completely false at the outset. What can be said here is that, provided some caution is exercised when applying the theory, it has proven to give very powerful results. What is unfortunate with this kind of asymptotics is that it seems to be very esoteric, on the one hand, while it looks suspect on the other for at least two reasons : the first is that in its simplest version, the only one reported here, it is based on $R_e^{-1/8}$ being much smaller than unity ; the second is that the higher $R_e$ the less plausible the laminar conditions seem to be. It is outside the scope of this paper to discuss this last issue, and we ask the reader to believe that laminar conditions can exist (obviously within a limited range). I only intend to debunk, if possible, the esoteric blurring which appears to have, from long, discouraged many scientists from investing in it. Let me explain the main goal assuming the reader is familiar with basic boundary layer theory and knows the corresponding hierarchy : computation of the inviscid flow first, followed by boundary layer computation as such then, finally (eventually) computations of corrections, generally of order $R_e^{-1/2}$, to both. The classical theory is based on what might be called an asymptotic "ansatz", namely that a thin (of order $R_e^{-1/2}$ thick) layer of fluid surrounds the surface of an undeformable smooth body on a $O(1)$ scale. Each time this asymptotic "ansatz" is broken, one should expect a failure of classical boundary layer theory.

Before discussing triple deck theory, I will touch on a more familiar subject : the instability theory of the laminar boundary layer. First of all a strong result has to be stated : Prandtl's boundary layer equations exhibit stability (Oleinik 1969, Nickel 1973). This looks a bit paradoxical, because instability of boundary layers is a well established fact, through so many experiments that have been repeated since the ones of Schubauer, Skramstad 1943. It has received a theoretical explanation, which agrees with well controlled experiments and which is firmly believed by scientists. For an account see Drazin, Reid 1982, especially §§ 25 to 28. The issue, raised by this paradox, has long been known and discussed even before the mathematical evidence given by Oleinik 1969. The paradox itself was (partially) resolved by Bouthier 1972, 1973, almost fifty years after the first attempt to predict instability, by Heisenberg 1924. The full asymptotic resolution, via the triple deck, waited seven years more the contributions of Smith 1979 and Bodonyi, Smith 1981. It is fair to mention an early attempt by Lanchon, Eckhaus 1964.

It is useful to describe first the classical theory of instability (of boundary layers) as it grew up from seminal ideas laid down by Heisenberg 1924. It has to be stressed that the first numerical results got by Tollmien 1929 relied on asymptotics and that the first, precise, full numerical solution was published by Kurtz, Crandall 1962. We consider the simplest case : a Blasius steady plane boundary layer on a flat plate for flow of an incompressible fluid. The perturbed stream function is sought under the form :

$$\psi = \mathrm{Re}\{\Phi(y)\ \mathrm{Exp}[i(\alpha x - \Omega t) + \sigma t]\}, \tag{3}$$

where $\alpha$, $\Omega$, $\sigma$ are real, $t$ is the time, $x$, $y$ are cartesian coordinates, with the plate at $y = 0$, $x > 0$. We remind the acquainted reader that $\Phi$ obeys the so-called Orr-Sommerfeld equation (look Drazin, Reid 1982, eq (25.12), which contains a Reynolds number $R$, which is not the previous one (written $R_e$). The difference arises from the length used : the displacement thickness for $R$, the longitudinal length for $R_e$. Here we simply need to know that $R = O(R_e^{1/2})$. The so-called neutral curve (defined by $\sigma = 0$ , and sketched on Fig. 1)

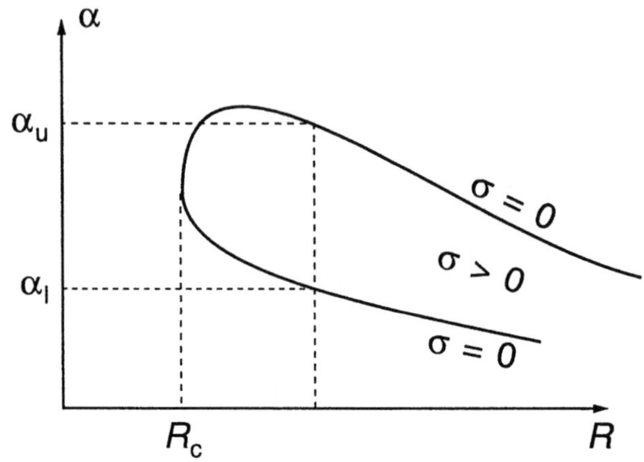


Fig. 1

delineates the region of temporal instability from the one where the laminar boundary layer is stable (to perturbations of infinitesimal amplitude). It has two branches which may be described, asymptotically, by explicit formulae :

$$\begin{cases} R^{1/3} = 0.396\ \alpha^{-4/3} + \ \dots \quad , \quad \Omega = 1.75\ \alpha^2 + \ \dots \ , \quad \text{Lower branch,} \\ R^{1/3} = 0.0275\ \alpha^{-10/3} + \ \dots \quad , \quad \Omega = 4.02\ \alpha^2 + \ \dots \ , \quad \text{Upper branch.} \end{cases} \tag{4}$$

The region of instability lies in between the two branches. These may be continued, numerically, from very high values of $R$ to moderate ones. When this is completed they join together at $R_c = 520$ and form a closed loop including the unstable region. We leave aside the upper branch and observe that, along the lower one, modes of perturbations evolve on a streamwise scale, the length of which is $O\left(R_e^{1/8}\right)$ times the length scale of the boundary layer thickness, and this is $O\left(R_e^{-3/8}\right)$ times the longitudinal length scale of the whole boundary layer. Two remarks are in order : first, the paradox which arises from the (advocated previously) stability of Prandtl's equations is avoided, at the expense of breaking the asymptotic "ansatz" which is basic to Prandtl's equation, because the longitudinal length scale is no longer $O(1)$ ; secondly we may state that the length scale $O\left(R_e^{-3/8}\right)$ is precisely the one which occurs in the simplest version of the triple deck. One may observe, incidently, that the time scale is $O\left(R_e^{-1/4}\right)$.

This reasoning might have been made very early, perhaps in the middle 1940, but it was not put to the point of discovering the triple-deck. Curiously enough, this powerful theory did not grow up from boundary layer instability but it was rather the latter who took profit of the former, and this occurred even ten years after the discovery of triple-deck by Stewartson, Williams 1969 and Neiland 1969. It is an easy exercise for the commentator to find an explanation "a posteriori". Triple-deck theory allowed to understand a fine point in boundary layer instability theory but it did not solve a completely mysterious problem. The issue at hand was to give an account of what were called nonparallel flow effects at the time. These are closely related to the boundary layer asymptotic "ansatz" which leads to the paradox of the stability of Prandtl's equations. The founders of the instability theory overcame this problem in a reasonable, but asymptotically inconsistent way. They studied whether damping or amplification occured by considering perturbation modes governed by the complete (linearized) Navier-Stokes equations, knowing that the flow they were perturbing did not satisfy these equations but rather Prandtl's ones. Concurrently, they assumed that the basic flow was translation invariant in the downstream direction, but this is only asymptotically (at high Reynolds numbers) correct. The process for overcoming this viciated circular logical deficiency was very long. To my knowledge it was achieved through the works of Lanchon, Eckhaus 1964, Bouthier 1972, 1973 and Smith 1979. It would be interesting to analyze the contribution of each of those three sets of papers in greater detail. I simply state that for me, the essential contribution was the last one, but it is fair to add that this major contribution was a rather natural output of triple-deck theory.

## 2.2 Towards the triple deck theory

### *2.2.1 The upstream influence mystery*

Now that the reader has been prepared to the $R_e^{-3/8}$ scaling, let us give an account of the historical process which led to the discovery of the triple-deck theory by Stewartson, Williams 1969 and, independently, by Neiland 1969. The problem at hand is completely different from the previous one, and arose from another paradox, which defied the perspicacity of all researchers, except Lighthill 1953, after a first unsuccessful attempt (Lighthill 1950). The paradox was due to an obvious discrepancy between theory and experiment, leaving such a gap between the two that it was a case of loosing confidence in the theory. Let us assume that a weak or moderately strong oblique shock wave impinges at a steady position on a rigid wall. Inviscid theory predicts specular reflection (with unequal angles unless the shock is of infinitesimally small strength). At high Reynolds numbers, the boundary layer is so thin ($O(R_e^{-1/2})$) that one cannot imagine a large effect of viscosity. But experimental data are irrefutable : the boundary layer is perturbed upstream of impingement for so many times the thickness of the boundary layer that the observations (reported in Chapman, Kuehn, Larson 1950) remained a mystery. Lighthill's resolution of the paradox in 1953 was satisfactory. But many scientists added a lot of work which did not contribute to raising the degree of understanding. All of this continued up to 1969 (discovery of the triple- deck theory) and was pursuing later. There is nothing to add in 1994 to the comment of Brown, Stewartson 1969 at the end of their review : "We make a final point about the recent attempts to obtain theories of wake flows and interaction problems. These methods lay tremendous emphasis on massive computation and ad hoc arguments to lead to their results, little attempt being made to justify the assumptions on mathematical grounds. While we recognize the need for such a computational effort in the end, it seems to us of great importance that a rational approach be adopted to make sure, for example, that terms neglected in the approximation really are smaller than those retained. Until this is done, and even now it is possible in part, it will be difficult to convince the detached and possibly skeptical reader of their value as an aid to understanding". The end of this comment has a special flavor when one realizes that the asymptotic solution of the paradox was published the same year, in particular by Stewartson himself (Stewartson, Williams 1969). The phenomenon observed was puzzling : outside the boundary layer, inviscid supersonic flow prevails, and it is well known that under such conditions, the perturbations created by the shock cannot run upstream. Within the boundary layer itself there is again no mechanism which can explain running upstream (this is a consequence of the parabolic character of Prandtl's equations). The basic hierarchy involved in boundary layer theory ruled out any possibility of the perturbation running upstream. Some scientists like Catherall, Mangler 1966 suspected that something was wrong with the hierarchy but the idea was difficult to include in a mathematical model. Of course Navier-Stokes equations exhibit ellipticity in space, which restores the possibility of perturbations running upstream. But given the high Reynolds numbers under consideration, the extent of this upstream running could not be expected, from past knowledge, to exceed a few boundary layer thicknesses. Yet, the observations were spectacular : for moderate shocks, the boundary layer was separated by up to fifty times its thickness upstream of the point of impingement. Scientists involved in this issue were completely puzzled. At the time, the concept of asymptotic structure was not well developped (the Lagerstrom, Cole, Trilling 1949 paper was known to only a few specialists, and the papers which founded the theory, namely Kaplun 1954, 1957, were not published). And especially, the very concept of breaking of an asymptotic structure was completely ignored. As a matter of fact this concept had been used, as we have seen, in stability theory, but why and how it was used was not at all understood.

*2.2.2 Lighthill's theory*

Lighthill 1953 had the idea of trying it without formalizing, asymptotically, the process, in order to obtain a tentative theoretical model containing, as a consequence, the observed phenomenon of usptream influence in shock-boundary layer interaction. His first argument was that the phenomenon should, in some sense, be basically inviscid. The argument was based on the observation that the boundary layer is subjected (in the phenomenon studied) to perturbations occurring on a longitudinal scale which was short compared to the scale of the boundary layer as a whole. The consequence extracted by Lighthill was that inertial terms could not be balanced by viscous ones. The argument of a parallel flow, probably borrowed from stability theory, was advocated. These premises led to a model of small inviscid disturbances in a pure parallel shear flow. In his 1950 paper, Lighthill considered that the perturbations were forced by the incident, infinitely weak, oblique shock and he obtained a reflected one, but no significant upstream influence emerged from the analysis. The parallel flow assumption allowed him to use Fourier transforms for the streamwise dependency, and this technical detail proved useful for the end of the story. This arose with the 1953 paper, which presented, among many other things, a full resolution of the paradox of upstream influence. Since the 1950 analysis was inviscid, the no-slip condition on the wall could not be enforced. A viscous analysis within a sublayer, which turned out to be $O\left(R_e^{-1/8}\right)$ thinner than the basic boundary layer (a scaling essential for the future triple-deck theory), gave the result that his previous, inviscid analysis was applicable not to the true wall itself, but to a fictitious one, displaced normally, by an amount of $O\left(R_e^{-5/8}\right)=O\left(R_e^{-1/2}.R_e^{-1/8}\right)$. Fitting together all the pieces, Lighthill found that eigenmodes of perturbations existed, with a pressure perturbation running like $\mathrm{Exp}\left(kR_e^{3/8}(x-x_0)\right)$ with a computed $O(1)$ constant $k$ and an arbitrary value of $x_0$. With a great deal of ingenuity, Lighthill had succeeded in constructing a theoretical model which included the possibility of spontaneous perturbations of the boundary layer, in the form of eigenmodes, that is perturbations apparently provoked by nothing. Turning back to the very weak impinging shock problem, the "nothing" could be explained but this was not basic for the future.

*2.2.3 The beginning of a new theory*

Lighthill's explanation of the upstream influence was a very deep result. It remained to turn it into a comprehensive theory, through use of systematic asymptotics. This necessitated sixteen years and was accomplished independently by Stewartson, Williams 1969 and Neiland 1969, both on this specific problem of upstream influence in supersonic boundary layers. Soon after, Stewartson 1969, followed by Messiter 1970, showed that a similar phenomenon occurred in the incompressible domain. It was then an easy matter to show, through use of the Prandtl-Glauert rule, that the whole subsonic domain could be reduced to the incompressible one. The transonic régime was treated by Messiter, Feo, Melnik 1971, and by Brilliant, Adamson 1974, while the hypersonic one was investigated a bit later and thoroughly covered by Lagrée 1992.

There are several good reviews of the early results achieved by the triple-deck theory ; we give only a few : Stewartson 1974, Smith 1982, Meyer 1983, this last one being short but illuminating. Further achievement are so extensive that an up-to-date review would be appropriate. The one by Nayfeh 1991 does not include the most recent, quite fascinating, achievements.

**2.3 Main asymptotic features of triple deck theory**

It is now time for me to describe the triple deck theory. I have already suggested that it is based on breaking the asymptotic structure which is at the root of Prandtl's boundary layer equations. This breaking is done by substituting another asymptotic structure for it and deriving a new mathematical model from it. I use the simplest situation : 2D, steady, incompressible flow over streamlined bodies, local $(x,y)$ coordinates, local $(u,v)$ velocity components and pressure $p$, with the wall at $y = 0$, all rendered dimensionless in some convenient way. The overall Reynolds number is $R_e$. Prandtl's asymptotic "ansatz" is that the flow is inviscid except when $(x,y)=\left(O(1),O\left(R_e^{-1/2}\right)\right), p=O(1)$. With no slip on the wall, and matching with the inviscid flow, this leads to Prandtl's so-called boundary layer theory, which is a mathematical model. I stress the fact that full asymptotic machinery was not needed by Prandtl to get his model, but it was undoubtedly essential to Van Dyke's (1962, 1964) higher order approximation, knowledge of which was crucial for estimating heat fluxes during reentry of space vehicles. I also stress that Prandtl's mathematical model became most efficient when it was complemented by a numerical scheme for computations, as in Flügge-Lotz, Blottner 1963. There are now many ways of breaking the asymptotic structure, but one is especially significant, because it leads to a new asymptotic structure which seems to be very rich and it is this one which leads to the triple deck. The first breaking is to place oneself near a particular position, say $x_0$, and set $x-x_0=O\left(R_e^{-3/8}\right)$. The specific order $R_e^{-3/8}$ is obtained by inspection, through trial and error, but we have already seen that it comes from examination of the instability theory, as known in the thirties, or from Lighthill's 1953 solution to the upstream influence paradox. Let us stress, anyway that Mauss (this volume) has indicated a logical way to proceed, in order to obtain it deductively. Once the asymptotic order of the longitudinal extent is obtained, one finds that a three-tier structure is needed normally to the wall. There are different equations depending on the order of $y$, namely $y=O\left(R_e^{-5/8}\right)$ for the so-called lower deck , $y=O\left(R_e^{-1/2}\right)$ for the main deck, and $y=O\left(R_e^{-3/8}\right)$ for the upper deck. This three-tier structure has suggested the name triple deck. But there exist situations in which more than three decks are necessary, so that the name "multi-structured boundary layers", which was suggested by Stewartson 1974, is preferable, even if the tradition has not retained it. Within the main deck, the perturbations can be written explicitly in terms of only one unknown function $A\left((x-x_0)R_e^{3/8}\right)$. The same situation prevails in the upper deck, but the solution requires continuation of an analytic function known in terms of $A(.)$ on the real axis, in the whole of the upper complex plane. The mathematical model proper concerns the lower deck with the asymptotic "ansatz" $(u,v,p-p_0)=O\left(R_e^{-1/8},R_e^{-3/8},R_e^{-1/4}\right)$. It consists of Prandtl's boundary layer equations (the same as for the original boundary layer theory, but with quite different scalings and interpretation) with no slip on the wall and the following condition at infinity from the wall :

$$R_e^{1/8}\,u \rightarrow U_B'(0)\left\{R_e^{5/8}y+A\left[(x-x_0)R_e^{3/8}\right]\right\} \ , \tag{5}$$

a quite unfamiliar one to the users of boundary layer theory. The mathematical model is completed by a functional relationship that we state, symbolicaly :

$$\Im\ \left\{(p(.)-p_0)R_e^{1/4}\ ;\ A(.)\right\} \quad . \tag{6}$$

where the "." stands for the reduced dummy variable $(x-x_0)R_e^{3/8}$, and the symbol $\Im$ does not stand for a function of the arguments but for a functional of these arguments. It does not matter what is the precise form of this relation. With (5) the expression $U_B\left(R_e^{1/2}y\right)$ denotes the streamwise velocity distribution which would exist in station $x_0$, if Prandtl's asymptotic structure were not broken. One very attractive feature of the model is its universality, all the specific features of the model being incorporated into the various scalings. As an example, let $R_B\left(R_e^{1/2}y\right), T_B\left(R_e^{1/2}y\right)$ and $\mu_B\left(R_e^{1/2}y\right)$ be the density, temperature and viscosity distributions in station $x_0$ for the basic incoming boundary layer ; the scaling for the velocity $u$ is then :

$$u=\beta^{-1/4}R_B^{\ 1/4}(\infty)R_B^{\ -1/2}(0)\mu_B^{1/4}(0)U_B^{1/2}(\infty)U_B'^{\ 1/4}(0)R_e^{-1/8}\ U, \tag{7}$$

and we have similar scalings for $v, x-x_0, y$ and $p-p_0$. The only notation which has not been defined in (7) is $\beta$ , namely :

$$\beta=\left|1-M_B^2(\infty)\right|^{1/2}, \tag{8}$$

where $M_B\left(R_e^{1/2}y\right)$ is the Mach number distribution in $x_0$ for the incoming boundary layer. When due care is taken with all the appropriate scalings, the problem issued from the triple deck asymptotic structure, does not contain any constants except possibly some which correspond to asymptotic inputs at $(x-x_0)R_e^{3/8} \to \pm\infty$. Apart from those inputs, which correspond to flexibility in adapting the model to various situations, the only choice occurs in (6), which has two different forms (very different indeed), depending on whether $M_B(\infty)$is below or above unity. An obvious consequence is that if one plots a large number of data issued from experiments (for instance pressure versus downstream distance to a specified point in the supersonic upstream influence problem), using the scaling issued from the triple deck, all the points should lie on a single curve. This was attempted by Stewartson, Williams 1969 (see their figure 9) with good success.

## 2.4 A few examples of significant achievements

The simplest triple deck model has been extended to three-dimensional unsteady flows by extending the scalings to time through $t-t_0 = O\left(R_e^{-1/4}\right)$ and to cross-stream coordinate through $z-z_0 = O\left(R_e^{-3/8}\right)$, the corresponding velocity component being $w = O\left(R_e^{-3/8}\right)$. Linearization of the three-dimensional unsteady triple deck leads readily to the asymptotic trends of linear instability theory and solves once and for all the paradox of nonparallelism, as shown by Smith 1979, Bodonyi, Smith 1981. A number of nonlinear extensions were explored by Smith and his team. We mention here only a few examples of this extensive work. Duck 1990 showed that in a supersonic flow, oblique eigenmodes are more amplified than purely two-dimensional ones, in the linear framework, while nonlinearity may lead to breakdown at a finite time (earlier for cross modes than for two-dimensional ones) ; Smith 1992, following previous work by Smith, Burgraf 1985, Smith, Doorly, Rothmayer 1990, who provided a fascinating picture of turbulent spots during transition to turbulence, work continued in Smith 1994. To be sure, this quite appealing piece of work is not based, strictly speaking, on the triple deck model, but it preserves the spirit of looking at asymptotic structures got from breaking previous ones. Beginners should not expect to read these papers without reading, previously, less sophisticated ones. In relation with a triple deck view of the nonlinear stability theory I would like to mention the so-called vortex-wave interaction theory (see Walton in this volume) as represented by Hall, Smith 1984, Hall, Smith 1991, and there are plenty of other noteworthy achievements, for example in transonic or hypersonic regimes.

The basic issues illuminated by the triple deck are so numerous that I want to select one more, concerning the so-called d'Alembert's paradox (the reader will find a full account in Stewartson 1981). Here, I merely want to recall, before entering the modern view of the subject, that the paradox was resolved by brillant contribution from Kutta 1902 and Joukowski 1912 (but oral tradition tells us that the French mathematician Villat had, about at the same time, the same idea ; the reader may find an allusion to this story in Villat 1972, at the end of page 1 and the beginning of page 2). I consider the simplest case of a profile, in a steady 2D incompressible flow, with a trailing edge making a small angle $\delta$. Let $x$ be the streamwise distance, putting the origin on the trailing edge itself, and let $C_L$ be the lift coefficient. We put $C_{L,KJ}$ to denote the value of $C_L$ which one computes by applying the Kutta-Joukowski condition. From inviscid theory one may express the streamwise pressure gradient, right ahead of the trailing edge, on either the windward or the leeward side of the profile as :

$$\frac{dp}{dx}=C_1\left(C_L-C_{L,KJ}\right)|x|^{-1/2}+C_2\delta|x|^{-1},\ x\uparrow 0,\ (x<0), \tag{9}$$

where $C_1$ and $C_2$ are constants that depend on the particular profile considered.According to the triple deck structure, centered on the trailing edge, one has :

$$\frac{dp}{dx}=ab^{-1}R_e^{\ 1/8}P'(X),\ x=bR_e^{\ -3/8}X,\ p-p_0=aR_e^{\ -1/4}P(X), \tag{10}$$

where two of the universal scalings appear, with $a$ and $b$ depending on the basic characteristics of the boundary layer, right on the trailing edge, before entering the triple deck structure. The reader may ask why a triple deck is inserted around the trailing edge. The reason is that Prandtl's asymptotic structure leads to a singularity at the trailing edge, on the downstream side, due to the discontinuous change in the boundary condition, when the flow leaves the wing to enter the wake. This singularity was studied by Goldstein 1930, and this early contribution is used in the form of asymptotic behavior of the solution of the triple deck problem, far downstream, on the scale of the triple deck. By comparing (9) and (10) we simply get :

$$X\to-\infty,\ P'(X)\approx C_1a^{-1}b^{1/2}R_e^{\ 1/16}\left(C_L-C_{L,KJ}\right)|X|^{-1/2}+C_2a^{-1}R_e^{\ 1/4}\delta\,|X|^{-1}. \tag{11}$$

This relation holds on the windward and the leeward sides of the profile. The first term on the right-hand side of (9) is associated with a departure from the so-called Kutta-Joukowski condition. In effect, the contribution to a solution to the paradox given at the beginning of the century was that the lift coefficient has to adjust itself to a value such that the severe singularity in the pressure gradient at the trailing edge ($|x|^{-1/2}$ singularity) disappears. This leads to the value $C_{L,KJ}$ for this lift coefficient. A moment's reflection suggests that (11) contains the solution of the d'Alembert's paradox. The proposal of Kutta and Joukowski (and Villat), that the lift coefficient adjusts itself to the value which eradicates the worst singularity, namely the value $C_{L,KJ}$ , is only true in the limit of infinite Reynolds number. At high but finite value of this number, there is a correction to the value of the lift coefficient :

$$C_L-C_{L,KJ}=Const\,R_e^{\ -1/16}. \tag{12}$$

The constant incorporates data from the inviscid theory of the Kutta-Joukowski flow around the profile, from the boundary layers, at the trailing edge, from both sides of the profile. It also incorporates data coming from the numerical solution of a universal triple deck problem. Contrary to the accepted Kutta-Joukowski scheme, the flow does not leave the profile right at the trailing edge, but, rather, there is a slight warping of it, with an inviscid stagnation point on the leeward side of the profile, at a distance $O\left(R_e^{\ -3/8}\right)$ ahead of the trailing edge. The second term on the right-hand side of (11) tells us that triple deck theory is able to account for the strong adverse pressure gradient $\left(|x|^{-1}\right)$ due to a noncusped trailing edge, provided the angle $\delta$

is, at most, on the order of $O(R_e^{-1/4})$. A lot of work has been done to go beyond these simple considerations. Here I shall just add a comment. The presence of the exponent 1/16 in (11) has two facets : one of them leads to argue that a theory based on $R_e^{-1/16}$ being small is hardly credible, while the other leads to argue that the correction to the Kutta-Joukowski condition, due to viscous effects, is far from being negligible. In the same vein one may consider that this 1/16 exponent means that one may go to rather high values of the lift coefficient without provoking stall, due to the impossibility of absorbing the singularity within the triple deck itself.

## 3. Asymptotics of branching in brittle fracture

### 3.1. The problem

A fascinating subject in fluid mechanics, involving a lot of asymptotics, is the dynamics of thin vortex filaments. I choose not to go into it here, simply referring the reader to the book of Ting, Klein 1991 and the work to be published by Klein, Knio 1994. The reason is that I want to report on what I consider as a breakthrough in a category of problems in solid mechanics with wide applications in industry, which, to my point of view, have a very loose analogy with the dynamics of slender vortex filaments. It is the problem of crack propagation in the theory of brittle fracture. I cannot enter into a detailed presentation of the subject of fracture. The interested reader can find a comprehensive, yet readily accessible, account of the field in Bui 1978, and, for those who cannot read French, in Freund 1990. I would like to recommend Maugin 1993 (especially Chapter 7) for an elevated point of view on the subject.

Now what is it about ? The reader should refer to Fig. 2 for a number of specific notations used below. A solid made of an isotropic, linearly elastic material, contains a crack supported by an area S, bounded by a curve T, called the tip of the crack.

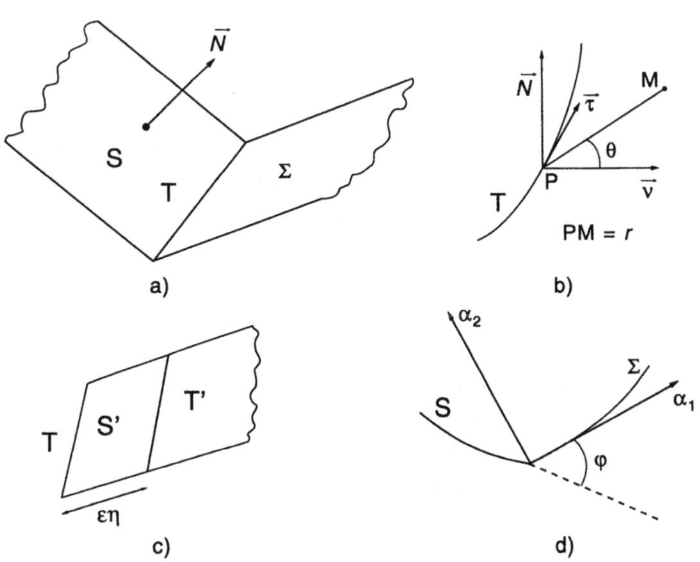


Fig. 2

All along S cohesion forces are broken, which means, in a mathematical model of the crack problem, that if $\vec{N}$ denotes the unit normal to S, the stress $\bar{\bar{\sigma}}.\vec{N}$ on both sides of S has to vanish. Of course $\bar{\bar{\sigma}}$ denotes the stress tensor, and the orientation of $\vec{N}$ has to be followed by continuity but the choice of this orientation does not matter. Concurrently, the displacement $\vec{u}$ may suffer a discontinuity $[|\vec{u}|]$ all along S, with the obvious condition that $[|\vec{u}|].\vec{N} \geq 0$, which means that the materials on either side of S may separate but are not allowed to penetrate one another. We remind the reader that $[|f|] = f^+ - f^-$, where $f^+$ (or $f^-$) is the limiting value of the field $f$ when approaching S from the downward (or upward) side when crossing S in the direction indicated by $\vec{N}$. One of the problems of brittle fracture is to predict the propagation of cracks, as the effect of various loads. The very loose analogy with vortex filaments lies in the fact that this propagation depends heavily on what happens to the material within a small tubular core surrounding the tip. One of the engineering issues has been to develop a local criterion (Griffith 1921, Irwin 1957, Barenblatt 1962) and to use it. This local criterion is expressed in terms of a vector attached to each point P of the tip. It is the so-called stress intensity factor (S.I.F.) $\vec{K}(s)$ ($s$ is the curvilinear abscissa along the tip).

### 3.2 The meaning of S.I.F.

In order to explain the meaning of the S.I.F., in the simplest way, imagine three coordinates $s, r, \theta$ attached to the tip curve. We already know $s$, the two others are polar coordinates in each plane normal to the tip at the corresponding point P($s$) (see Fig. 2). As a well known consequence of the theory of linear isotropic elasticity, one may state the following holds for any kind of body, any type of loads, any form of crack :

$$\operatorname*{Lim}_{r\downarrow 0} r^{1/2}\bar{\bar{\sigma}} = \sum_{p=I}^{III} \left\{ K_p(s)\, \bar{\bar{\Sigma}}_p^{(0)}(\theta) \; + \; r^{1/2} T_p(s) \bar{\bar{\Sigma}}_p^{(1/2)}(\theta) \right\} + \bar{\bar{O}}(r) \;, \tag{13}$$

where $K_p$ are the three components (in the local frame attached to the tip at P) of the vector $\vec{K}$, while $\bar{\bar{\Sigma}}_p^{(0)}(\theta)$ and $\bar{\bar{\Sigma}}_p^{(1/2)}(\theta)$ are functions of $\theta$ which are known from the theory as well as the three functions $T_p(s)$. The three components $K_p$ correspond to three modes of opening of the crack, as described by $[|\vec{u}|]$. Mode I (it is traditional to use roman representation of numbers) is the simplest : rotation of the two lips of the crack around the tangent to the tip, namely $[|\vec{u}|] \propto r^{1/2}\vec{N}+$..... Mode II corresponds to a shear with the two lips sliding with respect to each other in the plane tangent to the crack, and in the direction perpendicular to the tip, namely $[|\vec{u}|] \propto r^{1/2}\vec{\nu}+\ldots(\vec{\nu} = \vec{\tau} \wedge \vec{N})$, where $\vec{\tau}$ is the unit tangent to the tip (see Fig. 2). Finally, mode III is again a shear, but with slipping parallel to the tip, namely $[|\vec{u}|] \propto r^{1/2}\vec{\tau}+$...For our purpose the reader should be satisfied by accepting the following. Under a change of loading, the crack may go from S to another position (we assume that it corresponds to an extension) S', with a new tip T' ; provided we are able to compute the new distribution $\vec{K}'(s)$ of S.I.F along T', application of the so-called fracture criterion leads to a condition which allows, through a not so simple process, prediction of the precise form of the incremental extension of the crack from S to S'.

### 3.3 Mouchrif's formula for the variation of S.I.F.

The literature devoted to this topic is extremely abundant but in my opinion, a significant progress was made recently (Leblond 1989, Amestoy, Leblond 1992, Mouchrif 1994) and I would like to report on it because it relies on asymptotics. The most complete achievement, at the present time is Mouchrif 1994, but a joint publication with Leblond will appear. Before describing the results a word is necessary about the phenomenon of branching. The simplest situation occurs when the crack extension from S to S' occurs smoothly, and aside from simple geometries, the problem is, then, one of numerical computation. But such a smooth extension is not guaranteed : sometimes in the loading-extension process, branching may occur. This means that the planes tangent to S and S' along T (the tip of the initial crack before the incremental loading which provokes the branching) are distinct. When this happens, the extension exhibits a singularity in its curvature tensor along T . I am now ready to communicate the message contained in the work advocated before. It is necessary, first to specify the support $\Sigma$ of the crack extension (see Fig. 2.a). It is given by a piece of two-dimensional manifold, with the tip T as one of its boundaries, the others playing no role in the analysis. Concerning this manifold we only need to know it in a vicinity of T and this occurs through three functions of $s$ : the first, $\varphi(s)$, is the angle between the two tangent planes to S and S' along T (see Fig. 2.d which is a cut by a plane normal to T at P) ; the second is a kind of fractional 3/2 curvature $a(s)$, and the third is a true fully second-order curvature $C(s)$ (see Fig. 2.d where the curve $\Sigma$ is given by $\alpha_2 = a(s)\alpha_1{}^{3/2} + C(s)\alpha_1^2 + \ldots$). The extension itself is then made precise by giving its length $\varepsilon\eta\,(s)$ , where $\varepsilon$ is a small artificial parameter (see Fig. 2.c). The reason for the introduction of such a small parameter is that the process is incremental. This means that loading is incremented by an infinitesimal amount and that an infinitesimal extension is observed. The reader should expect a lot of asymptotics to arise in the analysis but I shall come back to this aspect later. Assume that we know the distribution of S.I.F. $\vec{K}(s)$ along T. Now, without changing the loading $L$ , we would like to know the new distribution $\vec{K}\,'(s)$ of S.I.F. along the new tip T ', after the extension. The reader should realize that this is a problem in pure elasticity, which has nothing to do with the physical problem of extension under a change of loading and corresponding destruction of cohesion forces. We may speak of a virtual extension as opposed to the real one, resulting from the application of the crack criterion when the loading is changed. The best way for me is to state the result given by Mouchrif 1994 after tedious calculation supported by a great deal of ingenuity, namely :

$$\vec{K}'(s) = \vec{\vec{F}}(\varphi).\vec{K}(s) + \varepsilon^{1/2}\left\{\vec{\vec{G}}(\varphi).\vec{T}(s) + a(s)\vec{\vec{H}}(\varphi).\vec{K}(s)\right\}\eta(s) + \varepsilon\vec{K}^{(1)}(s) + O\left(\varepsilon^{3/2}\right). \tag{14}$$

Some terms appearing in this formula have been explained, namely $\vec{T}(s)$ whose components $T_p(s)$ occur in (13). Then $\vec{\vec{F}}, \vec{\vec{G}}, \vec{\vec{H}}$ are matrix operators which depend on $\varphi(s)$ only. Mouchrif 1994 gives a formula for $\vec{K}^{(1)}(s)$ which is too lengthy to be reproduced here.

### 3.4 Comments on Mouchrif's formula

The only points that matter for us are : first, that the author has investigated the most general form that the equation can take ; second, that it contains an integral, evaluated in

Cauchy's principal value and that runs all along the tip T ; third, that the curvature $C(s)$, of the support $\Sigma$ of the extension, advocated previously, occurs only in that term $\vec{K}^{(1)}(s)$. Looking at (14) we should emphasize the hierarchy in which the three functions $\varphi$ (.), $a$(.), $C$(.) defining the support of the extension occur. One observes that $\varphi$ (.) occurs alone in the term independent of $\varepsilon$ while $\varphi$ (.) and $a$ (.) both occur, without $C$(.), in the term involving $\varepsilon^{1/2}$ ; finally $C$(.) occurs for the first time in the term proportional to $\varepsilon$ . One should also point out that concerning $O(1)$ and $O(\varepsilon^{1/2})$ parts, they are purely local. This means that the same value of $s$ occurs in the left-hand side and in the values of $\varphi(s)$, $a(s)$ which appear in the terms concerned of the right-hand side. These remarks are not innocuous, as one may see if one tries to answer the question : what to do with formula (14) ? When there is no branching, one can manage to compute the extension numerically without using (14), since one has simply to compute the function $\vec{K}'(s)$. Of course this is contained in $\vec{K}^{(1)}(s)$ and application of the local criterion of crack extension would equally well give the solution. Now, when there is branching, it is not sufficient to compute the function $\eta(s)$, which gives the length of the extension up to the factor $\varepsilon$ , numerically. One also has to compute the three functions $\varphi$ (.), $a$(.), $C$(.) describing the surface supporting the extension. Keeping in mind that the usual criterion of crack propagation is formulated in terms of the values of the S.I.F, it is plainly obvious, at least where $\varphi$ (.) and $a$ (.) are concerned, that (14) provides the necessary information. In order to determine $C$(.) one has to consider the term $O(\varepsilon)$ in (14) and this is why the Mouchrif 1994 work is really a breakthrough. Previously (and the whole work goes back to Amestoy 1987, in which he gives, at the end of his introduction, credit to Leblond for the main idea) there was no way of rationally attacking the phenomenon of branching in its most general aspect. Only restricted cases could be treated. Some skeptical readers might wonder why bother with $C(s)$ ? Anyway, when there is no branching, there is curvature, and the numerical process involving the solution of a problem of elasticity, coupled with application of the local criterion for crack propagation, provides the necessary ingredients for computing this curvature.This is quite true but the reader should notice that when there is no branching, the curvature is continuous, normal to the tip, whereas when there is branching, one should allow for the possibility of discontinuity in the curvature all along the tip from which branching occurs. As a matter of fact, Mouchrif 1994 states in Sec 1.6.b that the $O(\varepsilon)$ term in (14) contains sufficient information for determining both the length $\eta(s)$of the extension and its new curvature $C(s)$.

### 3.4 Asymptotic features of the theory

Now I would like to show that the derivation of (14) is quite a subtle proportioning of ingenuity and asymptotic technique. We select a special point P on the tip T (the one with abscissa $s$ which occurs in the left-hand side of (14)), we surround it with a sphere, centered in P, of radius $R$ (which we intend to let go to zero and which may be considered as small from the outset) see Fig. 3.

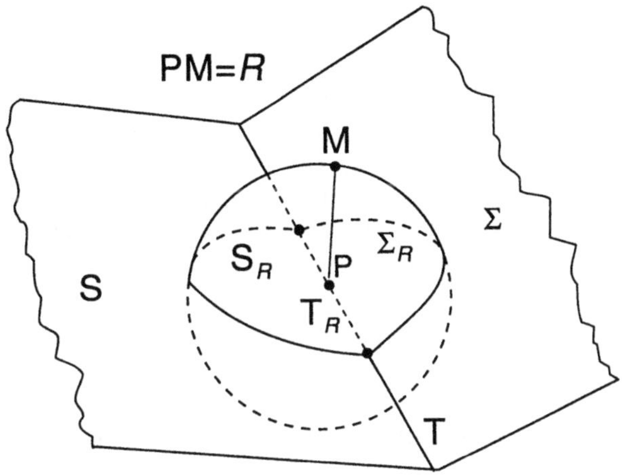


Fig. 3

We denote the parts of S, Σ and T contained inside this sphere as $S_R$, $\Sigma_R$ and $T_R$. We may recover all the fields corresponding to the original elasticity problem by specifying an appropriate stress distribution $\vec{\vec{\tau}}(R,.)$ all along the boundary of the sphere. Now, as a general consequence of elasticity theory, we can state :

$$\vec{K}\,'(\mathrm{P}) = \vec{L}(\mathrm{P}, R, \mathrm{S}_R, \Sigma_R, \eta\,(.), \vec{\vec{\tau}}\,(R,.)\;;\;\varepsilon)\;. \tag{15}$$

Before going ahead we require a bit of reflection on the part of the reader who should realize that, insofar as we are concerned with a pure problem of elasticity, independently of the problem of crack propagation, using a separate criterion, all the arguments in the functional $\vec{L}$, in (15), are completely free. We are allowed to play any game we want with these arguments. We (actually Mouchrif 1994, but the idea goes back to Amestoy 1987, and I have already said that this author gives credit to Leblond) will effectively play various games. For the present we just intend to play on the argument $\vec{\vec{\tau}}(R,.)$. In order to do so we imagine a special notation

$$\vec{L}\Big(\mathrm{P}, R, \mathrm{S}_R, \Sigma_R, \eta\,(.), \vec{\vec{\bar{\tau}}}\,(.)\;;\;\varepsilon\Big)\;, \tag{16}$$

for specifying that the stress distribution $\vec{\vec{\bar{\tau}}}\,(.)$ is an arbitrary function of the two polar coordinates specifying any point on the sphere, not necessarily the one which arises from the distribution of loads outside the sphere. The reader should understand that we have in no way forgotten argument $R$ : the sphere being specified, any stress distribution along that sphere is a function of only two polar coordinates. Now a crucial point in Mouchrif's argument is the following equivariance relation :

$$\vec{L}\Big(\mathrm{P}, \lambda R, \mathrm{S}_{\lambda R}, \Sigma_{\lambda R}, \eta\,(\lambda.), \vec{\vec{\bar{\tau}}}\,(.)\;;\;\lambda\varepsilon\Big) = \sqrt{\lambda}\,\vec{L}\Big(\mathrm{P}, R, \mathrm{S}_R, \Sigma_R, \eta\,(.), \vec{\vec{\bar{\tau}}}\,(.)\;;\;\varepsilon\Big)\;, \tag{17}$$

where $\lambda$ denotes any positive real number. This relation is self-explanatory, provided we specify the meanings of $\mathrm{S}_{\lambda R}, \Sigma_{\lambda R}$, and $\eta\,(\lambda.)$ . We first explain $\mathrm{S}_{\lambda R}$. The transposition to $\Sigma_{\lambda R}$

is obvious. We view $S_R$ as a manifold with two boundaries, one of which is located on T, the other being located on the sphere with radius $R$ centered in P. We get $S_{\lambda R}$ simply by submitting $S_R$ to the homothetic transformation $\vec{x}-\vec{x}_P \rightarrow \lambda(\vec{x}-\vec{x}_P)$, where $\vec{x}$ stands for the vector position of any point, $\vec{x}_P$ being that of P. The function $\eta\ (\lambda.)$ is deduced from $\eta\ (.)$ by a similar transformation that we explain now. Let $\sigma$ be the abscissa of a current point on T, $s$ being that of P. By applying the previously defined homothetic transformation to T, we obtain a curve $T_\lambda$ having P in common with T. The abscissae of corresponding points on T and $T_\lambda$ are related by $\sigma - s \rightarrow \lambda(\sigma - s)$ and we have simply set $\eta\ (\lambda.)$ for the function

$$\sigma \rightarrow \eta(s+\lambda(\sigma-s)) \ . \tag{18}$$

The proof of (17) is reduced to thinking about the problem of elasticity and we do not bother with it here. The second step in the analysis lies in assessing the following asymptotic expansion :

$$\begin{cases} \vec{L}\left(\mathrm{P},R,\mathrm{S}_R,\Sigma_R,\eta(.),\vec{\vec{\tau}}(.)\ ;\varepsilon\right) = \vec{L}^{(0)}\left(\mathrm{P},R,\mathrm{S}_R\Sigma_R,\eta(.)\ ;\ \vec{\vec{\tau}}(.)\right) \\ \qquad +\varepsilon^{1/2}\vec{L}^{(1/2)}(-----)+\varepsilon\,\vec{L}^{(1)}(-----)+O(\varepsilon^{3/2}) \end{cases} \ . \tag{19}$$

From this and (17) we readily deduce :

$$\begin{cases} \vec{L}^{(0)}\left(\mathrm{P},\lambda R,\mathrm{S}_{\lambda R},\Sigma_{\lambda R},\eta(\lambda.)\ ;\ \vec{\vec{\tau}}(.)\right) = \sqrt{\lambda}\,\vec{L}^{(0)}\left(\mathrm{P},R,\mathrm{S}_R\Sigma_R,\eta(.)\ ;\ \vec{\vec{\tau}}(.)\right) , \\ \qquad L^{(1/2)}(-(\lambda)) = L^{(1/2)}(-)\ ,\ L^{(1)}(-(\lambda)) = \lambda^{-1/2}L^{(1)}(-)\ , \end{cases} \tag{20}$$

where, on the second line, $-(\lambda)$ stands for the arguments in the left-hand side of the first line and — stands for the arguments on the right-hand side. Having achieved this, the final step consists in successfully performing the limiting process $R$ going to zero. This allows us to pick the most general form of formula (14), including the explicit form of $\vec{K}^{(1)}(s)$ whose derivation requires tedious calculations. This is not the whole story because one has yet to fully determine the matrix operators $\vec{\vec{F}}(\varphi),\vec{\vec{G}}(\varphi),\vec{\vec{H}}(\varphi)$ which are apparent in (14) and all the others (including nonlocal ones) which are hidden behind $\vec{K}^{(1)}(s)$. The most proper way to do so is to use matched asymptotic expansions (M.A.E.). Each operator occurring in (14) is obtained by solving some specific problem in plane elasticity (either plane deformation or an antiplane problem). When dealing with the $O(1)$ and $O(\varepsilon^{1/2})$ terms in (14), one has to solve only homogeneous problems, while nonhomogeneous ones arise at the level of $\vec{K}^{(1)}(s)$. In my personal opinion if M.A.E. had been applied in its systematic machinery from the outset, the technicalities would have hidden the simplicity of the basic argument, and possibly rendered the full discovery of (14) impossible. But such speculations are probably futile.

# 4. Asymtotics of Miscellanea

## 4.1 A problem of jet

Higuera, Martinez 1993 start from ad hoc modeling and use slight asymptotics. Consider a jet issuing from a hole in a plane at right angle and submitted to a weak, uniform, cross flow, parallel to that plane. The boundary layer in the cross flow is ignored as is the part of the jet very close to the hole. The main goal is to predict how the jet is bent and deformed by the cross flow. Ad hoc modeling is provided by Navier-Stokes incompressible flow equations with constant eddy viscosity, to account for turbulence. The choice of a convenient value is argumented. Asymptotics is introduced into the problem through a small parameter $\beta$ , which measures the slenderness of the jet, on the one hand, and the small ratio of the uniform cross velocity to the maximum axial velocity of the jet, on the other hand. The main unknowns used are $w$, the velocity component normal to the wall, and $\Omega$, the vorticity component in the same direction. The distance to the wall, made properly dimensionless, is $z$. The smallness of $\beta$ reduces the problem to two parabolic equations for $w$ and $\Omega$, without any second-order derivatives with respect to $z$. Actually, one has to add a two-dimensional Poisson equation for $u$, the velocity component parallel to the uniform cross flow velocity at infinity, and the continuity equation for obtaining the third velocity component, but that is a detail. Scrutinizing the problem, asymptotics reveals that $u$ has to be scaled with $1/\beta$ , in comparison with $w$, in order for the main part of bending of the jet to be captured numerically. Numerical computations are presented. They show spreading of the jet , formation of axial shear, laterally, as well as a kidney shape for its cross section, and formation of contrarotative vortices normal to the wall, which migrate laterally when $z$ grows. Numerical computations, which march in the $z$ direction have to be stopped when the jet approaches the boundary of the numerical grid. Asymptotics is used again to give plausible behavior for large values of $z$, but we refer the reader to the paper for that. We end this short presentation by mentioning another asymptotic analysis, for small $z$, which is of interest in itself and helps starting numerical marching in $z$. It amounts to an inner-outer expansion, involving the powers of $z$ as the gauge functions. Inner expansion corresponds to the core of the jet. Its leading approximation is the well-known similarity solution for a circular laminar jet, with constant axial momentum. Outer expansion mainly concerns the cross flow.

## 4.2 Problems in potential flow theory

### *4.2.1 Asymptotics versus numerics. A comparison*

Consider a slender body of revolution, with rounded edges, placed in an inviscid flow, uniform at infinity in the direction of the axis of the body. Assume that this body is defined by the law $\varepsilon^2 S(x)$, $|x| \leq 1$ , giving the areas of its sections, $\varepsilon$ being the small slenderness ratio. It is well known that the velocity potential can be obtained by distributing sources of intensity $f(x,\varepsilon)$ all along $-1 < x < 1$. This can be given by an expansion in powers of $\varepsilon^2$ including terms like $\varepsilon^{2q}(\text{Log}\,\varepsilon)^p$ and the like. Such an expansion, truncated after the $O(\varepsilon^6)$ term, leads to tractable formulae, involving $S$, $S' = \dfrac{dS}{dx}, S''$, $(SS'')'$, $(S(SS'')')'$ and various expressions formed with those terms and integrals. What is less well-known is that if the distribution of sources is limited to $-1<\alpha(\varepsilon)\leq x\leq\beta(\varepsilon)<+1$, and if $\alpha$ and $\beta$ are similarly expanded in powers of $\varepsilon^2$, one can account for the nonuniformity on the rounded noses. Concerning the pressure distribution on the body, Wong, Liu, Geer 1985 present a comparison between use of an asymptotic formula ($f$, $\alpha$, $\beta$ all truncated after $\varepsilon^6$) and a numerical solution involving 300 panels. Up to $\varepsilon=0.1$, the comparison is excellent and the asymptotics gives a better resolution near the noses. The computation time is in favor of asymptotics (2 against 14 seconds). The same method can be applied to a thin, cambered airfoil with rounded nose and sharp trailing edge, the airfoil being defined by $y=\varepsilon\,(f(x)\pm h(x)),|x|\leq 1$ . The asymptotics is again superior for small values of $\varepsilon$ . For a symmetric Joukowski airfoil one even knows the exact solution. For $\varepsilon=0.065$ at a 12 degree incidence, the error in the lift coefficient is 0.9 % while it is 5.5 % for the panel method. Here, asymptotics proceeds with even and odd powers of $\varepsilon$ , without logarithms ; it is truncated at $\varepsilon^2$, inclusive for the source distribution and $\varepsilon^4$, inclusive for the extremities of the segment over which the sources are distributed ; by comparison, 50 panels are used for the numerical method. Again, it is apparent that asymptotics is superior for the pressure distribution in the nose region.

*4.2.2 Numerics assisted by asymptotics*

The work of Geer, Andersen 1989 is quite different and, perhaps more promising. The idea is attractive and I would like to repeat the presentation that is given by the authors. Assume that one has to solve some operator equation for $U$, namely $L(U;\varepsilon)=0$, where $\varepsilon$ is a small parameter. Assume that one knows a regular asymptotic expansion $U=\sum_{n=0}^{N-1}\gamma_n(\varepsilon)u_n(x)+O(\gamma_N)$ for the solution. One uses this as a basis for a Bubnov-Galerkin numerical technique. The approximate solution is sought with the form $\tilde{U}=\sum_{n=0}^{N-1}\delta_n(\varepsilon)\,u_n(x)$. The $u_n(x)$ are delivered by asymptotic theory, while one attempts to find $\delta_n(\varepsilon)$ by using a numerical code. The method is simple : one uses the set of $N$ equations $\int_D L(\tilde{U};\varepsilon)v_p(x)dx=0$, $p=0,\ 1,\ 2...,\ N-1$ for the $N$ functions $\delta_n(\varepsilon)$, $n=0,\ 1,...,\ N-1$. One may choose $v_p(x)=u_p(x)$. The integral is computed over the entire domain of the variable (scalar or vector valued) $x$. The method is qualified as hybrid and provides a good example of skill combination of numerics and asymptotics. Various applications are presented. Here we select the application to potential flow around a slender body of revolution, under the same conditions as above. We have to be a little more specific about the numerical method. The velocity potential is again given as a distribution of sources along segment $\alpha\,(\varepsilon)\leq x\leq\beta\,(\varepsilon)$ of the axis of the body. The functions $\alpha\,(\varepsilon)$ and $\beta\,(\varepsilon)$ are known from their asymptotic expansions and what is looked for is the distribution of sources $f(x,\varepsilon)$, $\alpha\leq x\leq\beta$. Instead of using the asymptotic expansion for $f$, the authors change it to $\tilde{f}=\sum_{n=0}^{N-1}\delta_n(\varepsilon)f_n(x)$ where $f_n(x)$ come from asymptotics, while functions $\delta_n(\varepsilon)$, which do not constitue an asymptotic sequence of gauge functions, are obtained by the Bubnov-Galerkin procedure. Instead of writing that the slip condition is satisfied, pointwise, but asymptotically, the authors write that it is satisfied, integrated over the surface, with as many weighting functions as there are unknowns $\delta_n(\varepsilon)$, $n=0, 1,.. N-1$. They end up with a linear system :

$$\sum_{n=0}^{N-1} A_{np}(\varepsilon)\delta_p(\varepsilon)=B_n(\varepsilon) \quad , \quad n=0, 1..., N-1 \quad , \tag{21}$$

where $A_{np}(\varepsilon)$ and $B_n(\varepsilon)$ are known functions of $\varepsilon$ . The result is spectacular when applied to a prolate spheroid $S(x)=4x(1-x)$ (the reader should remember that the function $S(x)$ gives the distribution of areas of the cross-sections). With $N$=1, we know $f_0(x)=\frac{1}{4}S'(x)$ and it is found that $\delta_0=\varepsilon^2\left[\left(1-4\varepsilon^2\right)^{1/2}-2\varepsilon^2\, C(\varepsilon)\right]^{-1}$ (with $C(\varepsilon)=2\,\mathrm{Log}\left[\left(1+\left(1-4\varepsilon^2\right)^{1/2}\right)/2\varepsilon\right]$ ) which is obviously not a gauge function as is usually meant in asymptotic theory. The pressure distribution on the spheroid is known exactly and it is found that the hybrid method, with $N$=1, is not distinguishable from it for $\varepsilon=0.3$, while the asymptotitc solution, up to $N$=6, is some 8 % wrong. Of course this example provides an exceptional illustration but it shows how asymptotics may be efficient when skill use of it is made.

*4.2.3 A drastic reduction of numerical work*

The work of Barone, Caulk 1985 provides yet another example of asymptotics used in a combination with numerics. The original problem seems to be one in heat transfer but it seems simpler to explain it in electrostatic terms as done by the authors. Consider an interior domain $R$ whose external boundary $\partial R$ is at a given potential $\phi_\ell$. Inside the domain there are several slender, conducting, wires $W_\alpha$, $\alpha=1, 2,..N$, each being raised to a given potential $\phi_\alpha$. The problem is to find the potential $\phi$ within the domain outside the wires. The numerical method consists of finding a distribution of dipoles on the external surface $\partial R$ and on the surfaces of the wires $W_\alpha$. The distribution of sources, used in the representation of the solution, is given by data $\phi_\ell$ and $\phi_\alpha$. Asymptotics is introduced into the problem by analytically calculating the integral along each curve, section of any of the wires by any plane perpendicular to the axis of that wire. This integral can be carried out, assuming that the intensity of the dipole is constant along the curve considered, as a consequence of slenderness of the wires. The final problem is to find a surface distribution $D(x)$ of dipoles all along the external boundary $\partial R$ and a linear distribution $d_\alpha(\sigma)$ of dipole intensities all along a central curve $C_\alpha$ for each wire. Here $\sigma$ is the arc length along $C_\alpha$. Asymptotics has allowed us to go from surface density along each wire to linear density along each $C_\alpha$. The reduction in volume of the numerics is quite significant, and stiffness, due to slenderness of the wires, is avoided. Of course we must end with numerics using panels : two-dimensional along the external boundary, segments along the axis of each wire. The paper does not present applications involving the degree of generality expected, but it demonstrates the capabilities of the method, using several examples. Let us consider just one. The external boundary is rejected to infinity, while there are only two wires ; each one is a torus with a main radius $R$ and a circular section with radius $a$. The planes of the two wires are mutually perpendicular and the tori themselves are interlocked (each winds around the other and the closest distance is $2d$). There is only one datum : the potential $\phi_T$ of one torus, the potential of the other being $-\phi_T$ and the potential at infinity being zero. There is one unknown function $q(\psi)$, which is the linear density of dipoles corresponding to the torus whose potential is $\phi_T$, $\psi$ being the polar angle counted from the point on the main circle closest to the other torus. Each circle is divided into *2M* segments ; some evident symmetries are taken into account ; and we end with a system of *M+1* equations for *M+1* unknowns. Two tests are conducted. One consists in plotting $aq(\psi)/\phi_T$, for $M = 3$, 4, 6 and 12, in the case of $a/R = 0.05$, $d/R = 0.1$. The results for *M=3* and *M=12* are almost the same for $|\psi/\pi| > 0.2$, whereas, for $\psi = 0$, there are significant differences between *M=3, 4, 6*, but almost identity for *M=6* and *M=12*. Of course this is only a numerical check of internal consistency. Another is obtained by plotting $aq(0)/\phi_T$ as a function of $d/R$, with $d/a=2$. The result is almost a straight line down to $d/R = 0.01$. Extrapolating to $d/R = 0$ we find $aq(0)/\phi_T = 5.15$, which is the exact value for two cylindrical wires crossing at right angles.

### 4.3 Low Mach numbers

Tracey 1988 investigates the motion of an infinitely long, rigid circular cylinder in air, considered an inviscid, perfect gas. At negative times all is at rest and, at time *t=0*, the cylinder is put impulsively into motion with a velocity that is *M* times the speed of sound in the gas at rest. It is assumed that *M* is much smaller than unity. The purpose is to understand the complex problem arising from motion of the air, caused by translation of the cylinder, and deceleration of the body due to forces arising from unsteady, potential flow of air. Forces arising from dissipation (including vortex shedding) are neglected and the account I give here neglects the effect of gravity acting on the motion of the cylinder. This view of the problem fits very well with M.A.E. and leads to moderately sophisticated analysis, supported by some tedious calculations, but we need not bother with any details. The problem is attacked by expansion with respect to *M*, with gauge functions of the form $M^p(\mathrm{Log}\, M)^q$. A first expansion concerns short distances and short times (this means that distances are on the order of the cylinder radius and that time is on the order of that required for sound to travel over its). The leading approximation for the potential is a linear acoustic field produced by a dipole located on the axis of the body, whose strength is a function of time, found in closed but far from simple form. The cylinder velocity (pure translation) is also found in closed but not so simple form as a function of time, as an expansion in powers of *M*. The Log *M* does not appear in this part of the analysis. The author does not investigate formation of a shock because her main concern is with the slowing down of the cylinder by unsteady aerodynamic forces. One result may be mentioned here : the dimensionless velocity of the cylinder (after division by *M*) for large dimensionless times is

$$u = \frac{\alpha}{1+\alpha}\left\{1 - \frac{1}{(1+\alpha)t^2} + O\left(\frac{\mathrm{Log}\, t}{t^4}\right)\right\} + O(M) \ , \tag{22}$$

where $\alpha$ is the ratio of the cylinder density to the fluid density. We may roughly state that, by the end of the acoustic period of motion, the cylinder velocity has been reduced to $\alpha/(1+\alpha)$ of its initial value. For large value of $\alpha$ , this is very close to but less than unity as could be expected. For longer times (we use again *t* but with another scaling) on the order of the one required for the body to move by a distance equal to its diameter, two expansions are necessary, an inner one for distances comparable to the diameter, and an outer one for distances $O(M^{-1})$ times the diameter. Inner expansion involves $M^p$ and $M^p \operatorname{Log} M$ and is truncated at $M^3$ inclusive. Up to $M^2 \operatorname{Log} M$, the velocity potential obeys Laplace's equation for an incompressible potential flow. Higher order terms involve Poisson's equations. Terms involving $M^p \operatorname{Log} M$ are switchbacks, in the terminology of Lagerstrom 1988 (§ 2.5, page 118), and arise from matching with the outer expansion. The full potential in the outer field is $O(M)$ times smaller than in the inner field and is governed by a wave equation involving a nonlinear right-hand side, in two parts, one of them has $M^3$ in front of it as a multiplicative factor, the other has $M^6$. The expansion is worked out up to $O(M^3)$ inclusive, and matching between the inner and outer expansions is completed. Once the potential is found, we can compute the unsteady force on the cylinder, and write the dynamical equation giving its velocity as a function of time. We end this account with a suprisingly simple formula for the asymptotic ($t$ going to infinity) value of the velocity, namely :

$$u = \frac{\alpha}{1+\alpha} + \frac{M^2}{1+\alpha}\left\{\frac{1}{3}\left(\frac{\alpha}{2(1+\alpha)}\right)^3 - \frac{\alpha}{1+\alpha}\,\frac{1}{t^2} + k_2\right\} + o(M^2) \ , \qquad (23)$$

where, apparently, $k_2$ is an unknown constant.

I would like to take this opportunity for emphasizing the role of the Mach number in fluid mechanics. Chapter VIII of Zeytounian 1994 is devoted to this subject but a comprehensive review would be appropriate. The low Mach number case has a long story. The unsteady aspect has been renewed with Lighthill's theory of noise generated by turbulence in jets (Lighthill 1952, 1954), M.A.E. were brought into the subject by Lauvstad 1968, Crow 1970, Viviand 1970. Another aspect concerns an asymptotic derivation of the so-called Boussinesq model, for which there exists not so many papers, as, for example Bois 1991. Finally I mention Korobkin 1992 who studies the very tiny period of time concerning the beginning of the impact of a rigid rounded body with water.

### 4.4 An example from biomechanics

My next example is devoted to Lighthill's 1991, 1992 recent contribution to hearing sensitivity. It shows how far one may go towards scientific achievement by the clever use of asymptotics, provided it is not considered an end in itself but a tool. Of course, the account I give below is almost caricatural, and I mainly hope that it will excite readers' curiosity and incite them to read the references given and others mentioned in them. Two main points require an explanation : the first is that the ear, in mammals to be specific, acts as a very selective frequency analyzer ; the second is that very impressive experiments in vivo have demonstrated high sensitivity to very low intensity sound stimuli. I may add that this very high sensitivity is lost by death or with some drugs (of course I mean the sensitivity which is not conditioned by conscience). This second aspect which, according to Lighthill, is far from being understood, means that the ear is a highly nonlinear device, very peculiar because it enhances low excitation levels, so that nonlinearity is most effective at very low amplitudes. Purely mechanical devices do not exhibit this strange peculiarity, so that one suspects it may be due to some biophysical or biochemical process. What Lighthill tries to demonstrate (mainly in his

1991 paper) is that purely fluid mechanical phenomena may account for part of it. What mechanics, supported by asymptotics, may help in understanding is how sound stimuli are transmitted from the eardrum and middle ear through the cochlea to the auditory nervous system.

The cochlea is, roughly, a channel filled with a liquid whose physical properties are close to those of water. This channel is separated into two parts by the basilar membrane. The eardrum, through the mechanical device of the middle ear, excites vibrations in one of the two half-channels, by means of the oval window. These vibrations are transmitted to the other half-channel and all along the cochlea, through vibrations of the (basilar) membrane. Both experiments (in vivo because the property is lost on cadavers) and theory, using asymptotics at very high wave numbers for the latter, show the following. Waves at a given frequency run through smaller and smaller wavelengths when travelling along the cochlea from the base (adjacent to the middle ear) to the apex. At the same time they are amplified, then so strongly damped, that they pile up in a given spot on the membrane, each frequency corresponding to one particular spot. As the auditory nerve is ramified all along the basilar membrane, this may explain the very selective power of the cochlea as a frequency analyzer.

I shall attempt here to account very sketchily for the explanation, as given by Lighthill. Let $s(x)$ be the stiffness per unit length of the membrane and $h$ be its normal displacement, $x$ standing for the abscissa along it. For a pure tone at frequency $\omega$ , the total (kinetic plus potential) energy per unit length, averaged over a period can be expressed :

$$E=\frac{1}{2}s(x)<h^2>+\frac{1}{2}\left(A\rho k^{-1}+B\right)<\left(\frac{dh}{dt}\right)^2>=s(x)<h^2> \quad , \tag{24}$$

where the angle bracket $<.>$ stands for the average over one period. Use has been made (with the last part of the formula) of the well-known fact that energy is equally divided into its kinetic and potential parts. A lot of asymptotics is hidden behind (24). I mention only one point : we are interested in high wave numbers $k$ because the phenomenon of amplification followed by piling up occurs at high wave numbers. Then the vibrations of the liquid in the cochlea on the two sides of the membrane are located very close to that membrane, within a thin film whose thickness decreases as $O(k^{-1})$, and as consequence of the distribution of stiffness $s(x)$, $k$ increases downstream along the cochlea (downstream means following the direction of transmission of the acoustic tone). This explains why two terms appear in the inertial part of (24) : the first one, with $B$, accounts for the inertia of the membrane itself, and the second, with $A\rho\, k^{-1}$ accounts for the motion of the liquid with density $\rho$ . Looking at (24), one may guess that the wave system in the cochlea obeys the following approximate dispersion law, locally and asymptotically at high wave numbers :

$$(A\rho\, k^{-1}+B)\omega^2=s(x) \quad . \tag{25}$$

In order for asymptotics to be valid, we need $k^{-2}\left|\frac{dk}{dx}\right|<<1$. The analysis which follows is based on $A$ and $B$ being considered as independent on $x$, while $k$ varies widely. Setting

$\omega^2 B = s(x_r)$ yields :

$$k = \frac{A\rho\omega^2}{s(x) - s(x_r)} \tag{26}$$

In the ears of living mammals, $s(x)$ decreases considerably as $x$ increases, by about four orders of magnitude from the base to the apex. From the obvious condition that $k$ has to be positive, formula (26) is meaningless for $x > x_r$, but this does not matter because the transmission of the pure tone vibration stops at $x = x_r$.

For each pure tone at frequency $\omega$ , there is a given spot $x_r(\omega)$ such that $k$ grows without bound when $x$ approaches $x_r$ from the left (that is from the oval window to that given spot). The closer we are to that spot, the more appropriate the high wave number asymptotics which sustains the theory. Let $c_g$ defined by :

$$c_g = \frac{\partial \omega}{\partial k} = \frac{(s(x) - s(x_r))^2}{2A\rho\,\omega\, s(x)} \quad , \tag{27}$$

be the group velocity of a wave packet. High wave number asymptotics tells us that the energy per unit length, $E$ varies along the cochlea in such a way that :

$$c_g\, E\ \mathrm{Exp}\left\{\int^x \frac{D(\xi,\omega)}{c_g(\xi,\omega)} d\xi\right\} = const \quad , \tag{28}$$

where $D$ is the amount of energy dissipated by viscosity within the liquid filling the cochlea, per unit length. Asymptotics leads to :

$$D = C\left(\frac{\nu}{\omega}\right)^{1/2} k^2 c_g \quad , \tag{29}$$

where $\nu$ is the kinematic viscosity of the liquid and $C$ is a dimensionless constant. Introducing two quantities with the length dimension *i.e* :

$$\ell_1 = \frac{s(x_r)}{A\rho\,\omega^2} \quad , \quad \ell_2 = -\frac{s(x_r)}{s'(x_r)} \quad , \quad \left(s' = \frac{ds}{dx}\right) \quad , \tag{30}$$

we can rewrite (28) under the form :

$$c_g E \ \mathrm{Exp}\left\{ C\left(\frac{\nu}{\omega}\right)^{1/2} \frac{\ell_2^2}{\ell_1^2 (x_r - x)} \right\} = const \quad . \tag{31}$$

We are now ready for discussing the role of frequency analyzer played by the cochlea. Consider a given frequency $\omega$ ; when the corresponding wave rides along the basilar membrane, the wave number $k$, according to (26), grows without limit when the spot $x = x_r(\omega)$ is approached from the left. According to (27) the group velocity $c_g$ simultaneously decreases rapidly to zero. Due to the very low viscosity, the exponential factor in (31) is close to one, except in a very tiny neighborhood of $x_r$, so that the first effect of the rapid decrease in $c_g$ is a rapid growth of $E$ such that $< h^2 >^{1/2} (x_r - x)$ is roughly constant. Then, when $x$ is very close to $x_r$, the exponential factor grows very rapidly, and its growth is overwhelming in comparison to the effect of the decrease of $c_g$, so that the waves at that particular frequency $\omega$ , pile up quite suddenly at $x = x_r(\omega)$. The first function of the cochlea is thus explained brillantly and in a fairly plausible way in purely mechanical terms by Lighthill. The second aspect of enhanced sensitivity to low level excitation cannot be accounted by pure mechanical considerations. Lighthill's 1992 shows that a nonoscillatory flow normal to the membrane, excited by the so-called acoustic streaming phenomenon may be evaluated quantitatively. How this outflow finds its way, in the scala media, through the organ of Corti, and especially the outer hair cells, producing a feedback which amplifies low amplitude vibrations, is probably to be found with the aid of biophysics, but that is another story.

### 4.5. Microwave heating : asymptotics' assistance to numerics

Kriegsman 1993 uses multiple scaling to build a model of microwave heating in a three-dimensional medium. This model has to be solved numerically in two steps, but the model itself is issued from asymptotics. Let me try to explain this in words, with a minimum of equations. The ultimate goal is to predict heating in a material as a function of time. This is a more or less standard problem of nonlinear heat conduction in a three-dimensional domain, with a Robin condition on the boundary, and a heat source in the interior. The Robin condition arises from heat losses through the boundary, the heating is caused both by ohmic resistance and by polarization. The heat sources are operating at the very high frequencies of microwaves. They are quadratic in the fields, and the high frequency modulation of those heat sources can be averaged by the multiple scaling technique. The ratio of times sustaining this asymptotics is fantastically small $(10^{-10})$. The computation of the averaged heat source requires solving a problem of electromagnetic field scattering at given frequency. This is a quasi-steady, three-dimensional problem, linear, with coefficients which vary in time and space (it is their variation in time which produces the quasi-steady character, instead of steadiness). When the heat source has been computed from the solution of the electromagnetic scattering problem, one has to solve an unsteady three-dimensional problem in heat conduction. Both computations, the electromagnetic one and the heat conduction one, are coupled. This is due to the fact that the coefficients occuring in the scattering problem depend on the temperature. However, from time-marching, the computations themselves are decoupled in practice to a large extent. The storage space needed corresponds roughly to two complex vector fields (electric and magnetic) and one scalar, the temperature. It may be useful here to write a few equations, to give the reader an idea of the complexity of the final equations. Let us denote as $\vec{e}$ and $\vec{h}$ the complex amplitudes of the scattered electric and magnetic fields, and as $T$ the (real) temperature, all rendered conveniently dimensionless. Time and space are also dimensionless. For the scattering electromagnetic problem, there are a couple of equations

$$\vec{\nabla} \wedge \vec{e} = i\vec{h} \quad , \quad \vec{\nabla} \wedge \vec{h} = (\alpha + i\beta)\, \vec{e} + \vec{f} \ , \tag{32}$$

where $\alpha, \beta$ and $\vec{f}$ are zero outside the heated material $\Omega$. They are known functions of the temperature. We don't care about the transmission conditions on the boundary of the heated material, the computations must run over all space. For the temperature we have the problem

$$\begin{cases} \dfrac{\partial \mathrm{T}}{\partial t} = \vec{\nabla}.(k\vec{\nabla}\mathrm{T}) + Q & \text{, inside } \Omega \ , \\ k\dfrac{\partial \mathrm{T}}{\partial n} + B\mathrm{T} = 0 & \text{, on the boundary of } \Omega \ , \end{cases} \tag{33}$$

where $Q$ is given by

$$Q = Y(t)\gamma\left|\vec{e} + \vec{u}\ \mathrm{Exp}(i\vec{k}.\vec{x})\right|^2 \ , \tag{34}$$

$Y(t)$ being the Heaviside unit step function ; $\vec{u}$ stands for the unit vector giving the polarization of the planar microwave input, with wave number $\vec{k}$. Finally, $\gamma$ is a given function of the temperature. The Heavisisde $Y(t)$ function appears through multiple scaling, and accounts for the beginning of the microwave input. The asymptotics is rather elementary but crucial. Without its support, the numerics would be unsolvable (the asymptotics allows us to skip $10^{10}$ time steps). This is not the whole story with asymptotics. In a number of applications $B$ is very small (on the order of $10^{-4}$) and another asymptotic game may be played. We refer interested readers to the original paper.

## 5. Asymptotics of some spurious phenomena

### 5.1 Snap-through and combustion

Reiss 1980 uses a rather sophisticated "ansatz" which is not, strictly speaking, an asymptotic expansion, for explaining dynamic snap buckling of an elastic arch. Kassoy 1982 tries to asses this "ansatz" by using it on a mathematical model of combustion and discovers that Reiss, by failing to use a true asymptotic expansion, missed a spurious phenomenon. Buckling is described, according to Reiss, by a more or less ad hoc but reasonable model, consisting of a second order, nonlinear, ordinary differential equation, for the amplitude of the deformation of the arch, from some initial equilibrium position, namely

$$\frac{d^2A}{dt^2} = -\alpha A + 3hA^2 - A^3 + q \ , \tag{35}$$

where $A$ is the amplitude, $t$ is the time, $\alpha$ and $h$ are coefficients which depend on the arch, and $q$ is an external weight acting laterally. By equating the right-hand side to zero, one gets $A$ as a function of $q$, and this describes the equilibrium positions of the deformed arch. The curve $A(q)$ has the typical disposition which leads to hysteresis.

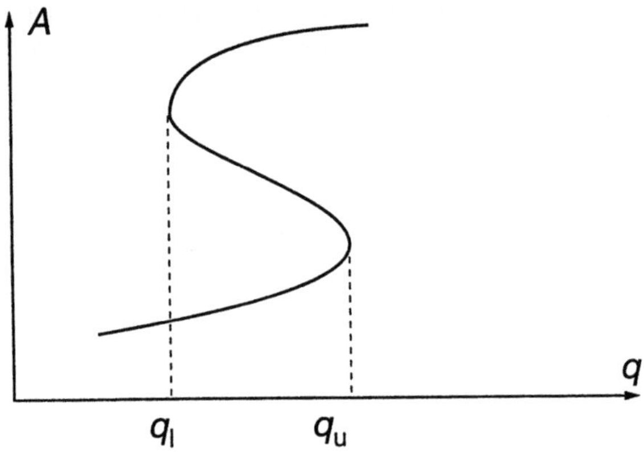


Fig. 4

On Fig. 4 the curve has three branches : the upper and lower ones correspond to stable behaviors, the intermediate to unstable one. A snap occurs when $q$ reaches $q_u$, increasing, and one tries to increase $q$ even more ; the equilibrium then jumps suddenly from the first, lower stable branch, to the upper, again stable, branch. Reiss attemps to capture the dynamics of snapping by using the dynamic equation (35). He sets $A(t) = A_u + V(t)$ and studies the dynamics under the initial conditions $V(0) = \varepsilon$ , $V'(0) = 0$. Asymptotics is introduced into the mathematical model by $\varepsilon$ which describes a small deviation from equilibrium, without velocity, at the initial time, $q$ being maintained at the value $q_u$. An explicit solution can be written in terms of Jacobian elliptic functions ; it contains $\varepsilon$ , and the problem is to analyze this solution when $\varepsilon$ is small. From the known solution, the author suggests the following formula :

$$V(t,\varepsilon) = \frac{P(\varepsilon) \ + \ Q(\varepsilon)\ \phi[\omega(\varepsilon)t]}{R(\varepsilon) \ + \ S(\varepsilon)\ \phi[\omega(\varepsilon)t]} \ , \tag{36}$$

where $P$, $Q$, $R$, $S$ are power series in $\varepsilon$ , while $\omega$ contains fractional powers, and $\phi$ (.) is a periodic function. Under the condition chosen ($\varepsilon$ <0), the solution is a cycle with period $O\left(|\varepsilon|^{-1/2}\right)$ but the amplitude of the oscillations is not $O(|\varepsilon|)$ , as would be suggested by the initial data. A long part of the cycle is spent with $O(|\varepsilon|)$ departure from the equilibrium value, but a short part of it, with duration $O(|\varepsilon|)$ , consists of a jump near the second equilibrium value. Reiss succeeds in constructing asymptotic expansions of *P, Q, R, S,* $\omega$ , but it should be noted that it is not, in the usual meaning of the term, an asymptotic expansion of the solution. The numerator of (36) is $O(|\varepsilon|)$ whereas the denominator is $O(1)$ except for short phases of the reduced time $\omega(\varepsilon)\ t = \tau$, during which it is also $O(|\varepsilon|)$ , so that the solution becomes $O(1)$, a fact interpreted as a jump near the second equilibrium. Reiss qualifies his approach as one of "rational function approximation".

The problem is to be sure that such a technique captures all the fine details of the phenomenon. He applies it to the following mathematical model of combustion :

$$y'^2 = y^2(1-y),\ \ y(0)=\varepsilon \tag{37}$$

The steady solution $y=0$ corresponds to the absence of a reaction and the steady solution $y=1$ corresponds to complete combustion of all the combustible material. The rational function method leads to an asymptotic representation of the solution that we now describe in words. As long as $\varepsilon\ \ t$ is smaller than one, the solution is a smooth, slow rise from zero, $y$ remaining $O(\varepsilon\ )$. But when $t$ is greater than $|\varepsilon|^{-1}$, the solution is a rapid (the phenomenon is described with $\varepsilon\, t = 1+\tau, \tau = O(1)$ ) growth to $y$ near unity an then an asymptotic approach to $y=1$. Plotting the result of an exact solution and comparing it with the approximate one reveals that, even at $\varepsilon\ =0.02$, there is a significant error, from $t = 0.6\varepsilon^{-1}$ to $t = 1.06\varepsilon^{-1}$ , during which the major part of the increase of $y$ arises. Kassoy 1982 reassessed Reiss's solution for this combustion model. He proceeded in two steps. The first one is an asymptotic analysis, when $\varepsilon$ goes to zero, of the exact analytic solution given by Reiss. He found three time phases. During the first one, the proper scaling is $t = \tau / \varepsilon$, and this phase lasts from $\tau = 0$ to $\tau = 1 + O(\varepsilon|\mathrm{Log}\ \varepsilon|)$. A second phase, of ignition, is obtained by the scaling $t = \varepsilon^{-1} + \sigma\ \mathrm{Log}(\varepsilon^{-1} - 1)$, and it lasts from $\sigma = -\infty$ to $\sigma = 1\_$ (the infinities have to be understood in the sense of M.A.E). It is during this phase that Reiss's rational function approximation is not accurate. Finally, a final phase corresponds to the scaling $t = \varepsilon^{-1} + \mathrm{Log}(\varepsilon^{-1} - 1) + s$ and $s$ goes from $s = -\infty$ to $s = +\infty$ (the infinities being, again, understood in the sense of M.A.E). During this third phase as well as during the first one, Reiss's rational function approximation is accurate. In a second part of his work, probably the more interesting one, Kassoy 1982 succeeded in constructing a three-layer asymptotic expansion without using this knowledge of the solution and he recovered what I have described. These two publications are illuminating for some spurious problems that can arise with asymptotics. Applying asymptotics without caution can lead to missing some fine details, even with significant errors. The reader might ask whether the dynamic snapping described by Reiss 1980 is erroneous. The answer is probably not. The reasoning is simple : the rational function approximation, for the arch problem, already has the form given by the exact solution, whereas, for the combustion problem, the rational function approximation is the one used for the approximate solving of a transcendental equation. For the approximate solving of the exact equation, Reiss substituted the approximate solving of another equation whose accuracy had not been carefully evaluated.

## 5.2 Spurious logarithms

### *5.2.1 Presentation of a class of problems*

A phenomenon which is well known to specialists, but is generally disconcerting to beginners, is the appearance of gauge functions, in the asymptotic solution, which are in no way suggested by the definition of the problem. A classical example is the occurrence of a small parameter $\varepsilon$ in the statement of the problem when Log $\varepsilon$ occurs in the asymptotic expansion of the solution. There is a famous example in fluid mechanics, namely a two-dimensional steady flow at a low Reynolds number around a fixed rigid body. This topic is discussed in general terms by Lagerstrom 1988 in § III.3.5 and more extensively on a model in § II.2.5 ; and the problem of low Reynolds number flow is specifically treated by Van Dyke 1975 Chapter VIII. Here we will discuss a different topic where the same type of spurious expansion occurs, based on some recent papers : Ward, Henshaw, Keller 1993, Ward, Keller 1993, Lange, Weinitschke 1994. The last paper can be considered an extension of Lange, Weinitschke 1991 that we commented in Section 1. The three papers referred to above are not detailed separately and we have changed the notations ; Ward, Keller 1993 is wider in scope and we neglect many aspects contained in it. In order to fix the setting we consider a domain $D$ with boudary $\partial D$, and the following nonlinear eigenvalue problem :

$$\Delta T+\lambda F(x,T)=0 \ , \ x\in D \ ; \ B(x)T=0 \ , \ x\in\partial D \ , \tag{38}$$

where $B(x)T=\alpha(x)\dfrac{\partial T}{\partial n}+\beta(x)T$, is the Robin operator. In this formula, $\dfrac{\partial}{\partial n}$ is the exterior normal derivative. The linear eigenvalue problem :

$$\Delta u+\lambda f(x)u=0, \ x\in D \ ; \ B(x)u=0, \ x\in\partial D \ , \tag{39}$$

is also considered. A relation between (38) and (39) occurs when $T(x\,;\,\lambda)$ is a solution of (38) such that $\lambda$ is a real eigenvalue of

$$\Delta\theta+\lambda\frac{\partial F}{\partial T}(x,T(x\,;\lambda))\ \theta=0, \ x\in D \ ; \ B(x)\theta=0, \ x\in\partial D. \tag{40}$$

This gives a way to detect a so-called limit-point that we discuss in Section 1. Asymptotics, in these problems, occurs if small cavities are inserted within domain $D$. Assume that a cavity $D_j(\varepsilon)$ surrounds $x_j$ and that $D_j(\varepsilon)$ is homothetic (with a ratio $\varepsilon$ and center $x_j$) with $D_j(1)$. The first equation of (38) (39) (40) is then valid in $D^*$ which is $D$ from which all $D_j(\varepsilon)$ have been deleted. The second equation in (38) (39) (40) must be supplemented by

$$\varepsilon\frac{\partial u}{\partial n}+\kappa u=0 \ , \ x\in\partial D_j(\varepsilon) \ , \tag{41}$$

where $\partial\ D_j(\varepsilon)$ is the boundary of $D_j(\varepsilon)$ and the normal derivative is internal to the cavity.

*5.2.2 A general perturbation formula*

We start discussing the linear problem (39) and its perturbation by (41). The dimension $n$ of the embedding space has to be taken into account. For $n \geq 3$ (we need not worry about the meaning of $n > 3$), any eigenvalue of (39) is modified by a correction $O(\varepsilon^{n-2})$ , when one skips to (41). The corresponding eigenfunction is corrected by a term $O(\varepsilon^{n-2})$ , except in a neighborhood of each $x_j$ where it is subjected to an $O(1)$ correction in a domain $\|x - x_j\| = O(\varepsilon)$ . The result is sufficiently simple to allow us to give it :

$$\lambda(\varepsilon) = \lambda_0 + (n-2)\varepsilon^{n-2}\left(2\pi^{n/2} / \Gamma(n/2)\right)\sum_j C_j |u_0(x_j)|^2 + O(\varepsilon^{n-1}) \tag{42}$$

Here $\Gamma(.)$ is the so-called Gamma function, $\lambda_0$ and $u_0(.)$ are respectively an eigenvalue and an eigenfunction of (39) and $C_j$ depend on the form of each cavity, specified by each $D_j(1)$, and on $\kappa$ ($\kappa_j$ if there is a different $\kappa$ in (41) for each cavity). Higher order terms may be given (see Ward, Keller 1993) but we do not pursue because (42) suggests that the two-dimensional case (*n=2*) is much more interesting for our purpose. As a matter of fact, letting *n=2*, at least formally, in (42), we guess that the basic small parameter will be $(|\text{Log}\ \varepsilon|)^{-1}$, and we will now account for what occurs then. The first question to ask is why there is such a spectacular effect of the dimension ? Of course (42) gives a first answer because, for fixed $\varepsilon << 1, (n-2)(\Gamma(n/2))^{-1}\varepsilon^{n-2}$ becomes larger and larger when $n$ decreases and suggests that the power of $\varepsilon$ should be replaced by a power of $(|\text{Log}\ \varepsilon|)^{-1}$ when $n = 2$. But this is not the answer we want. Put in another way we can say that (42) shows that the weight of the cavity is decreasing when $n$ increases. A pioneering work in this field is that of Sanchez-Palencia 1985 (see also Sanchez-Hubert, Sanchez-Palencia 1989, Chapter VIII) with a slightly different problem. In this work the reader will see that perturbation formulae like (42) are not the whole matter and that other effects, on the nature of the spectrum, can occur when $\varepsilon \downarrow 0$. But this goes beyond my purpose. Going back to the previous question we can rephrase it, but we will formulate the new question later. Near each cavity the eigenfunction is perturbed by $O(1)$ . The reason is obvious : if $u_0(x_j)$ does not vanish, which is the general situation, (41) fails by $O(1)$ (unless $\kappa$ is small). Then it is not hard to suspect that $u_0(x)$ has to be modified by an $O(1)$ quantity when $x$ is near $x_j$. Just think in terms of gravity field due to a continuous distribution of mass. Assume that this distribution is perturbed by a very heavy mass concentrated on a very small domain, of radius $\varepsilon$ , so that the total mass corresponding to the perturbation is of the same order as the unperturbed total mass. You may ask over what distance the gravity field is significantly perturbed near the concentrated mass ? But this is too vague to obtain a precise answer. Let us put it another way : at what rate does the perturbed gravity potential decay when $\|y\| = \varepsilon^{-1}\|x - x_j\|$ goes to infinity ?. The answer is then clear : the decay occurs at the rate $O\left(\|y\|^{-(n-2)}\right)$, so that, when $\varepsilon\|x - x_j\|$ is $O(1)$, the perturbation in potential is $O(\varepsilon^{n-2})$, which agrees qualitatively with (42). When *n=2*, the perturbation no longer decays but grows like

$O(\mathrm{Log}\|y\|)$, so that when $\varepsilon\|x-x_j\|$ is $O(1)$, the perturbation in potential is $O\left(|\mathrm{Log}\,\varepsilon|^{-1}\right)$. Of course, this is in no way a proof and the full argument is much more subtile, but I hope that some readers will find this remark illuminating and that those acquainted with M.A.E. (Matched Asymptotic Expansions) will find in it one of the many facets of matching.

*5.2.3 The special case of dimension two. The spurious logarithms*

We now come to the more specific case of dimension $n=2$. It was proved by Ward, Henshaw, Keller 1993, that any eigenvalue to (39) (41), with only one cavity, issued from an eigenvalue $\lambda_0$ to (39) alone, may be expressed as an asymptotic expansion :

$$\lambda(\varepsilon)=\lambda_0+\sum_{P=1}^{\infty}\lambda_P\left(\mathrm{Log}\left[\frac{1}{\varepsilon\;|d(\kappa,D_1(1))|}\right]\right)^p+O(\varepsilon\;|\mathrm{Log}\;\varepsilon\;|)\,. \tag{43}$$

But rather than trying to explain how to get this formula, I prefer to comment on it. First the $\lambda_p$ do not depend either on the shape of the cavity, represented by $D_1(1)$, or on $\kappa$, but on the location of its center. Second, an expansion involving powers of $|\mathrm{Log}\,\varepsilon|^{-1}$ is very poor and this may be confirmed by comparison with an exact solution. On the other hand, $d(\kappa,D_1)$ depends on $\kappa$ and on the shape of the cavity, *i.e.* on $D_1(1)$, but does not change when the cavity is arbitrarily rotated around its center. We give an example of poorness of the expansion. Assume $D$ is a circular disk with a radius of unity, the cavity a concentric disk with radius $\varepsilon, \kappa=+\infty$, $\beta/\alpha=+\infty$. For the lowest eigenvalue, the error in (43), truncated after $p=2$, is 0.3 % for $\varepsilon=0.01$, but increases to 4.9 % for $\varepsilon=0.125$. As a comparison, when $\kappa=0$, (43) is replaced by an expansion involving powers of $\varepsilon$ and the error in the corresponding formula, truncated after $O(\varepsilon^2)$ is 0.2 % at $\varepsilon$ =0.15. The reason why the equivalent of (43) contains powers of $\varepsilon$, instead of powers of $|\mathrm{Log}\,\varepsilon|^{-1}$ in this case, is that condition (41) is not violated at $O(1)$ by the eigenfunction of (39). If we consider, in dimension $n=3$, two concentric spheres with radii of unity and $\varepsilon$, (42) again contains powers of $\varepsilon$, instead of powers of $|\mathrm{Log}\,\varepsilon|^{-1}$. Truncation after $O(\varepsilon^2)$ leads, for $\kappa=+\infty$, $\beta/\alpha=+\infty$, to an error which is 0.71 % at $\varepsilon=0.125$.

*5.2.4 Further comments*

The reader may wonder why the $\lambda_p$ in (43) do not depend on the details pertaining to the shape of the cavity. We present an argument, which is not an explanation, but may throw some light on the matter. Let $\lambda$, $u$ be an eigenvalue, eigenfunction couple for (39) (41) and $\lambda_0, u_0$ the corresponding couple for (39) unperturbed. Let $\Gamma(\eta)$ be a circle with radius $\eta$, centered on the center of the cavity, and let $\Delta(\eta)$ be the part of the domain $D$, which is exterior to $\Gamma(\eta)$. One has :

$$(\lambda-\lambda_0)\int_{\Delta(\eta)} f(x)u(x)u_0(x)dS_x = \int_{\Gamma(\eta)}\left(u_0\frac{\partial\, u}{\partial\, n}-u\frac{\partial\, u_0}{\partial\, n}\right)d\sigma_x\ , \tag{44}$$

with self-explanatory extra notations. It is hard to go farther with (44) without entering into the machinery of M.A.E, but one may simply observe that the detailed shape of the cavity does not enter in (44) and $\kappa$ does not either. One more comment about (43) truncated after *p=2.* The domain *D* is a circle with a radius of unity ; the function *f* is identically equal to unity ; $\kappa=\infty, \beta/\alpha=\infty$, but there are five cavities. One is centered on the center of the main domain *D* and the four others are regularly distributed at distance $\frac{1}{2}$ from the center. Even for $\varepsilon=0.025$ and for the lowest eigenvalue, the error in (43), truncated after *p=2,* is 28 %, which is unexpectedly high. But scrutinizing the surface plot of the eigenfunction given by Ward, Henshaw, Keller 1993 shows that there are rather strong interactions between the perturbations created by the various holes and this may explain why the asymptotic representation of (43) is so poor. Of course these comments do not lead to much enthusiasm for asymptotics. But it is safe to keep in mind that, on several occasions, there are severe limitations, and it is safer to have arguments which allow such kinds of limitations to be suspected. Stated this way, even this example provides a message of hope, that we borrow again from Ward, Henshaw, Keller 1993.

We consider (39) perturbed by (41), and assume that all the cavities are identical, with the same value of $\kappa$ on each of them. Then, by setting a numerical problem, defined on the whole domain, ignoring the cavities, at the cost of introducing a variable $z=\varepsilon\, d(\kappa, D_1(1))$, associated to the center of each cavity, we can find a function $\lambda^*(z)$, which, when expanded in powers of $|\text{Log}\,\varepsilon|^{-1}$, reproduces the whole of expansion (43), truncated as indicated. Here $d(\kappa, D_1(1))$ has the same meaning as in (43). This result is so unexpectedly rewarding that it requires a special comment. But, before that, we present the accuracy taking the example discussed a few lines above. The error in $\lambda^*(\varepsilon\, d)$, with respect to the exact value, is 0.25 % at $\varepsilon=0.025$ , it is 1 % at $\varepsilon=0.05$ and 4 % at $\varepsilon=0.1$ . Why does this prove to be possible ? The main reason is that the so-called inner expansion, near each cavity, may be summed up to all logarithmic terms. This sum is well defined by a specific mathematical problem which contains, in its definition, a multiplicative factor $A_j(\varepsilon\, d(\kappa, D_1(1))$, with one such factor for each cavity. These factors, then, appear in the problem, defined on the whole domain and they appear linearly. The authors of this noteworthy paper do not indicate how to proceed with cavities differing in shapes and in the values of $\kappa$ . My opinion, which is not based on a careful study, is that in quite a similar way we should be able to build a function $\lambda^*(z_1..,z_N)$, where $z_j=\varepsilon\, d(\kappa, D_j(1))$ corresponds to the *jth* cavity, $j=1,\ 2,..,\ N$. Then the multiplicative factors $A_j$ would not depend on the same argument $\varepsilon\ d$ but one such argument for each corresponding cavity. If my conjecture turns out to be correct a slight modification of the numerical code will give the same kind of answer as with the case of identical cavities.

Now suppose that we consider the nonlinear eigenvalue problem (38), possibly perturbed by a condition like (41). The critical value $\lambda_c$ that may lead to runaway is obtained by solving the linear eigenvalue problem (40), possibly perturbed by a condition like (41). In dimension

$n$=2 we again find an expansion like (43), with the same poor accuracy as before, but the same process of summation occurs. We give just one result. The domain $D$ is a circle with a radius of unity and we have $F(x,T)=e^T$ in (38), $\beta/\alpha=\infty$ in the definition of the operator $B$ (see the line after (38)), there is just one elliptical cavity, centered in the center of the circle $D$, with 6.25 as the ratio of its axes and $\kappa=1$. The error in the critical value $\lambda_c$ is 0.2 % (or 1.2 %) at $\varepsilon=0.025$ and 0.3 % (or 2.4 %) at $\varepsilon=0.1$ when the summed formula like $\lambda^*_c(\varepsilon\, d(\kappa,D_1(1)))$ (see a few lines above) is used in comparison to a full numerical solution (or when an asymptotic expansion in powers of $|\mathrm{Log}(\varepsilon\, d)|^{-1}$, truncated after the term going as $|\mathrm{Log}(\varepsilon\, d)|^{-2}$ is used).

**5.3 Spurious oscillations**

Pernarowski, Miura, Kevorkian 1992 consider a model built from biophysics to predict an observed spurious phenomenon. It consists of

$$\frac{d^2u}{dt^2}+F(u)\frac{du}{dt}+G(u,c)=\varepsilon H(u,c),\quad \frac{dc}{dt}=\varepsilon(\beta\, h(u)-c)\ . \tag{45}$$

The scalar $u$ is directly related to the electrical potential of a membrane, whose calcium content is $c$ ; $\varepsilon$ is quite a small parameter and $\beta$ is a control parameter, which is $O(1)$ . Observations shows the existence of periodic oscillations (limit cycle) which apparently behave strangely. The concentration $c$ oscillations are made of almost straight segments, with very rapid changes from ascending to descending phases during one period. During the ascending phase there are very tiny oscillations superposed on an almost straight ascension. No such tiny oscillations occur during the descending phase of $c$. The behavior of $u$ is even stranger. When $c$ decreases smoothly, $u$ also increases smoothly. This phase of the cycle ends with an apparently explosive growth of $u$. This growth stops when $c$ increases and gives way to oscillations of an order of one amplitude. These oscillations continue almost up to the end of the ascending phase of $c$, then $u$ drops by one order, very rapidly, just to start a new cycle. Full numerical solutions agree with this behavior, and the purpose of the paper is to derive it from a combination of asymptotics, qualitative integration, and some numerics. It seems to be quite instructive as an example of the versatility of asymptotics, when exercised with ingenuity.

The phenomenon of rapid oscillations in $u$ when $c$ increases is the so-called burst of electrical activity (B.E.A). The analysis is based on an asymptotic description of the two main phases of the cycle ($c$ decreasing, followed by $c$ increasing), and a description of the transitions in between, as well as gluing together all the pieces. The so-called active phase ($c$ increasing) is described through a straightforward expansion with respect to $\varepsilon$ . The leading approximation is $c$=*Const*, while $u$ obeys a Lineard type of equation. Of course $c$ is constant on a tiny time scale and varies on a large time scale. Numerical simulations show that $F(u)\, du/dt$ may be considered small during each active phase so that the oscillations are governed, with a good approximation, by :

$$\frac{d^2u}{dt^2}+G(u,c)=0\ , \tag{46}$$

with $c$ almost constant during each oscillation, whose duration is $O(1)$. The smooth (as opposed to active) phase evolves on a time scale $O(\varepsilon^{-1})$ so that, with $u\left(\frac{\tilde{t}}{\varepsilon}\ ;\ \varepsilon\right)=U(\tilde{t},\varepsilon)$, $c\left(\frac{\tilde{t}}{\varepsilon}\ ;\ \varepsilon\right)=C(\tilde{t},\varepsilon)$, one finds (45) again, but $\frac{d}{dt}$ is replaced by $\varepsilon\frac{d}{d\tilde{t}}$ .

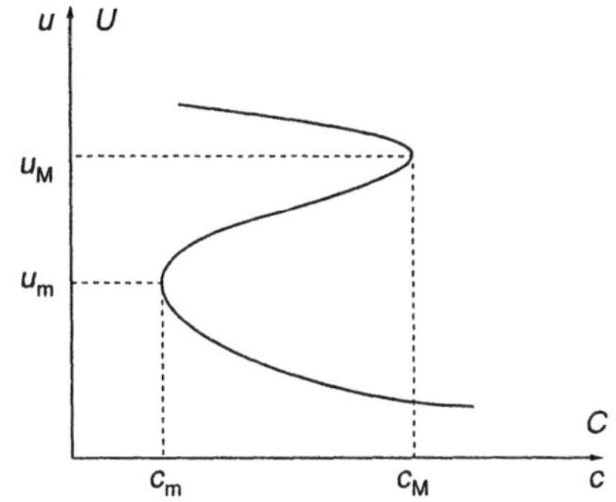


Fig. 5

The leading approximation follows $G(U,C)=0$. Of course, the analysis is based on the nature of the function $G(.,.)$. It is such that $U$ as a function of $C$, arising from $G(U,C)=0$, has three branches, like a cubic. We use indifferently capital or lower case letters for $u$ and $c$. We set $(u_m,c_m)$ and $(u_M,c_M)$ for the two turning points (see Fig. 5). The slow-smooth phase of the cycle corresponds to a part of $G=0$ where $u$ increases while $c$ decreases, time increasing. Two problems remain : first, the beginning and end of the smooth-slow and oscillating-fast phases of the cycle, second, the transition from one phase to the other. The transition, from smooth-slow to oscillating-fast, occurs at $c=c_m$, while the other transition occurs at $c=c_e>c_m$. During the transition itself $u$ obeys one of the equations :

$$\frac{du}{dt}=-T_n(u)\ G(u,c_e)\ \ ;\ \ \frac{du}{dt}=-T_n(u)\ \ G(u,c_m)\ . \tag{47}$$

Here $T_n(.)$ is a well defined function which occurs in the definition of the function $F(.)$. The value $c_m$ corresponds to the minimum of $c$ on $G(u,c)=0$, as the reader may see on Fig. 5. This arises because the smooth-slow phase has to follow $G=0$ on a part of it where $u$ is a decreasing function of $c$. Such a phase ends at the beginning of a transition phase during which $c=const$, and it is tempting to choose, the value for which $c$ is a minimum along the adequate part of $G=0$ as the appropriate value of $c$.

In order to describe the cycle in the phase space $(u,\dot{u},c)$, one has to compute $c_e$, the value of the concentration at which the cycle escapes from the oscillating-fast phase. During the oscillating-fast phase one has (46) as a leading approximation, with $c$ changing from one oscillation to the other. The problem is one of bifurcation. Let $V(u,c)$ be such that $\frac{\partial V}{\partial u}=G$, and consider the oscillations during the oscillating-fast phase. They are governed approximately by :

$$\frac{1}{2}\left(\frac{du}{dt}\right)^2 + V(u,c) = 0 \quad , \tag{48}$$

but there is a slight distortion, due to the term *F(u) du/dt* in (45), which is small. Due to the distortion, and depending on the value of *c*, the solution distorted, either spirals towards the solution of (48), which runs from the saddle point at $t=-\infty$ to the same saddle point at $t=+\infty$, a so-called homoclinic orbit, or it spirals outward. The value of $c_e$ is the one which separates the two behaviors. It can be computed to leading approximation with respect to ε .

### 5.4 Some spurious WKB asymptotics

A well-known and useful set of asymptotic issues goes around the so-called Liouville-Green, alias WKB (J), alias stationnary phase or steepest descent derivation of asymptotic expansions, which are of much help in a number of problems. One may find an account of it in various publications but I shall mention only Bender, Orszag 1978 (especially § 6.5 and 6.6, and Chapter 10) and Olver 1974, Chapter 6. The prototype is the search for an asymptotic representation of appropriate solutions to :

$$\varepsilon^2 \frac{d^2u}{dt^2} = f(t)u \ , \tag{49}$$

when ε goes to zero. A strong impetus for this search was given by the correspondence between classical and quantum mechanics, when De Broglie wave length is very close to the scale of the potential barrier. The reader may find a concise account of the asymptotics of this quantum mechanical problem in Landau, Lifshitz 1959, Chapter VII. The first use of it for a problem in fluid mechanics was by one of the founders of quantum mechanics, namely Heisenberg 1924, à propos of the hydrodynamic stability of viscous shear flow. It is likely that most fluid dynamics specialists of my generation learned the WKB method with this problem and an account of it may be found in Drazin, Reid 1982, Chapters 4 and 5. The papers devoted to this topic are so numerous that the whole of the present paper would be necessary simply for listing them.

A spurious phenomenon which occurs with (49) is concerned with two related questions : the so-called turning point problem, arising near the zeroes of *f*(.), and the Stokes' phenomenon which is encountered when crossing Stokes' lines in the complex *t* plane. Of course, this means that we deal with an *f*(.) which is analytic. We need to start with the statement that, away from such spurious lines, one may state two independent solutions of (49), namely :

$$u = U^{\pm}(F)\left[f(t)\right]^{-1/4} \operatorname{Exp}\left\{\pm F(t)/\varepsilon\right\} \ , \quad F(t) = \int^t \left[f(\tau)\right]^{1/2} d\tau \tag{50}$$

In these formulae, the integral with respect to τ , is taken along a proper path in the complex plane, ending at the point *t*. Finally, the function $U^{\pm}(F)$ may be found, often most efficiently, by an asymptotic expansion :

$$U^{\pm}(F) \approx \sum_{n=0}^{\infty} \varepsilon^n U_n^{\pm}(F) \quad , \tag{51}$$

where the series on the right-hand side diverges, and (51) has to be interpreted as an asymptotic equivalence.

Of the two solutions occurring in (50), the one such that $\mathrm{Re}(\pm F(t)) > 0$ (or $< 0$), is said to be dominant (or recessive). The theory of (49) is heavily burdened by the fact that both solutions in (50) cannot be used at the same time. The reason is that the |recessive| / |dominant| ratio is exponentially small, being of the order of $\mathrm{Exp}\left(-2\varepsilon^{-1}|\mathrm{Re}\,F|\right)$, so that when both dominant and recessive representations must imperatively be taken into account in the solution of a problem, the truncation error of the asymptotic series (51) of the dominant part in any order, is very much larger than the leading term of the recessive contribution. The Stokes' phenomenon occurs when one crosses a so-called Stokes' line, along which $\mathrm{Re}\,F(t) = 0$, so that, when crossing such a line, the dominant or recessive characters are exchanged for one and the same solution. The phenomenon is well documented with the case $f(t)=t$, which leads to the Airy functions (see Olver 1974 § 4.4 or 11.8) and is extensively used in the theory of hydrodynamic stability. Unfortunately, under such circumstances, one has to know the full solution, and the answer to some physical problems is conditioned by knowledge of the continuation of a dominant solution whereas, after crossing a Stokes' line, it becomes recessive. The problem is a very difficult one indeed, and we cannot give any precise account here. I shall just suggest some reading. Dingle 1973 (for example § 1.3.7) was probably the first to deal with the asymptotic equivalence (51) by trying to replace it by an equality :

$$U^{\pm}(F;\varepsilon) = \sum_{n=0}^{N(\varepsilon)} \varepsilon^n U_n^{\pm}(F) + R_N^{\pm}(F;\varepsilon) \quad . \tag{52}$$

The main idea is that the truncation integer $N(\varepsilon)$ depends on $\varepsilon$ . Simultaneously, one relegates the difficulty in trying to find a new asymptotic approximation for the remainder $R_N(F;\varepsilon)$. This has a story that cannot be dealt with here. I simply refer to Chapter 4 of Olver 1974, to the very recent paper of Jones 1994 (for which a limited knowledge of non-standard analysis is required, but see my Section 6) and to Berry, Howls 1990, Howls 1992.

## 6. Flashes on some mathematical aspects

### 6.1 Rigorous asymptotics : a few examples

Up to a certain point, asymptotics has gained the status of a mathematical theory. The phenomenon is far from being new. While a great number of major scientific achievements obtained through use of asymptotics are only heuristically founded, rigorous mathematics is getting involved in an increasing number of results. Although the main part of the present review is not at all concerned with mathematically rigorous matters, I feel necessary to touch lightly on it. Eckhaus 1979 is a good introduction for beginners (especially Chapter 6) but it is simpler, for me, to discuss rapidly the main topic of Chang, Howes 1984. It is devoted to the scalar equation :

$$\varepsilon \frac{d^2y}{dt^2} = f\left(t, y, \frac{dy}{dt}\right) \quad , \quad a < t < b \quad , \tag{53}$$

with various boundary conditions at $t=a$ and $t=b$. Chapter VIII of that book discusses an extension to systems of ordinary differential equations. The heart of the work is concerned with the point of giving mathematically precise conditions under which a solution of

$$f\left(t, y, \frac{dy}{dt}\right) = 0 \quad , \tag{54}$$

conveniently complemented by boundary or interior layers, gives an asymptotically correct solution of the original boundary value problem associated with (53) when $\varepsilon$ goes to zero. The results are very powerful for a simple reason. The conditions stated for getting mathematically proven results are formulated on the solutions concerned of (54), and on $f$ as a function of its three arguments, considered as being independent of each other. Technicalities preclude entering into any detail, but what I have already said is sufficient to give the reader a desire to get more acquainted with the subject. The main shortcoming in these results is that the error estimates provided are generally poor (they are limited to the leading order approximation), a price to be paid for the rigor and relative simplicity of the mathematical tools used. As a matter of fact it seems relevant, for the purpose of the present review, to give a hint on this aspect. It is based on so-called differential inequalities, originally used in this kind of work by Nagumo 1939 and Jackson 1968. The main underlying idea is simple enough to be reported here. Assume that we know two functions $\alpha(t)$ and $\beta(t)$, $t \in (a,b)$, of class $C^2$, such that

$$\begin{cases} \alpha(t) \leq \beta(t) \quad , \quad a \leq t \leq b \ , \\ \alpha(a) \leq A \leq \beta(a) \quad , \quad \alpha(b) \leq B \leq \beta(b) \quad , \\ \alpha'' \geq f(t, \alpha, \alpha') \quad , \quad \beta'' \leq f(t, \beta, \beta') \quad , \end{cases} \tag{55}$$

then, one may state under some conditions to be formulated concerning $f$, $\alpha$, $\beta$, that there exists one solution to (53), with $y(a) = A$, $y(b) = B$, and that this solution is such that

$$\alpha(t) \leq y \leq \beta(t) \quad . \tag{56}$$

The result is obtained by a combination of ingenuity in choosing the comparison functions $\alpha(.)$ and $\beta(.)$ and proper use of the maximum principle, a topic reported in Protter, Weinberger 1967. We will end this very short account by mentioning that differential inequalities and an extension to systems, named invariant regions, are the subject of the books Bernfeld, Lakshmikantham 1974 and Schroder 1980. The same subject is used in the mathematical theory of Prandtl's boundary layer equations, as in Nickel 1973. We need mention also that the maximum principle is the main tool used in Chapter 6 of Eckhaus 1979.

M.A.E. for elliptic problems, mainly linear ones, leads sometimes to a complete theory, with the use of functional analysis in Sobolev spaces. A rather early result is the one of Rigolot

1972 who succeeded in developping a high order formal M.A.E. theory of three-dimensional steady linear isotropic elasticity in slender cylindrical bodies. The result is a formal justification of the engineering theory of rods which is readily accessible in Landau, Lifshitz 1959 a (§ 16 to 21 inclusive) or in a more comprehensive way in Antman 1972. I should stress that Rigolot's expansion can be pursued to any order desired, and, of course, far beyond the level which justifies the engineering theory. In the same paper Rigolot was able to give an estimate of the error in the expansion truncated to a specified order, in comparison with the exact solution. He uses the technique of correctors which is systematically investigated in Lions 1973, as a tool for getting expansion with controlled error obtained through use of a priori estimates in Sobolev spaces and compactness arguments for limiting processes. Roughly one may state that a corrector is what has to be added to an outer expansion in order to get a composite one. The corrector is particularly simple to obtain when the inner solution is asymptotic to its limiting value at infinity in an exponential way.

Multiple scale expansion has led to a complete mathematical theory in the case of the so-called periodic homogeneization. The formal theory was given for the first time by Sanchez-Palencia 1970 and soon after this same author raised it to the status of a rigorous mathematical theory with controlled error estimates (see Sanchez-Palencia 1980, especially Chapters 5 to 7 inclusive and references therein). A comprehensive account with an extension to random homogeneization is given in Bensoussan, Lions, Papanicolaou 1978. The mathematical tool with homogeneization is much more sophisticated than differential inequalities and rests on the so-called G-convergence which goes back at least to Spagnolo 1967, 1968.

### 6.2 Solutions got with diverging series

There are a lot of works devoted to series expansions, computed to a very high order, with the assistance of a computer. At the same time there is much interest in extending the domain of convergence of these expansions, when possible. A comprehensive review is given by Van Dyke 1984. We add a specific example worked out by Dadfar, Geer, Andersen 1984. By extensions of power series they succeeded in studying the limit cycle of the Van Der Pol equation

$$\frac{d^2u}{dt^2}+\varepsilon(u^2-1)\frac{du}{dt}+u=0 \ , \tag{57}$$

from $\varepsilon=0$ up to $\varepsilon=+\infty$. The computer performed the expansion up to $\varepsilon^{163}$. An interesting point is that the technique of extension, used a recasting of the series, from one with respect to $\varepsilon$ into another with respect to $w(\varepsilon,t)$. Such a procedure was used to grasp the whole of the phase plane, instead of simply either the amplitude or the period of the limit cycle, as a function of $\varepsilon$. Phase plots close to the ones obtained through numerical computations by Zonneveld 1966 were obtained using a simpler transformation than this $t$ dependent one, $\tilde{w}=\tilde{w}(\varepsilon)$, suggested by it, but independent of $t$, up to $\varepsilon=3.5$. On the other hand, the extension suggested by the expansion of either the period or the amplitude yielded spurious oscillations along certain phases of the cycle, from $\varepsilon=1.6$ onwards. The idea of looking at the transformation from $\varepsilon$ to $w(\varepsilon,t)$ was given by examining the location of the singularity closest to the origin, as a function of $t$. In effect, it was shown that near the phase of the cycle which develops spurious oscillations, the singularity moves closer towards the origin than it did concerning the expansion for either the period or the amplitude.

This short incursion into a subject concerned with specific applications, and not especially oriented towards mathematics, provides me with an introduction to the second example of the mathematically oriented topics I want to examine. It deals with deep results obtained recently (see Ramis 1993, Sect 3.1) on rather old ideas due to Maillet 1903 and Gevrey 1918. Assume that $G(x,Y_0,Y_1,..,Y_N)$ is an analytic function of all its $N$+2 complex arguments. Assume that with

$$y \approx \sum_{n=0}^{\infty} a_n x^n \quad , \tag{58}$$

one has found a formal (not necessarily convergent) power series, which, when substituted into $G(x,y,y'y'',..,y^{(N)})$, gives a formal power series, identically equal to zero. The question arises whether the right-hand side of (58) may be summed, in some sense, to give a function $f(x)$ and whether such a function may be differentiated $N$ times, such that $G\left(x,f,\frac{df}{dx},\dots,\frac{d^N f}{dx^N}\right)=0$. Gevrey 1918 was concerned with the nature of formal Taylor power series, of solutions of partial differential equations. The prototype is Poisson's equation :

$$\frac{\partial^2 u}{\partial x^2}+\frac{\partial^2 u}{\partial y^2}=f(x,y) \ , \tag{59}$$

for which it is known that whenever $f$ is analytic with respect to both its arguments, solutions of (59) are also analytic. Gevrey 1918 was concerned with the extension to

$$f(x,y)=\sum_{m_1\geq 0}\sum_{m_2\geq 0} a_{m_1,m_2} x^{m_1} y^{m_2} \quad , \tag{60}$$

the series being a purely formal one. For this purpose, he created the concept of a Gevrey class. Here $a_{m_1,m_2}$ is defined by :

$$\frac{\partial^{m_1+m_2} f}{\partial x^{m_1}\partial y^{m_2}} = m_1!\,m_2!\,a_{m_1,m_2} \quad , \tag{61}$$

and $f$ is said to be of the Gevrey class $(\alpha_1,\alpha_2)$ if there exist real positive numbers $R_1,R_2,M$, such that

$$\left|a_{m_1,m_2}\right| \leq M\frac{(m_1!)^{\alpha_1-1}(m_2!)^{\alpha_2-1}}{R_1^{m_1}\ \ R_2^{m_2}} \quad . \tag{62}$$

Among many other results, Gevrey was able to prove that any solution of (59), in the vicinity of $(x,y)=(0,0)$, is of the same Gevrey class as $f$. We can now go back to the differential equation $G(x,y,y',..,y^{(N)})=0$. Maillet 1903 succeeded in getting a formal power series "solution" of this equation with the meaning of "solution" that we explained after (58). He proved the existence of a positive real *k* such that (58) may be considered as issued from the Taylor series expansion around *x=0* of a function of the Gevrey class *k*. Ramis 1993 attributed to Sibuya the proof of a theorem which may roughly be stated as follows. Given $G(.,.,....,.)$, one can find the smaller real *k* such that the formal "solution" (58) is of Gevrey class *k* . Furthermore, there exists a process of summation of (58) leading to a function *f*(*x*) which is indeed a solution of $G(x,f,f',..,f^{(N)})=0$. Such a powerful result should incite fluid mechanicists mainly interested in asymptotics to pay attention at what is published in this field. Let me just add a word about the paper by Ramis, Sibuya 1989, which is concerned with solutions in the complex domain of *x*, near *x=0*, of the equation $x^r \dfrac{dY}{dx} = \phi(x,Y)$ where *Y* is an *N*-dimensional complex vector, while $\phi$ is analytic in all its arguments (the rather strong requirement of analyticity may be weakened, at least as far as the dependency on *x* is concerned). Such a kind of singularity has been of interest to many researchers and the most comprehensive treatment, at least to my knowledge, is to be found in Wasow's 1965 book. It seems to me that the referenced work goes much farther than Wasow but I feel difficult to summarize briefly these results.

### 6.3 Asymptotics and non-standard analysis

The reason why I have chosen to give a short account of this topic is that, although it deals with the foundation of mathematics, in quite an abstract way, yet it seems to me that it has some impact on asymptotics as I shall try to show, at least superficially. In most applications of asymptotics, the root of the method is to use a constructive process for building, more or less recursively, a function which is a candidate for some solution to a definite problem. We use a very simple example : the expected solution is a function $f(x,\varepsilon)$, *x* being a real variable and $\varepsilon$ a small parameter. If the approximation is obtained by M.A.E., it has the form, say, $F\left(x,\dfrac{x-a}{\varepsilon};\varepsilon\right)$ where *a* is some specific real number. Starting from a mathematical model which contains only the function *f*, the variable *x* and the parameter $\varepsilon$ , the M.A.E. technique generates a function $F(x,y;\varepsilon)$ with an extra real variable *y* which should, in the process of construction, be considered as a completely independent variable. We simply emphasize that both the effect of stretching on differential operators, and matching, are essential ingredients for the success of the operation. We also emphasize that the process works because the extended function *F* is sought not as a precisely definite one but as a formal asymptotic expansion. If the approximation is obtained by multiple scaling expansion, the substitution for *y* is not $(x-a)/\varepsilon$ but rather something like $\varphi(x)/\varepsilon$ , so that *y*, instead of playing a role mainly when *x* is close to *a*, gets into the approximation for, generally, the whole range of values of the variable *x* concerned from the outset. Here, secularity or compatibility condition play the role of matching. We do not proceed further along these lines because we only intended to focus the attention of the reader on this process of extension. Now, other processes of extension are encountered in problems issued from mechanics, and there exists a suitable mathematical theory for many of them. As an example, consider shell theory. Here, two extensions occur. One is to grasp the whole of the shell by gluing together small pieces of it, each piece being mapped on an area of the plane. The mathematical structure for this is the theory of manifolds. Of course, scientists may well do good work on shells without mastering manifolds, but anyone who has an even elementary knowledge of both is convinced that knowing the manifold theory is a significant help to understanding and sometimes much more than simply understanding. The second example comes from differential calculus on manifolds, and there is a mathematical structure for that, namely the machinery of fiber spaces, which consists in adding to each point of the manifold, as an extra element, its tangent plane as a vector space, named a fiber over the point considered.

What about asymptotics in that matter ? The point is that dealing with asymptotic expansions relies on yet another extension, and that the mathematical structure for it is, precisely, non-standard analysis. The basic operation with the construction of an asymptotic expansion is scaling by a conveniently chosen gauge function of the small parameter $\varepsilon$ and going repeatedly to a limit. An asymptotic expansion with respect to $\varepsilon$ is an expression

$$\sum_{n=0}^{N} \varphi_n(\varepsilon) F_n(.) \quad , \tag{63}$$

where the "." is set for a set of arguments possibly including the extended variables (which have been alluded to before and which, when considered in such a way, non longer contain $\varepsilon$ ). Of course, at the end of the process, some real number (presumably small with respect to unity) has to be substituted for $\varepsilon$ , but, before that, the asymptotic expansion (63) is associated with the limiting process :

$$\lim_{\varepsilon \downarrow 0} = \left\{ \varepsilon \downarrow 0 \text{ while all the arguments of } F \text{ other than } \varepsilon \text{ are held fixed} \right\} . \tag{64}$$

Non-standard analysis has created a whole mathematical machinery for dealing with (64) (and, of course, other more sophisticated things). It might be called calculus with infinitesimal quantities or, more precisely here, infinitesimal real numbers. The parameter $\varepsilon$ is in effect not a real number at all, but something else, which is called an infinitesimal non-standard real (of course $\varepsilon^{-1}$ is infinitesimally great). We insist on the fact that $\varepsilon$ has to be considered as something quite different from the usual reals. The reader should think about that, and try to become familar with the very idea, independently of how it might be realized. In a first step he does not need to understand how this is possible but simply to admit that a meaning may be given to the sentence which follows. "The object $\varepsilon$ and all the $\varphi_n(\varepsilon)$ considered are infinitesimals (we consider that $\lim_{\varepsilon \downarrow 0} \varphi_n(\varepsilon) = 0,\ \lim_{\varepsilon \downarrow 0} (\varphi_{n+1}(\varepsilon) / \varphi_n(\varepsilon) = 0)^{"}$. Then to say that (63) is an asymptotic expansion is expressed by saying that "for any $n \in (0,1,..,N)$, the function $\varphi_n(\varepsilon)$ is infinitesimally small when $\varepsilon$ is, and $\varphi_{n+p}(\varepsilon) / \varphi_n(\varepsilon)$ is also infinitesimally small, for any integer $p>0$, such that $n+p \leq N$ ". Again, for any such $n$, the function of the infinitesimal $\varepsilon$ , defined by $\left( \sum_{q=0}^{n-1} \varphi_q(\varepsilon) F_q(.) / \varphi_n(\varepsilon) \right)$ is infinitesimal for any fixed value of the set of arguments, other than $\varepsilon$ , in $F$. This, apparently innocuous substitution of infinitesimals to the whole limiting process, is a rather large step towards creating a mathematical language especially designed for economy and short-cutting. Of course, there is no magic at all in that. Infinitesimals have driven away the limits because these limits are included in the process of building the infinitesimals themselves. It is in the very process of building the infinitesimals that the whole machinery of non-standard analysis comes in, at least for the purpose considered here.

It is time now to dive the reader for a short incursion into the machinery of infinitesimals. Let

$$\left\{\varepsilon_n, n \geq n_0\ ;\ \lim_{n\to\infty} \varepsilon_n = 0\right\} \quad , \tag{65}$$

be a sequence of positive real numbers going to zero. We would like to say that such a sequence defines an infinitesimal, say $\varepsilon$ . A small amount of thinking suggests that taking into account the whole sequence (65) is too much and leads to something which is useless. Let $F$ be any set of integers greater than $n_0$ but finite. Consider two sequences of the type (65), namely $\left\{\varepsilon_n^{(1)}\right\}, \left\{\varepsilon_n^{(2)}\right\}$. One intuitively expects them to lead to the same infinitesimal if $\varepsilon_n^{(1)} = \varepsilon_n^{(2)}$, for any $n \geq n_0$, such that $n \in F$(which is any finite set of integers). It is possible to develop this remark, but we prefer to leave the reader at this step and urge him to read, for example, Chapters 1 and 2 of Davis 1977. We simply state here that it is possible to define an equivalence relation on the set of sequences (65). There is then a one-to-one correspondence between equivalence classes of sequences (65) according to this relation, and infinitesimals, which we are discussing. In order to carry over the whole machinery of asymptotics, we need be able to define gauge functions $\varepsilon \to \varphi(\varepsilon)$, which map infinitesimals into infinitesimals. To achieve this, the best way is to learn non-standard analysis, but we may however give a hint. Let a be any (standard) real, and let the infinitesimal $\varepsilon$ be represented by the sequence (65). Then, the sequence

$$\left\{a.\varepsilon_n, n \geq n_0; \lim_{n\to\infty} \varepsilon_n = 0\right\} \quad , \tag{66}$$

is obviously a representative of the infinitesimal $a.\varepsilon$ . In the same way, the sequence *(a>0)* :

$$\left\{\varepsilon_n^{\,a}, n \geq n_0; \lim_{n\to\infty} \varepsilon_n = 0\right\} \quad , \tag{67}$$

is a reprentative of $\varepsilon^a$. Of course, this argument is not at all a proof, simply a hint that maps $\varphi(.)$, from infinitesimals into infinitesimals can be defined, and that these maps $\varphi(\varepsilon)$ play the role of gauge functions in asymptotic theory.

In order to persuade the reader, we leave asymptotic expansions in order to give the definition of continuity (of a real valued function of a real variable), in the words of non-standard analysis. "The function $f$ (.) is continuous at $x_0$ if and only if $f(x) - f(x_0)$ is infinitesimal when $x - x_0$ is itself infinitesimal". Of course, continuity is such an elementary notion that the gain is slight. But we think it is nevertheless perceptible that, and perhaps the reader will be ready to accept this, in more intricate questions, the gain may be significant.

Let us come back to asymptotics in order to translate the rule of matching in M.A.E. into non-standard analysis. We assume that $F_N(x\ ;\ \varepsilon)$ and $\tilde{F}_M(\tilde{x}\ ;\ \varepsilon)$ , with $\tilde{x} = x/\varepsilon$, are outer and inner expansions respectively (to given orders, indicated by the indexes $N$ and $M$ ). We say that $F_N$ and $\tilde{F}_M$ match to the order given by the infinitesimal $\zeta(\varepsilon)$, on the overlap ray defined by the infinitesimal $\gamma(\varepsilon)$ whenever the function of the infinitesimal $\varepsilon$ (with $y$ fixed) :

$$\frac{F_N\big(\gamma(\varepsilon)y\ ;\ \varepsilon\big)-\tilde{F}_M\left(\dfrac{\gamma(\varepsilon)y}{\varepsilon}\ ;\ \varepsilon\right)}{\zeta(\varepsilon)} \tag{68}$$

is itself infinitesimal, for any standard $y$. One may skip from an overlap ray to an overlap domain by choosing $\gamma(\varepsilon)$ in a class of gauge functions.

I end this too short account by suggesting that it is indeed a moderate investment, but a rewarding one, to first read Chapters 1 and 2 of Davis's 1977 book, and then a few papers, namely : Hartong 1981,1983, Diener 1979, Diener 1980, Benoit 1981, Benoit, Lobry 1982, Ingleton 1982, Rogers 1982, Tall 1982, Hoskins 1982, Cutland 1982, Cartier 1981, Zvonkin, Shubin 1984, Brons, Bar-Eli 1994.

## 7. Conclusion

Mathematics and, more specifically, calculus, ranging from elementary to quite advanced one, supported by numerical simulation will, in the future, play a prominent role in technological progress. It is likely that more will be accomplished in the next few decades than has been achieved during the past few. Even biotechnology or economics and sociology will rely, at least partly, on mathematical modeling. So far as one accepts such a prospective view on the role of mathematics and calculus, on those human activities, and if one adheres to the opinion that they are important, among many other ones, one must think about the scope of asymptotics within the range of mathematical modeling. My opinion is that, presently, it appears more limited than it really is, but that, progressively it will gain the status of a full component. I have tried in the preceding pages to show how wide its spectrum is. I am a fluid dynamics specialist, addressing himself to other fluid dynamics specialists, so that the demonstration is limited, but it might be rendered much more comprehensive. On the other hand, I hope that the reader has understood that one may use asymptotics with a degree of technicality which ranges from quite elementary calculus to sophisticated mathematics. This depends on the nature of the problem and on the amount of refinement that one requires for the solution. Another point is that scientists interested in applications of their works to technology cannot hope to work exclusively with asymptotics. Of course, asymptotics, as a part of mathematics, or calculus, may be considered a scientific activity by itself, and it has to be treated as such when a high degree of sophistication is needed to progress. But this will concern very few scientists indeed. Most of them will have to use asymptotics as a tool among others. The big problem with asymptotics is to be convinced that the effort that has to be put into it in order to be able to work efficiently, is worth while the rewards it brings. Probably something has to be done at the level of education. Everyone interested in science and technology feels that some amount of training in calculus is necessary, even for those who will devote themselves almost exclusively to experimental research. Of course, calculus is very wide in scope, so that a choice has to be made, among the large number of techniques that one might be tempted to teach. Nevertheless, I express the view that asymptotics is basic for so many scientists that it should be part of their background, at least on an elementary level. For a great number of scientists it is more a matter of receptivity than one of actual knowledge.

Going further that this point is in the realm of communication. Any scientist who is experienced in asymptotics should make it a point of duty to bring his contribution to scientific communication up to a higher level than is usually expected from scientists. With sustained effort along these lines, I hope that asymptotics will enter in the background of scientifically educated people, looking as natural as matrices, ordinary differential equations, plane wave solutions and the like look now. Meanwhile, it has to be put in the hands of young scientists who wish to make it more and more efficient, and who do not spare their time where applications to technological problems and communication with other scientists, are concerned.

I am very grateful to Ms. Joan Potter for her careful reading of an early version of the manuscript and her contribution to improving my poor English. She is not responsible for any spelling or syntax errors I may have introduced in the final version. I am also much grateful to the editors for allowing me to write this paper and for very useful suggestions about it.

## References


Adou, J. (1993) : "Modélisation et stabilité non linéaire d'un régulateur piloté de gaz", *Entropie N° 177, pp. 39-42*

Amestoy, M. (1987) : " Propagation de fissures en élasticité plane", *Thesis University Pierre and Marie Curie*

Amestoy, M., Leblond, J.B. (1992) : "Crack paths in plane situation. II detailed form of the expansion of the stress intensity factors", *Intern. J. Solids Struct. 29, pp. 465-501*

Antman, S. (1972) : "The theory of rods", *In Handbuch der Physik, Vol VI a/2, Mechanics of Solids, pp. 641-703*

Barone, M.R., Caulk, D.A. (1985) : "Special Boundary Integral Equations for Approximate Solution of Potential Problems in Three-dimensional Regions with slender cavities of Circular Cross-section", *IMA J. Appl. Math. 35, pp. 311-325*

Barenblatt, G.I. (1962) : "The mathematical theory of equilibrium cracks in brittle fracture", *Adv. Appl. Mech. 7, pp. 55-125*

Bender, C.M., Orszag, S.A. (1978) : "Advanced mathematical methods for scientists and engineeers", *International student edition, Mc Graw Hill*

Benoît, E. (1981) : "Equations différentielles : relation entrée sortie", *C.R.A.S. Paris A 293, pp. 293-296.*

Benoît, E., Lobry, C. (1982) : "Les canards de $\mathbb{R}^3$", *C.R.A.S. Paris A 294, pp. 483-488*

Bensoussan, A., Lions, J.L., Papanicolaou, G. (1978) : Asymptotic analysis of periodic structure", *Studies in mathematics and its applications, Vol 5, North-Holland.*

Bernfeld, S.R., Lakshmikantham, V. (1974) : "An introduction to nonlinear boundary value problems", *Academic Press*

Berry, M.V., Howles, C.J. (1990) : "Hyperasymptotics", *Proc Roy Soc London A 430, pp. 653-657*

Bleistein, M., Handelsman, R.A. (1975) : "Asymptotic expansions of integrals", *Holt, Rinehart, Winston*

Bodonyi, R.J., Smith, F.T. (1981) : "The upper branch stability of the Blasius boundary layer including non-parallel flow effects", *Proc. Roy. Soc. London A 375, pp. 65-92*

Bois, P.A. (1991) : " Asymptotic aspects of the Boussinesq approximations for gases and liquids", *Geophys. Astro. Fluid Dyn. 58, pp. 45-55*

Bouthier, M. (1972,1973) : "Stabilité linéaire des écoulements presque parallèles", "Stabilité linéaire des écoulements presque parallèles Partie II : La couche limite de Blasius", *J. Mécanique 11, pp. 599-621 ; 12, pp. 75-95*

Brilliant, H.M. , Adamson, T.C. (1974) : "Shock wave boundary layer interactions in laminar transonic flow", *A.I.A.A. J. 12, pp. 323-329*

Brons, M., Bar-Eli, K. (1994) : "Asymptotic analysis of canards in the E.O.E equations and the role of the inflection line", *Proc. Roy. Soc. London A 445, pp. 305-322*

Brown, S.N., Stewartson, K. (1969) : "Laminar separation", *Ann. Rev. Fluid Mech. 1, pp. 45-72*

Bui, H.D. (1978) : "Mécanique de la rupture fragile", *Masson*

Cartier, P. (1981) : "Perturbations singulières des équations différentielles ordinaires et analyse non standard", *Séminaire Bourbaki 34ème Annéé 1981/1982, N° 580, pp. 580.01 to 580.24*

Catherall, D., Mangler, K.W. (1966) : "The integration of the two-dimensional laminar boundary-layer equations past the point of vanishing skin friction", *J. Fluid Mech. 26, pp. 163-182*

Chang, K.W., Howes, F.A. (1984) : "Non linear singular perturbation phenomena : theory and applications", *Applied Mathematical Sciences, Vol 56, Springer*

Chapman, D., Kuehn, G.D., Larson, H.K. (1950) : "Investigation of separated flow in supersonic and subsonic streams with emphasis on transition", *NACA Rept 1356*

Cole, J.D. (1968) : "Perturbation methods in applied mathematics", *Blaisdell Publishing Company*

Crocco, L. (1972) : "Coordinates perturbations and multiple scale in gas dynamics", *Phil. Trans. R. Soc. A 272, pp. 274-301 + 699*

Crow, S.C. (1970) : "Aérodynamic sound emission as a singular perturbation problem", *Studies Appl. Math. 49, pp. 21-44*

Cutland, N. (1982) : "Infinitesimal methods in measure theory, probability theory and stochastic analysis", *Bulletin I.M.A. 18, pp. 52-57*

Dadfar, M.D., Geer, J., Andersen, C.M. (1984) : "Perturbation analysis of the limit cycle of the free Van der Pol equation", *SIAM J. Appl. Math. 44, pp. 881-895*

Davis, M. (1977) : "Applied non standard analysis", *Wiley*

Diener, F. (1979) : "Famille d'équations à cycle limite unique", *C.R.A.S. Paris A, 289, pp. 571-574*

Diener, F. (1980) : "Deux nouveaux "phénomènes canards"", *C.R.A.S. Paris A, 290, pp. 541-544*

Dingle, R.B. (1973) : "Asymptotic expansion : Their derivation and interpretation", *Academic Press*

Drazin, P.G., Reid, W. (1982) : "Hydrodynamic stability", *Cambridge Univ. Press, 2nd Paperback Edition*

Duck, P.W. (1990) : "Triple deck flow over unsteady surface disturbances : the three-dimensional developpment of Tollmien-Schlichting waves", *Comp. and Fluids 18, pp. 1-34*

Eckhaus, W. (1979) : "Asymptotic analysis of singular perturbations", *Studies in Math. Appl., Vol 9, North-Holland*

Favret, F., Cornil, J.P., Deneuve, F., Guiraud, J.P. (1990) : "Stabilité des détendeurs de gaz", *107ème Congrès du gaz, Paris 18-21 Sept 1990, Tome 2, pp. 169-195*

Flugge-Lotz, I., Blottner, F.G. (1963) : "Finite difference computation of the boundary layer with displacement thickness interaction", *J. Mécanique 2, pp. 397-423*

Freund, L.B. (1990) : "Dynamic fracture mechanics", *Cambridge Univ. Press*

Geer, J.F., Andersen, C.M. (1989) : "A hybrid perturbation Galerkin technique with application to slender body theory", *SIAM J. Appl. Math. 49, pp. 344-361*

Gevrey, M. (1918) : "La nature analytique des solutions des équations aux dérivées partielles", *Ann. Sci. Ec. Norm. Sup. Paris 25, pp. 129-190*

Goldstein, S. (1930) : "Concerning some solutions of the boundary layer equations in hydrodynamics", *Proc. Camb. Phil. Soc. 26, pp. 1-30*

Griffith, A.A. (1921) : *Phil. Trans. Roy. Soc. London A, 221*

Guiraud, J.P., Zeytounian, R. (1977) : "A double scale investigation of the asymptotic structure of rolled-up vortex sheets", *J. Fluid. Mech. 79, pp. 93-112*

Hall, P.G., Smith, F.T. (1984) : "The effect of non parallelism, three-dimensionality, and mode interaction in non linear boundary-layer stability", *Studies in Appl. Math. 70, pp. 91-120*

Hall, P.G., Smith, F.T. (1991) : "On strongly non linear vortex wave interactions in boundary layer transition", *J. Fluid Mech. 227, pp. 644-666*

Harris, S.E., Crighton, D.G. (1994) : "Solitons, solitary waves, and Voidage disturbances in fluidized beds", *J. Fluid Mech. 266, pp. 243-276*

Hartong, J. (1981) : "Le moiré", *Advances in Applied Mathematics, 2, pp. 24-75*

Hartong, J. (1983) : "L'analyse non standard", *La Recherche 14, N° 148, pp. 1194-1201*

Heisenberg, W. (1924) : "Über stabilität im Turbulenz von Flussigkeitströmen", *Ann. Phys. Leipzig 74, pp. 577-627*

Higuera, F.J., Martinez, M. (1993) : "An incompressible jet in a weak cross flow", *J. Fluid Mech. 249, pp. 73-97*

Hoskins, R.F. (1982) : "Infinitesimal, non-standard analysis and generalized functions", *Bulletin I.M.A. 18, pp. 49-51*

Howls, C.J. (1992) : "Hyperasymptotics for integrals with finite end points", *Proc. Roy. Soc. London A 439, pp. 373-396*

Ingleton, A. W. (1982) : "An introduction to non-standard analysis", *Bulletin I.M.A. 18, pp. 34-37*

Irwin, G.R. (1957) : *J. Appl. Mech. 24, p. 361*

Jackson, L.K. (1968) : "Subfunctions and second order ordinary differential inequalities", *Adv. in Math. 2, pp. 308-363*

Jones, D.S. (1994) : "Asymptotic remainders", *S.I.A.M. J. Appl. Math. 25, pp. 474-490*

Joukowski, N. (1912) : "Uber die Konturen der Drachenflieger", *Z .Flugtech. 1, p. 281 and 2, p. 81*

Kaplun, S. (1954) : "The role of coordinate systems in boundary layer theory", *Z.A.M.P. 5, pp. 111-135*

Kaplun, S. (1957) : "The role of coordinate systems in boundary-layer theory", *J. Math. Mech. 6, pp. 595-603*

Kassoy, D.R. (1982) : "A note on asymptotic methods for jump phenomena", *S.I.A.M. J. Appl. Mat.h 42, pp. 926-932*

Kevorkian, J., Cole, J.D. (1981) : "Perturbations methods in applied mathematics"

Klein, R. , Knio, O.M. (1994) : "Asymptotic vorticity structure and numerical simulation of slender vortex filaments", *Submitted J. Fluid. Mech.*

Korobkin, A. (1992) : "Blunt-body impact on a compressible liquid surface", *J. Fluid. Mech.. 244, pp. 437-453*

Kriegsmann, G.A. (1993) : "Microwave heating of dispersive media", *S.I.A.M. J. Appl. Math. 53, pp 655-669*

Kurz, E.F., Crandall, S.H. (1962) : "Computer aided analysis of hydrodynamic stability", *J. Math. Phys. 44, pp. 264-279*

Kutta, M.W. (1902) : "Auftriebskräfte in strömenden Flussigkeiten", *Ill. Aero. Mitt. 6, p. 133*

Lagerstrom, P.A., Cole, J.D., Trilling, L. (1949) : "Problems in the theory of viscous compressible fluids", *Caltech Report Pasadena*

Lagerstrom, P.A., Casten, R.G. (1972) : "Basic concepts in singular perturbation techniques", *S.I.A.M. Review 14, pp. 63-120*

Lagerstrom, P.A. (1988) : "Matched asymptotic expansions. Ideas and techniques", *Applied Mathematical Sciences, Vol 76, Springer*

Lagrée, P.Y. (1992) : "Structures interactives fluide parfait - couche limite en hypersonique, variations sur le thème de la triple couche", *Thesis University Paris VI*

Lanchon, H., Eckhaus, W. (1964) : "Sur l'analyse de la stabilité des écoulements faiblement divergents", *J. Mécanique 3, pp. 445-459*

Landahl, M. (1981) : "Mathematical modelling in fluid mechanics", *Z.A.M.M. 61, supplement T 9-T14*

Landau, L.D., Lifshitz, E.M. (1959) : "Quantum mechanics. Non-relativistic theory", *Vol 3 of Course of Theoretical Physics", Pergamon*

Landau, L.D., Lifshitz, E.M. (1959 a) : "Theory of elasticity", *Vol 7 of Course of Theoretical Physics", Pergamon*

Lange, C.G., Weinitschke, H.J. (1991) : "Singular perturbations of limit points with application to tubular chemical reactors", *Studies in Appl. Math. 84, pp. 7-42*

Lange, C.G., Weinitschke, H.J. (1994) : "Singular perturbations of elliptic problems on domains with holes", *Studies in Appl. Math. 92, pp. 55-93*

Lauvstad, V.R. (1968) : "On non uniform Mach number expansion of the Navier-Stokes equations and its relation to aerodynamically generated sound", *J. Sound Vib. 7, pp. 90-105*

Leblond, J.B. (1989) : "Crack paths in plane situations. I General form of the expansion of the stress intensity factor", *Int. J. Solids Struct. 25, pp. 1311-1325*

Lighthill, M.J. (1950) : "Reflection at a laminar boundary layer of a weak steady disturbance to a supersonic stream, neglecting viscosity and heat conduction", *Quart. J. Mech-Appl. Math. 3, pp. 303-325*

Lighthill, M.J. (1952,1954) : "On sound generated aerodynamically - I & II", *Proc. Roy. Soc. London 211, pp. 564-587, 222, pp. 1-32*

Lighthill, M.J. (1953) : "On boundary layers and upstream influence. II Supersonic flow without separation", *Proc. Roy. Soc. London A 217, pp. 478-507*

Lighthill, M.J. (1991) : "Biomechanics of hearing sensitivity", *J. Vibration Acoust. 113, pp. 1-13*

Lighthill, M.J. (1992) : "Acoustic streaming of the ear itself", *J. Fluid Mech. 239, pp. 551-606*

Lions, J.L. (1973) : "Perturbations singulières dans les problèmes aux limites", *Lecture Notes in Mathematics, Vol 323, Springer*

Maillet, E. (1903) : "Sur les séries divergentes et les équations différentielles", *Ann. Sci. Ec. Norm. Sup. Paris 10, pp. 487-518*

Maugin, G.A. (1993) : "Materials inhomogeneity in elasticity", *Chapman & Hall*

Messiter, A.F. (1970) : "Boundary layer near the trailing edge of a flat plate", *S.I.A.M. J. Appl. Math 18, pp. 241-257*

Messiter, A.F., Feo, A., Melnik, R.E. (1971) : "Shock wave strength for separation of a laminar boundary layer at transonic speeds", *A.I.A.A. J. 9, pp. 1197-1198*

Meyer, R.E. , Wilson, L. (1973) : "Streching and the calculus of negligibility", *S.IA.M. J. Appl. Math.. 25, pp. 388-402*

Meyer, R.E. (1983) : "A view of the triple deck", *S.I.A.M Jal. Appl. Math 43, pp. 639-663*

Mouchrif, S.E. (1994) : "Trajets de propagation de fissures en mécanique linéaire tridimensionnelle de la rupture fragile", *Thesis University Paris 6*

Nagumo, M. (1939) : "Uber die differentialgleichung $y'' = f(x,y,y')$", *Proc. Phys. Math. Soc. Japan 19, pp. 861-866*

Nayfeh, A.H. (1991) : "Triple deck structure", *Comp. Fluids 20, pp. 269-292*

Neiland, V.Ia. (1969) : "On the theory of laminar separation in supersonic flow", *Mech. Liquid-Gaz, N° 4 (in Russian) pp. 53-57*

Nickel, K. (1973) : "Prandtl's boundary-layer theory from the point of view of a mathematician", *Ann. Rev. Fluid Mechanics* *5*, *pp. 405-428*

Oleinik, O.A. (1969) : "Mathematical problems of boundary layer theory", *Lectures notes University of Mineapolis Press*

Olver, F.W.J. (1974) : "Asymptotics and special functions", *Academic Press*

Pernarowski, M., Miura, R.M., Kevorkian, J. (1992) : "Perturbations techniques for models of bursting of electrical activity in pancreatic β-cells", *S.I.A.M. J. Appl. Math.* *52*, *pp. 1627-1650*

Protter, M.H., Weinberger, H.F. (1967) : "Maximum principle in differential equations", *Prentice Hall*

Ramis, J.P., Sibuya, Y. (1989) : "Hukuhara domains and fundamental existence and uniqueness theorems for asymptotic solutions of the Gevrey type", *Asymptotic analysis* *2*, *pp. 39-94*

Ramis, J.P. (1993) : "Séries divergentes et théories asymptotiques", *Bulletin Soc. Math. France supplément,* *121*, *pp. 1-74*

Reiss, E.L. (1980) : "A new asymptotic method for jump phenomena", *S.I.A.M. J. Appl. Math.* *39*, *pp. 440-455*

Rigolot, A. (1972) : "Sur une théorie asymptotique des poutres", *J. Mécanique* *11*, *pp. 673-703*

Rogers, L. (1982) : "Infinitesimals, Continuity and Meaning", *Bull. I.M.A.* *18*, *pp. 38-42*

Sanchez-Palencia, E. (1970) : "Solutions périodiques par rapport aux variables d'espace et applications", *C.R.A.S. Paris* *271*, *pp. 1129-1132*

Sanchez-Palencia, E. (1980) : "Non homogeneous media and vibration theory", *Lectures Notes in Physics, Vol. 127, Springer*

Sanchez-Palencia, E. (1985) : "Perturbations of eigenvalues in thermoelasticity and vibration of systems with concentrated mass", *pp. 346-368* dans Ciarlet, P.G., Roseau, M. : "Trends and applications of pure mathematics to mechanics", *Lecture Notes in Physics, N° 195*

Sanchez-Hubert, J., Sanchez-Palencia, E. (1989) : "Vibration and coupling of continuous systems, asymptotic method", *Springer*

Schröder, J. (1980) : "Operator inequalities", *Academic Press*

Schubauer, G.B., Skramstad, H.K. (1943) : "Laminar boundary layer oscillations and stability of laminar flow. Original in Nat. Bureau-stand. Res. Paper 1772.", *Republished as J. Aeron. Soc.* *14*, *pp. 69-78, 1947*

Smith, F.T. (1979) : "On the non-parallel flow stability of the Blasius boundary layer", *Proc. Roy. Soc. London A* *366*, *pp. 573-589*

Smith, F.T. (1982) : "On the high Reynolds number theory of laminar flows", *I.M.A. J. Appl. Math.* *28*, *pp. 207-281*

Smith, F.T., Burgraf, O.R. (1985) : "On the development of large sized disturbances in boundary layers", *Proc. Roy. Soc. London A* *399*, *pp. 25-55*

Smith, F.T., Doorly, D.J., Rothmayer, A.P. (1990) : "On displacement thickness wall layer and mid-flow scales in turbulent boundary-layers, and slugs of vorticity in channel and pipe flows", Proc. *Roy. Soc. London A* *428*, *pp. 255-281*

Smith, F.T. (1992) : "On non linear effects near the wing tip of an evolving boundary layer spot", *Phil. Trans. Roy. Soc. London A* *340*, *pp. 131-165*

Smith, F.T. (1994) : "On global dynamics of spots : a theoretical approach", *J. Engng. Math.* *28*, *pp. 73-91*

Spagnolo, S. (1967) : "Sul limiti delle soluzioni di problemi di Cauchy relativi all'equazione del calore", *Ann. Scuola Norm. Sup. Pisa* *21*, *pp. 657-699*

Spagnolo, S. (1968) : "Sulla convergenza di soluzioni di equazioni paraboliche ed ellitiche"", *Ann. Scuola Norm. Sup. Pisa* *22*, *pp. 577-597*

Stewartson, K., Williams, P.G. (1969) : "Self-induced separation", *Proc. Roy. Soc. London A* *312*, *pp. 181-206*

Stewartson, K. (1969) : "On the flow near the trailing edge of a flat plate II", *Mathematica* *16*, *pp. 106-121*

Stewartson, K. (1974) : "Multistructured boundary-layers", *Adv. Applied. Mech.* *14*, *pp. 145-239*

Stewartson, K. (1981) : "D'Alembert's paradox", *S.I.A.M. Review* *23*, *pp. 308-343*

Tall, D. (1982) : "Elementary Axioms and Pictures for Infinitesimal Calculus", *Bulletin I.M.A.* *18*,, *pp. 43-48*

Ting, L., Klein, R. (1991) : "Viscous vortical flows", *Springer*

Tollmien, W. (1929) : "Über die enstehtung der turbulenz", *Mitl. Nachr. Ges. Wiss. Göttingen, Math-Phys. Klasse 21-44*

Tracey, E.K. (1988) : "The initial and ultimate motions of an impulsively started cylinder in a compressible fluid", *Proc. Roy. Soc. London A* *418*, *pp. 209-228*

Van Dyke, M.D. (1962,1964) : "Higher approximations in boundary layer theory : I. General analysis, II. Application to leading edge, III. Parabola in uniform stream", *J. Fluid Mech.* *14*, *pp. 161-177 ;* *14*, *pp. 481-495 ;* *19*, *pp. 145-159*

Van Dyke, M.D. (1975) : "Perturbation methods in fluid mechanics", *Parabolic Press*

Van Dyke, M.D. (1984) : "Computer extended series", *Ann. Rev. Fluid Mech.* *16*, *pp. 287-309*

Villat, H. (1972) : "As luck would have it - a few mathematical reflections", *Ann. Rev. Fluid Mech.* *4*, *pp. 1-5*

Viviand, H. (1970) : Etude des écoulements instationnaires à faible nombre de Mach avec application au bruit aérodynamique", *J. Mécanique 4, pp. 573-599*

Ward, M.J., Keller, J.B. (1993) : "Strong localized perturbations of eigenvalue problems", *S.I.A.M. J. Appl. Math. 53, pp. 770-798*

Ward, M.J., Henshaw, W.D., Keller, J.B. (1993) : "Summing logarithmic expansions for singularly perturbed eigenvalue problems", *S.I.A.M. J. Appl. Math. 53, pp. 799-828*

Wason, W. (1965) : "Asymptotic expansions for ordinary differential equations", *Interscience*

Wong, T.C., Liu, C.H., Geer, J.F. (1985) : "Comparison of uniform perturbation and numerical solutions for some potential flows past slender bodies", *Comp. Fluids 13, pp. 271-283*

Zeytounian, R. Kh., Guiraud, J.P. (1987) : "Asymptotic modelling of flow : a few examples", *Atti Peloritana dei Pericolanti, classe I, 45, pp. 103-122*

Zeytounian, R. Kh. (1994) : "Modélisation asymptotique en mécanique des fluides newtoniens", *Mathématiques et Applications, Vol. 15, Springer*

Zonneveld, J.A. (1966) : "Periodic solutions of the Van der Pol equation", *Indag. Math. 28, pp. 620-622*

Zvonkin, A.K., Shubin, M.A. (1984) : "Non-standard analysis and singular perturbations of ordinary differential equations", *Russian Math. Surveys 39, pp. 69-131*

MIX
Papier aus verantwortungsvollen Quellen
Paper from responsible sources
FSC® C105338

If you have any concerns about our products,
you can contact us on
**ProductSafety@springernature.com**

In case Publisher is established outside the EU,
the EU authorized representative is:
**Springer Nature Customer Service Center GmbH**
**Europaplatz 3, 69115 Heidelberg, Germany**

Printed by Libri Plureos GmbH
in Hamburg, Germany